Silica and Clay Dispersed Polymer Nanocomposites

Related titles

Progress in Rubber Nanocomposites
(ISBN: 978-0-08-100409-8)

Micro and Nanofibrillar Composites (MFCs and NFCs) from Polymer Blends
(ISBN: 978-0-08-101991-7)

Polymer-based Nanocomposites for Energy and Environmental Applications
(ISBN: 978-0-08-101910-8)

Woodhead Publishing Series in Composites Science and Engineering

Silica and Clay Dispersed Polymer Nanocomposites

Preparation, Properties and Applications

Edited by

Md. Rezaur Rahman

WP
WOODHEAD PUBLISHING
An imprint of Elsevier

ELSEVIER

Woodhead Publishing is an imprint of Elsevier
The Officers' Mess Business Centre, Royston Road, Duxford, CB22 4QH, United Kingdom
50 Hampshire Street, 5th Floor, Cambridge, MA 02139, United States
The Boulevard, Langford Lane, Kidlington, OX5 1GB, United Kingdom

Notices

Knowledge and best practice in this field are constantly changing. As new research and experience broaden our understanding, changes in research methods, professional practices, or medical treatment may become necessary.

Practitioners and researchers must always rely on their own experience and knowledge in evaluating and using any information, methods, compounds, or experiments described herein. In using such information or methods they should be mindful of their own safety and the safety of others, including parties for whom they have a professional responsibility.

To the fullest extent of the law, neither the Publisher nor the authors, contributors, or editors, assume any liability for any injury and/or damage to persons or property as a matter of products liability, negligence or otherwise, or from any use or operation of any methods, products, instructions, or ideas contained in the material herein.

Library of Congress Cataloging-in-Publication Data
A catalog record for this book is available from the Library of Congress

British Library Cataloguing-in-Publication Data
A catalogue record for this book is available from the British Library

ISBN: 978-0-08-102129-3 (print)
ISBN: 978-0-08-102130-9 (online)

Publisher: Deans Matthew
Acquisition Editor: Gwen Jones
Editorial Project Manager: Peter Jardim
Production Project Manager: Joy Christel Neumarin Honest Thangiah
Cover Designer: Mark Rogers

Typeset by SPi Global, India

Transferred to Digital Printing in 2018

Contents

Contributors ix
Preface xi

1 Introduction and reinforcing potential of silica and various clay dispersed nanocomposites **1**
Md. Rezaur Rahman, Josephine Lai Chang Hui, Sinin bin Hamdan
1.1 Introduction **1**
1.2 Problem statement **2**
1.3 Scope of the study **3**
1.4 Objectives of the study **3**
1.5 Literature review **3**
1.6 Conclusion **15**
Acknowledgments **16**
References **16**

2 Preparation and characterizations of silica and various clay dispersed nanocomposites **25**
Md. Rezaur Rahman, Josephine Lai Chang Hui, Sinin bin Hamdan
2.1 Materials **25**
2.2 Preparations of nanocomposites **26**
2.3 Fourier transform infrared spectroscopy analysis **28**
2.4 X-ray fluorescence analysis **28**
2.5 Scanning Electron microscopy analysis **29**
2.6 Adsorption isotherm **29**
2.7 Tensile testing **30**
2.8 Thermogravimetric analysis **30**
2.9 Moisture absorption test **30**
2.10 Conclusion **31**
Acknowledgments **31**

3 Impact of nanoclay on physicomechanical, morphological, optical, BET isotherm, and thermal analysis of polyvinyl alcohol/fumed silica/clay nanocomposites **33**
Md. Rezaur Rahman, Josephine Lai Chang Hui, Sinin bin Hamdan
3.1 Introduction **33**
3.2 Results and discussion **34**
3.3 Conclusion **42**

Acknowledgments 42
References 42

4 **Polyvinyl alcohol/silica/clay nanocomposites: effect of clay on surface morphology, electrical and thermo-mechanical properties** **45**
Md. Rezaur Rahman, Josephine Lai Chang Hui, Sinin bin Hamdan
4.1 Introduction **45**
4.2 Results and discussion **46**
4.3 Conclusion **55**
Acknowledgments **55**
References **55**

5 **Nanoclay dispersed phenol formaldehyde/fumed silica nanocomposites: effect of diverse clays on physicomechanical and thermal properties** **59**
Md. Rezaur Rahman, Josephine Lai Chang Hui, Sinin bin Hamdan
5.1 Introduction **59**
5.2 Results and discussion **60**
5.3 Summary **69**
Acknowledgments **69**
References **69**

6 **Study on physical, mechanical, morphological and thermal properties of styrene-*co*-glycidyl methacrylate/fumed silica/clay nanocomposites** **71**
Md. Rezaur Rahman, Sinin bin Hamdan
6.1 Introduction **71**
6.2 Results and discussion **72**
6.3 Summary **82**
References **82**

7 **Physico-mechanical and thermal properties of clay/fumed silica diffuse polylactic acid nanocomposites** **87**
Md. Rezaur Rahman, Josephine Lai Chang Hui
7.1 Introduction **87**
7.2 Results and discussion **88**
7.3 Summary **105**
References **105**

8 **Biomedical and packaging application of silica and various clay dispersed nanocomposites** **109**
Akshay Kakar, Elammaran Jayamani, Muhammad Khusairy Bin Bakri, Md. Rezaur Rahman
8.1 Introduction **109**
8.2 Structure and characteristics of layered silicates and clays **111**
8.3 Organic modification of layered silicates **113**

8.4 Nanocomposite structure **115**
8.5 Nanocomposite structural characterization **116**
8.6 Nanocomposite mechanical properties **117**
8.7 Toughness and strain **122**
8.8 Flame retardance **126**
8.9 Nanocomposites: Advantages and applications **129**
References **132**

9 Durability and sustainability of the silica and clay and its nanocomposites 137
Akshay Kakar, Elammaran Jayamani, Muhammad Khusairy Bin Bakri, Md. Rezaur Rahman
9.1 Introduction **137**
9.2 The role of nanofillers on nanocomposites **141**
9.3 Clay **141**
9.4 Silica **143**
9.5 Durability and sustainability of clay nanocomposites and silica nanocomposites **144**
9.6 Challenges and applications **146**
9.7 Summary **150**
References **150**
Further reading **157**

10 Low-density polyethylene/silica nanocomposites foams: Relationship between chemical composition, particle dispersion, cellular structure and physical, mechanical, dynamic mechanical, electrical, and thermal properties 159
Md. Rezaur Rahman, Md. Faruk Hossen, Sinin bin Hamdan
10.1 Introduction **159**
10.2 Chemical composition and physical properties **160**
10.3 Mechanical properties **164**
10.4 Thermal properties **176**
10.5 Scanning electron microscopy (SEM) study **188**
10.6 Cellular structure of nanocomposite foams **194**
10.7 Electrical properties (DC current) **196**
References **196**
Further reading **198**

11 The Effect of clay dispersion on polypropylene nanocomposites: Physico-mechanical, thermal, morphological, and optical properties 201
Md. Rezaur Rahman, Sinin bin Hamdan, Md. Faruk Hossen
11.1 Overview of the nanocomposites **201**
11.2 Possibilities and prospects in nanocomposites **205**
11.3 Preparation of nanocomposites **208**
References **239**
Further reading **256**

12 Improvement of epoxy nanocomposites on physical, morphology, and mechanical properties as well as fracture behavior with the addition of mesoporous silica/nano-silica **259**

Muhammad Khusairy Bin Bakri, Elammaran Jayamani, Md. Rezaur Rahman, Akshay Kakar

12.1 Introduction **259**
12.2 Nanocomposites and nanofillers **261**
12.3 Epoxy **262**
12.4 Mesoporous silica/nano-silica **263**
12.5 Physical and mechanical properties **264**
12.6 Fracture behavior and morphological properties **269**
12.7 Summary **274**
References **275**
Further reading **280**

Index **281**

Contributors

Muhammad Khusairy Bin Bakri Faculty of Engineering, Computing and Science, Swinburne University of Technology Sarawak Campus, Kuching, Sarawak, Malaysia

Sinin bin Hamdan Department of Chemical Engineering and Energy Sustainability, Faculty of Engineering, University Malaysia Sarawak (UNIMAS), Kota Samarahan, Sarawak, Malaysia

Md Faruk Hossen Department of Chemical Engineering and Energy Sustainability, Faculty of Engineering, University Malaysia Sarawak (UNIMAS), Kota Samarahan, Malaysia

Josephine Lai Chang Hui Department of Chemical Engineering and Energy Sustainability, Faculty of Engineering, University Malaysia Sarawak (UNIMAS), Kota Samarahan, Sarawak, Malaysia

Elammaran Jayamani Faculty of Engineering, Computing and Science, Swinburne University of Technology Sarawak Campus, Kuching, Sarawak, Malaysia

Akshay Kakar Faculty of Engineering, Computing and Science, Swinburne University of Technology Sarawak Campus, Kuching, Sarawak, Malaysia

Md Rezaur Rahman Department of Chemical Engineering and Energy Sustainability, Faculty of Engineering, University Malaysia Sarawak (UNIMAS), Kota Samarahan, Sarawak, Malaysia

Preface

Recently, nanoclay-dispersed polymeric materials have begun to exhibit better utilization in diverse applications. Nanoclay's utility as an inorganic filler results from exfoliation or dispersion at the nanoscale into polymeric matrices, enhancing and improving nanocomposites' properties by adding slight amounts of clay due to its high specific area. Current studies of silica and clay-dispersed polymer nanocomposites' preparation, properties, and applications report the demonstration of better reinforcing potential in elastomeric polymer matrices.

The proposed book is focused on clay-dispersed elastomeric nanocomposites' properties and applications. It will also include an introduction and discussion of the reinforcing potential of various clay- and monomer-dispersed elastomeric nanocomposites. Readers will find complete information about the preparation and characterization of various clay-dispersed elastomeric nanocomposites, polyvinyl alcohol/fumed silica/clay effects, phenol formaldehyde/fumed silica nanocomposites, styrene-co-glycidyl methacrylate/fumed silica/clay nanocomposites, clay/fumed silica diffuse polylactic acid nanocomposites, biomedical and packaging application of nanocomposites, durability and sustainability of the silica and clay and its nanocomposites, low-density polyethylene/silica nanocomposite foams, and the effect of clay dispersion on polypropylene nanocomposites and epoxy nanocomposites.

I am thankful to all my co-authors who contributed book chapters and provided their valuable ideas and knowledge to this edited book. I attempted to gather all the information of my co-authors from the same fields in elastomeric polymer nanocomposites and finally produce this project, which will hopefully become a success. I greatly appreciate my co-authors' support in bringing our idea to reality.

I thank the Woodhead Publishing team (Cambridgeshire, United Kingdom) for their substantial cooperation at every stage of the book production.

Md. Rezaur Rahman
Kota Samarahan, Malaysia

Introduction and reinforcing potential of silica and various clay dispersed nanocomposites

1

Md. Rezaur Rahman, Josephine Lai Chang Hui, Sinin bin Hamdan
Department of Chemical Engineering and Energy Sustainability, Faculty of Engineering, University Malaysia Sarawak (UNIMAS), Kota Samarahan, Sarawak, Malaysia

1.1 Introduction

Polymers are chain substances that consist of bonding of similar monomer units in a molecular state (Chu, 2010). Polymer nanocomposites are fabricated by incorporation of nanosized fillers or a reinforcing agent. The production of polymer nanocomposites has gained attention from both researchers and manufacturers due to its outstanding performance. Polymer nanocomposites are important in nanoscience due to the significant improvement in physicomechanical, thermal, and barrier properties compared to microcomposites or conventional composites (Pukánszky, 2005). However, the application of the polymer nanocomposites is limited to certain fields. In polymer nanocomposites, it is difficult to achieve uniform dispersion between the filler and matrix, as it forms agglomerations easily (Chu, 2010). To obtain excellent properties of nanocomposites, preparation and the structural relationships between the polymer matrix and fillers should be highly studied (Pukánszky, 2005).

Silica, fumed silica, and clays are referred to as reinforcing agents. Silica and fumed silica are widely used as filler embedded in a polymer matrix to improve the mechanical and morphological properties (Conradi, 2013). Clay is widely applied onto polymer matrix, as the hydrophobic clay enhances the strong interaction with hydrophilic polymer matrix (Roelofs & Berben, 2006). The clay with weak bondings between the layers expands the clay galleries and enhances the intercalation in the polymer nanocomposites. Clay is used as a filler material in small amounts due to its availability, low cost, and high surface-to-volume ratio (Greesh, 2011). It leads to a better reinforcement due to its large interface and evenly distributed nanosized particles. The nanofillers are dispersed in polymer matrix to form polymer nanocomposites. Polymer matrix is the continuous constituent that is found in higher quantities in the nanocomposites. It helps to bind and distribute the fillers uniformly in the nanocomposites (Thomas, Kuruvilla, Malhotra, Goda, & Sreekela, 2012, Chapter 1).

A polymer such as PVA functions as polymer matrix. It is a hydrophilic polymer that has high solubility in aqueous solution with low degrees of hydrolysis (Kokabi, Sirousazar, & Hassan, 2007). PVA performs better when crosslinking with fillers; it

Silica and Clay Dispersed Polymer Nanocomposites. https://doi.org/10.1016/B978-0-08-102129-3.00001-4

has advantages such as compatibility, excellent film foaming, it is nontoxic, and a simple chemical structure (Rosi, Iskandar, & Abdullah, 2014). It is widely applied in adhesives, paints, plastics, coatings, and others. PVA nanocomposites are applied in pharmaceutical and biomedical tools due to their outstanding properties (Tanigami, Yano, Yamaura, & Matsuzawa, 1995). However, PVA has limitations, such as thermal reversibility and low mechanical strength (Rosi et al., 2014).

PF is another common polymer produced by polycondensation of phenol and formaldehyde in acidic or basic catalyst at low molecular weight that is used as a polymer matrix in polymer nanocomposites (Edoga & Kovo, 2006). PF is one of the thermosetting polymers that can be applied as a bonding agent and binders in coatings, as PF has good mechanical and thermal properties. However, PF has limited usage due to its brittleness and poor adhesion properties (Inamdar & Rathod, 2013).

ST-*co*-GMA is a polymer matrix that undergoes polymerization with the introduction of an initiator (Ma, Shi, & Song, 2014). This polymer resin can be applied in paper strength additive (Zhang & Tanaka, 2001). ST-*co*-GMA matrix has good mechanical and thermal properties as well as better water resistance from acids and deionized water (Garg & Srivastaba, 2014). However, ST-*co*-GMA is unable to be widely applied due to its high toxicity and low polymerization temperature (El-Demerdash, Sadik, & El-Maghraby, 2008).

PLA is another polymer matrix. It is a thermoplastic with high strength that can be easily modified by simple hydrolysis of the ester bond without any enzyme (Garlotta, 2002). PLA is used as packaging material for construction, tissue regeneration, and scaffolds for bone reconstruction (Ray, 2012). Recently, PLA has been applied in green plastics. PLA nanocomposites can also be applied in biomedical fields, as they have bioadsorbable characteristics (Lim, Kim, Han, Khan, & Seo, 2014).

In this study, selected polymers such as PVA, PF, ST-*co*-GMA, and PLA are used as polymer matrix, while silica/fumed silica and clay are chosen as fillers. The fabricated nanocomposites are characterized by Fourier-transform infrared spectroscopy (FTIR), X-ray fluorescence (XRF), scanning electron microscopy (SEM), adsorption isotherm, tensile testing, thermogravimetric analysis (TGA), and moisture absorption test.

1.2 Problem statement

Nowadays, polymer nanocomposites are materials with high demand due to their unique characteristics. However, the pure polymer matrices cannot be applied widely due to the limitations of their properties and environmental concerns. Pure polymers are hydrophobic in nature, and their drawbacks include low thermal resistance, low mechanical properties, and poor adhesion. Due to the drawbacks of pure polymers, single filler has been incorporated into the pure polymers to synthesize polymer nanocomposites. The incorporation of single filler has gradually improved either thermal or mechanical properties of the polymer nanocomposites. However, the physical and morphological properties of the single filler polymer nanocomposites have not been improved. To overcome the aforementioned problem, dual fillers such as silica/fumed silica (si/fsi) and clay are used as reinforcing agents to be incorporated into polymer matrices to enhance their physical, morphological, mechanical, and thermal properties. In this study, PVA, PF, ST-*co*-GMA, and PLA matrices were used to fabricate the nanocomposites.

1.3 Scope of the study

Due to their significant properties, especially their biodegradability and eco-friendliness, the research of polymer nanocomposites is of growing interest in the development of new materials. It is speculated that the hydrophilic nature of the pure polymers affects the resultant physical, mechanical, and thermal properties. According to the literature review and the aforementioned problem statement, there was no research reported on PVA-fumed silica-clay, PVA-silica-clay, PF-fumed silica-clay, ST-*co*-GMA-fumed silica-clay, and PLA-fumed silica-clay nanocomposites. Due to the research gap, the researcher is responsible for finding the appropriate solution.

1.4 Objectives of the study

The specific objectives of this study are:

a) To manufacture the polymer nanocomposites with various fillers loading.
b) To evaluate the physical, mechanical, morphological, and thermal properties of polymer nanocomposites.
c) To investigate the compatibility of dual fillers with various polymer matrices in the nanocomposites.
d) To compare the fabricated polymer nanocomposites with the existing nanocomposite materials.

1.5 Literature review

1.5.1 Clays and clay minerals

Clays are the main constituents of fine-grained rocks that exist in mudstones and shales from marine sediments and soils. Most clays are formed due to the weathering and secondary sedimentary processes (Bilotti, 2009).

Clay minerals are referred to as layered silicates due to their stacked structure of 1 nm silicate sheets with a variable basal distance. They belong to the structural family known as phyllosilicates. The fundamental building units of phyllosilicates consist of tetrahedral silica (Si) and octahedral alumina (Al). The tetrahedral sheets are composed of individual tetrahedrons where every silicon atom is equidistant from four oxygen or hydroxyls to balance the structure. They are arranged in a hexagonal pattern. Octahedral sheets are composed of individual octahedrons that share edges, which are composed of oxygen and hydroxyl anion groups coordinated by cations like

Al, Mg, Fe^{3+}, Fe^{2+}, and others. These sheets are held together by electrostatic and Van der Waals forces (Bilotti, 2009).

The clay minerals can be divided into two types, 1:1 and 2:1 phyllosilicates (Ray & Okamoto, 2003). 1:1 phyllosilicates are defined as the "nonswelling" clays such as kaolinite. This clay mineral contains single unit sheets of tetrahedral silica between octahedral alumina sheets (Yaris & Cross, 2002). Each layer is free of charge due to the absence of isomorphic substitution in the octahedron or tetrahedron sheet. The layers are held together by hydrogen bonding between the hydroxyl groups in the octahedral sheets and oxygen in the tetrahedral sheets of the adjacent layers. 2:1 phyllosilicates are known as "swelling" clays, and are composed of one central octahedral sheet fitted in between two tetrahedral sheets and condensed into one unit layer. Most of the clays are 2:1 phyllosilicates (montmorillonite (MMT), saponite) (Yaris & Cross, 2002). This layered silicate is very useful for production of nanocomposites.

1.5.2 Modifications of clay to enhance the properties of nanocomposites

There are two conditions that are considered for preparing polymer-silica or polymer-clay nanocomposites. The first condition is the incompatibility of the high hydrophilicity of silicate layers with the hydrophobic polymers. The second condition is the degree of penetration of polymer into the silicate interlayer. This is due to the narrow basal spacing of the silicate layers that lead to the weak physical interactions between organic and inorganic phases (Ray & Okamoto, 2003). Therefore, clay surfaces should be modified to render the layered silicates to be miscible with a polymer matrix. There are several common processes to modify clays, such as reaction with silane compounds and ion exchange with organic cations (Herrera, Putaux, David, & Bourgeat-Lami, 2006).

1.5.2.1 Modification of clay by Silane reaction

Clay surface modification through silane reaction has been extensively investigated. From the chemical structure of clay, silanol bonds are formed due to the reaction of the hydroxyl groups at the edges of the clay platelets (Ogawa, Okumoto, & Kuroda, 1998). Covalent bonds are formed between clay filler and polymer matrix. Silane is grafted at the external surface of the clay. It is highly dependent on accessible surface area. There is no change in the interlayer distance. There are two basic steps for silane grafting process. The first step is that the silane molecules interact into the clay interlayer. The second step is the condensation reaction that takes place between these silane molecules and clay surface (He et al., 2013).

H^+-titanosilicate (formed by proton exchange of Na^+-titanosilicate) derivatives are found to produce H^+-titanosilicate/dodecylamine/tetraethylorthosilicate (DDA/TEOS) intercalation compounds in DDA-TEOS solution. The basal spacing of these derivatives is about 4.16–4.32 nm with uniform pore size of 2.8–3.4 nm. DDA acts as base catalyst during the hydrolysis of TEOS. The rapid hydrolysis of TEOS in water controls TEOS outflow from interlayers and contributes to the formation of firm silica-pillars. The results prove that intercalation and silylation processes are more effective due to the evaporation of solvents (Park, 2012).

1.5.2.2 Modification of clay by ion-exchange with organic cations

Another simple method to modify clay's surface is the traditional cation-exchange modification. This modification involves the substitution of the structural clay cations with organic cations. Small cationic molecules can replace the exchangeable cations present on the clay surface due to the loose bonding of cations to the clay surface. The cation-exchange method with alkylammonium ions is an easy process to make mineral fillers become hydrophobic, and is effective in enhancing the properties of nanocomposites (Ahmad et al., 2009). The surface polarity of the filler will not be altered. The intergallery distance will be expanded, which increases the probability of polymers penetrating into the galleries (Zaharri, Othamn, & Mohd Ishak, 2013).

According to Singla, Mehta, and Upadhyay (2012), MMT has been successfully modified using organic cationic surfactants. The intercalation between the clay and organic cationic surfactants, such as phosphonium-MMT, helps increase the basal spacing. This modified MMT can be introduced into polymer matrix to form nanocomposites, which renders a higher level of dispersion. The mechanical and thermal properties can be improved using modified MMT. The structure of clay depends on the molecular size of organic onium and its orientation in the clay galleries after modification. Ion exchange with sodium cations showed a different arrangement of onium ions within the clay galleries. Different arrangements of the ions are possible due to the charge density of the clay and the ionic surfactant. In general, the longer the surfactant chain length, the higher the charge density of the clay (LeBaron, Wang, & Pinnavaia, 1999).

1.5.3 Structure of nanocomposites

A nanocomposite is defined as a composite material in which at least one dimension of the component is in the nanometer size scale (<100 nm). The three different types of nanocomposites morphologies that are generally obtained include phase separated systems, intercalated systems, and exfoliated systems. Different systems are formed due to the nature of the components (clay and polymer matrix) and preparation methods (Camargo, Satyanarayana, & Wypych, 2009).

Phase separated nanocomposites are obtained when the polymer is unable to intercalate between the silicate sheets, whose properties stay in the same range as that of traditional microcomposites due to the hydroxylated edge-edge interaction (Mallikarjuna, 2009). Thus, the *d*-spacing of clay will not change and cause the nanocomposites to have poor mechanical properties. For intercalated nanocomposites, the polymer chains are inserted in the interlayer gallery. Which increases the interlayer distance. This leads to the formation of a well ordered multilayered morphology, with alternating polymeric and inorganic layers (Olad, 2011). The mechanical and thermal properties of intercalated nanocomposites are significantly improved compared to phase separated nanocomposites (Chow, 2008). Exfoliated nanocomposites are also referred to as delaminated nanocomposites. The individual clay layers are separated and uniformly distributed in a continuous polymer matrix (Alexandre & Dubois, 2000). Overall, intercalated nanocomposites exhibit better homogeneity than exfoliated nanocomposites (Chen, 2004).

1.5.4 Polymer nanocomposites

Polymer nanocomposites are defined as composites in nanoscale that are filled by at least one dimension of the dispersed particles (Greesh, 2011). Currently, both academic and industry field are interested in the development of polymer nanocomposites due to the enhanced properties and better performance compared to both conventional polymer composites and pure polymers. Another interesting feature of polymer nanocomposites is that the incorporation of small amount of filler will lead to significant changes in the polymer nanocomposites. The filler is the main factor that affects the final properties and surface morphology of the polymer nanocomposites.

1.5.4.1 PVA-silica/fumed silica nanocomposites

PVA is a biocompatible synthetic polymer that acts as a matrix, which could be applied in biomedical fields (Rescignano et al., 2014). Silica is filler that is commonly applied in PVA to enhance the properties of the nanocomposites. Son et al. (2015) showed that 30 μm of silica nanoparticles coupled well with the PVA to reduce accumulation in oil recovery. The precipitated silica was one of the reinforcing fillers that were introduced into PVA gel composites to enhance the mechanical properties (Zhang & Le, 2014). In addition, the increasing surface area of the silica led to the better reinforcement due to the interaction between the polymer and filler (Bhattacharya & Chaudhry, 2013).

1.5.4.2 PVA-clay nanocomposites

PVA is an outstanding host polymer matrix for composite films (Thomas et al., 2012). Filler such as hydrophilic natural sodium- MMT (Na^+-MMT) performed as a crosslinking agent in PVA nanocomposite hydrogels (PVA-MMT) and improved the physical, mechanical and thermal properties, as well as surface morphology. The PVA-(NA^+-MMT) nanocomposites were suitable for biomedical application (Karimi & Wan Daud, 2015). The water barrier of PVA-MMT nanocomposites had been greatly improved due to the incorporation of clay, which acted as reinforcing agent between polymer matrix and filler (Sirousazar, Kokabi, & Hassan, 2011). In addition, PVA hybrids achieved polymer crystallization at higher temperature with the addition of MMT (Fonseca, Strawhecker, Manias, & Harrison, 2000).

1.5.4.3 PF-fumed silica nanocomposites

PF resin is used as polymer matrix in the composites. The addition of fillers into the resin greatly improves the mechanical behavior of the composites (Chauhan, Patnaik, Kaith, Satapathy, & Dwivedy, 2009). According to Wei et al. (2015), nanosilica, which acted as a reinforcing agent, was introduced into sisal fiber-PF composites to strengthen friction properties. Additionally, in situ polymerized PF foam nanocomposites had improved mechanical and thermal properties due to the introduction of silica as the strong reinforcing agent (Li et al., 2015). Furthermore, nucleating agents such as fumed silica improved the compression strength of the phenolic resin foams, which were fabricated through microwave technology. Silica performing as a reinforcing filler and surface compatibilizer enhanced the properties of the phenolic-silica-nitrile composites (Achary & Ramaswamy, 1998).

1.5.4.4 PF-clay nanocomposites

Phenolic resin is a good thermoset matrix resin that provides strong interfacial bonding with fillers, which enhances the mechanical strength of the composites (Fink, 2013). 5 wt% of nanoclay (acting as crosslinking agent) was introduced into PF resin, which produced higher tensile strength and Young's modulus (Ku, Yap, Lee, & Trada, 2012). According to Kang, Hong, and Kim (2010), the addition of organoclay that functioned as compatibilizer improved both morphology and mechanical properties of the melt-mixing fabricated phenolic resin. The organic MMT (OMMT) reinforced PF nanocomposites and improved the mechanical and thermal properties (Gao, Jiang, & Su, 2010).

1.5.4.5 ST-co-GMA-fumed silica nanocomposites

ST, also known as vinyl benzene, has high molecular weight thermoplastic resin produced by free radical polymerization reaction (Ash, 2004). GMA is a monomer that is used in various industrial applications such as coatings and adhesives (Yoshida, 2012). Silica acts as a reinforcing agent that enhances the dispersion and exfoliation in ST-related rubber tread compounds (Bao, Tao, & Flanigan, 2015). ST-butadiene-GMA terpolymer improves the compatibility with the addition of silica into the rubber compounds (Kim et al., 2015). Silica also plays a role of a coupling agent that enhances the physical and mechanical properties of ST-butadiene rubber (Zhao et al., 2012).

1.5.4.6 ST-co-GMA-clay nanocomposites

ST monomer is one of the copolymers with excellent physical strength that prevent the copolymers from oxidation degradation or hydrolysis (Dorfner, 1972). GMA is a common monomer that is readily biodegradable and hydrolyzed (OECD SIDS, 2002). According to Jlassi et al. (2015), clay plays the role of a reinforcing agent that was incorporated into poly(GMA)-clay nanocomposites through free radical polymerization. Clay is also a coupling agent that enhances the mechanical properties of carbon black-filled ST butadiene rubber composites (Ma et al., 2014). Clay acting as a crosslinking agent was well dispersed into poly(GMA) nanocomposites through free radical polymerization (Bayramoglu, Senkali, & Arica, 2013).

1.5.4.7 PLA-fumed silica nanocomposites

PLA is a polymer that is used as polymer matrix in nanocomposites (Auras, Lim, Selke, & Tsuji, 2010). Silica is one of the fillers regularly introduced into PLA nanocomposites. It functioned as nucleation agent to improve the mechanical properties as well as the surface morphology (Myoung, Im, & Kim, 2015). Strong covalent bonds between PLA, poly(methyl methacrylate) (PMMA) and silica were formed which improved the heat-resistance properties. This was due to the silica that acted as coupling agent in the polymer matrix. Furthermore, fumed silica was widely used as reinforcing filler in thermoplastic polymers to improve the physical properties of the composites (Lightsey, Kneiling, Long, & Kolbert, 1999).

1.5.4.8 PLA-clay nanocomposites

PLA is a polymer that is fabricated from lactic acid, which is widely applied in biodegradable products (Oksman, Skrifvars, & Selin, 2003). The addition of nanoclay increased the mechanical and morphological properties of PLA-thermoplastic starch. Nanoclay enhanced the covalent bond between the polymer matrix and fillers (Ayana, Suin, & Khatua, 2014). The addition of organoclay that acted as compatibilizer was introduced into PLA-natural rubber nanocomposites. This compatibilizer had improved the bonding adhesion that confirmed the compatibility of organoclay with the polymer matrix (Martinez-Tong et al., 2014). Based on the research by Bitinis et al. (2012), MMTs, which acted as compatibilizers on PLA nanocomposites, had improved the permeability properties. Moreover, clay acting as a nucleation site had facilitated the crystallization process, thus resulting in smaller crystallites and lower crystallinity (Liao, Nawaby, & Naguib, 2012).

1.5.5 The changes in properties of silica and clay nanocomposites

1.5.5.1 Physical properties of silica/fumed silica nanocomposites

FTIR detected an increase in the amount of carbonyl groups with the increments of silica loading in PVA nanocomposites. This was due to the hydrogen bonding interactions between hydroxyl groups on the nanoparticle surfaces and carbonyl groups of PVA (Lin et al., 2015). FTIR analysis also showed the strong bonding between hydroxyl and silanol groups in PVA-MMT nanocomposite hydrogels (Karimi & Wan Daud, 2015). Zuo, Li, Yu, Ma, and Ruan (2013) showed that the absorption peak at $1097\,cm^{-1}$ represented the peak of Si-O-Si structure in the modified phenolic resin. In addition, the broader peak at $3600\text{-}3000\,cm^{-1}$ represented the presence of alcohol and phenolic groups due to the crosslinking with silica (SiO_2) and zinc sulfate ($ZnSO_4$) (Streckova et al., 2012). Broader FTIR peaks could be detected due to the strong formation of covalent bonds between the epoxy groups of ST-butadiene-GMA and silanol groups on the silica surface (Kim et al., 2015). According to Grama et al. (2014), FTIR results showed that the uniform dispersion of silica had improved the biocompatibility of poly(GMA) microspheres. The modified fumed silica was successfully incorporated to enhance the segmental mobility and plasticity of PLA polymer (Lai, Chen, Han, Yu, & Lai, 2013). Melt-compounded PLA-silica nanocomposites presented the well intercalation of hydroxyl groups between silica surfaces and PLA chain-end (Wen et al., 2011).

From the elemental aspect, the disappearance of hydroxyl peak and formation of PVA and silica bonding were clearly observed in the XRF data. The bonding of the hybrids was indicated by the Si-O-C peak in the spectra (Pirzada, Arvidson, Saquing, Shah, & Khan, 2012). For phenolic resin-silica nanocomposites, elemental analysis showed high amounts of silica groups with equivalent carbon and nitrogen groups in the nanocomposites (Yu, Wei, Lv, Liu, & Meng, 2012). XRF results showed that Si acting as a reinforcing agent increased the Si:Ca ratio particles in the modified calcium carbonate ($CaCO_3$)-PLA nanocomposites (Nekhamanurak, Patanathabutr, & Hongsriphan, 2012).

On the other hand, silica shows well intercalation with PVA matrix to prevent the aggregation formed on the surface of PVA-silica nanocomposites (Asghari, Hatami, & Ahmadipour, 2015). Strong chemical bonding between the organic and inorganic phases are formed when silica nanoparticles (<10 nm) were well-dispersed into the PVA-maleic acid-silica hybrid membrane (Xie et al., 2011). PF microcapsules with the addition of silica showed regular spherical shape and smooth surface without any agglomerations. In addition, a well-ordered mesostructured PF resin-silica-Pluronic F127 composite showed uniform morphology due to the strong intercalation between the hydrophobic PF resin and fillers (Lin, Chien, Tang, & Lin, 2006). ST-butadiene-GMA-silica compounds reduced the agglomeration compared to the conventional rubbers (Kim et al., 2015). According to Grama et al. (2014), the silica improved the surface morphology when well dispersed into the poly(GMA) microspheres. According to Wu, Zhang, Yang, Liu, and Yang (2014), silica nanoparticles were uniformly dispersed into the modified PLA matrix. The presence of nanosilica into PLA foams improved the uniformity of the surface morphology (Ji et al., 2013).

From adsorption isotherm analysis, PVA-silica-(3-mercaptopropyl) trimethoxysilane (MPTMS15%) membrane had higher surface area and pore volume compared to pristine PVA (Keshtkar, Irani, & Moosavian, 2013). The addition of PVA content decreased the surface area and pore volume of PVA-silica composites. This proved that mesoporous PVA-silica composites membrane had higher adsorption capacities than pure PVA membrane (Asghari et al., 2015). The smaller diameter of phenolic resin-microporous carbon ultrafine fibers reduced water adsorption, and thus improved the specific surface area (Bai, Huang, & Kang, 2014). The addition of silica into resorcinol-formaldehyde composite gels exhibited high Brunauer-Emmett-Teller (BET) specific surface area and pore volume of $251\,m^2/g$ and $0.965\,cm^3/g$ respectively (Kong et al., 2013). GMA grafted ethylene-*n*-octene copolymer acting as a compatibilizer had improved the specific surface area of modified nanosilica-filled polypropylene composites (Studzinski, Jeziorska, Szadkowska, & Zielecka, 2014). According to Robertson et al. (2011), the incorporation of silica improved the specific surface area of ST-butadiene rubber. Javier, Francisco, Carlos, Alma, and Saul (2013) revealed that the grafted silica onto PLA matrix blocked the mesopores in the nanocomposites, and thus the surface area and pore volume were reduced. The specific surface area of PLA-silica hybrid microspheres was calculated by BET equation, and higher specific surface area increased the nitrogen desorption within the pores in the microspheres (Wang, Wei, Xia, Chen, & He, 2012).

1.5.5.2 Physical properties of clay nanocomposites

According to Sirousazar et al. (2011), the reduction of hydroxyl groups proved the interaction of hydroxyl groups of PVA with Na^+-MMT silicate layers in the nanocomposites. Sengwa, Choudhary, and Sankhla (2010) showed the strong hydrogen bonding between PVA matrix and poly(ethylene oxide) (PEO) in the presence of 0.5 wt% exfoliated nanosheets of MMT. This proved the compatibility of MMT with PVA-PEO blend nanocomposites. On the other hand, the FTIR analysis showed the degradation of pure boron-containing phenol-formaldehyde resin (BPFR) due to fracture of methylene (Lu et al., 2014). According to Rangari et al. (2006), the strong chemical structure of bonding between the Cloisite clay and phenolic foam was detected due to the well dispersion of clay nanoparticles into the nanocomposites. FTIR study of Devi and Maji (2012) confirmed the presence of clay had improved the intercalation of ST-acrylonitrile copolymer with clay in the presence of GMA. Based on the FTIR analysis of He et al. (2013), the incorporation of clay had improved the compatibility on polyamide 6-maleated ST-ethylene-butadiaene-styrene copolymers through melt compounding. FTIR analysis proved that higher clay dispersion and exfoliation levels are achieved when Cloisite clay was added into the melt-compounded PLA nanocomposites (Maio, Garofalo, Scarfato, & Incarnato, 2015). According to Jiang, Zhang, and Wolcott (2007), the FTIR analysis proved the strong interaction between the semicrystalline PLA and MMT platelets, which reduced the microvoids in the composites.

Elemental analysis of PVA-silica nanocomposites showed that the vinyl silica nanoparticles had been successfully formed through covalent bonding between silica nanoparticles and PVA matrix (Jia, Li, Cheng, Zhang, & Zhang, 2007). XRF was used to determine the types and percentages of elements of total phenolics in leaves (Sykorova et al., 2009). According to Feng and Deng (2004), XRF under helium atmosphere was useful in detecting sodium ions in the resole type of PF resin. The elemental analysis of Hoidy, Ahmad, Al-Mulia, and Ibrahim (2010) proved that the amounts of intercalant in the modified clay based on carbon and nitrogen were almost equivalent. This confirmed the cation exchange capacity of the clay on PLA nanocomposites.

The well interaction between PVA matrix and modified layered double hydroxides (MLDH) nanofiller through hydrogen bonding improved the surface morphology of the nanocomposites (Mallakpour & Dinari, 2014). SEM morphology showed the spherical shape and smooth surface of resorcinol-based novel PF microcapsules, which proved the well intercalation between polymer matrix and filler (Hedaco, Mahulikar, & Gite, 2013). The introduction of clay was evenly dispersed in the urea formaldehyde matrix, which proved the positive interaction between matrix and filler particles as well as the effective reinforcement of the fillers (Mamantha & Jagadish, 2013). The clay improved the surface morphology of the nanocomposites by well intercalation between the poly(GMA) and clay through free radical polymerization (Jlassi et al., 2015). From the SEM analysis of Devi and Maji (2012), in the presence of GMA, the ST-acrylonitrile copolymer was well intercalated with clay at lower loadings. The dispersion of the small amount of clay platelets into PLA matrix enhanced large interfacial areas as well as improved tensile strength (Lai, Wu, Lin, & Don, 2014). Additionally, Joncryl-based clay platelets dispersed into PLA matrix more significantly compared to OMMT. According to Das et al. (2010), the SEM micrographs showed clear fibrillar pattern of solution-casted PLA-clay nanocomposites.

BET adsorption isotherm analysis showed that lower clay loading introduced in the PVA-MMT nanocomposite hydrogel would significantly improve the network mesh size and average molecular weight of polymer chains between crosslinks (Sirousazar et al., 2011). The BET adsorption isotherm analysis proved that the surface area of the blends of PVA-linear low density polyethylene (LLDPE) was about $310\,m^2/g$ (Vidya, 2012). The adsorption of the PF-mesoporous carbon fibers was determined by the pore size distribution as well as the exposed surface area (Teng, Qiao, Li, & Bera, 2012). The surface area and pore volume of phenol-mixed pillared MMT, which were suitable to be used as oxidation catalysts, increased significantly (Kurian & Sugunan, 2006). According to Singh and Ghosh (2014), the acrylonitrile-butadiene-ST nanocomposites had higher specific surface area due to the reinforcement of clay. Addition of clay onto GMA resin increased the BET surface area, which enhanced the removal of ions from aqueous solutions compared to clay-free resin (Donia, Atia, & Rashad, 2011). The blended PLA with about 5 vol% of clay provided larger surface areas so that the nanocomposites could be applied in biomedical application and green plastic (Ray, 2012). In addition, the BET surface area of phenolic resin-silica nanocomposites that undergo nitrogen adsorption-desorption isotherm at 77 K was found to be $46\,m^2/g$, and pore diameter was categorized as mesoporous (Yu et al., 2012).

1.5.5.3 Mechanical properties of silica/fumed silica nanocomposites

The tensile strength and tensile modulus of the silica nanoparticle-reinforced PVA cast sheets improved remarkably. This improvement was observed as the strong interface minimized interfacial sliding between the polymer and the silica nanoparticle in the elastic range (Bhattacharya & Chaudhry, 2013). PVA-nanosilica films improved the mechanical properties significantly at 0.3 wt% fumed silica (Wang et al., 2012).

According to Mirzapour, Asadollahi, Baghshaei, and Akbari (2014), the addition of 3 wt% of nanosilica into carbon fiber-phenolic nanocomposites that undergo compression molding increased their bending strength up to 13%. By introducing the silica into PF resin, the hardness of the resole polymer nanocomposites was greatly improved (Streckova et al., 2012).

With the introduction of silica into ST rubber tread compounds, the mechanical properties were greatly improved (Bao et al., 2015). According to He et al. (2013), the melt-compounded polyamide 6-maleated-ST-ethylene-butadiene-ST copolymer composites slightly improved the tensile strength and tensile modulus. Silica helped to improve the tensile strength with well dispersion of silica into ST-butadiene-rubber composites (Lei & Chow, 2011).

Addition of rice husk ash (RHA) into PLA-silica composites did not improve the mechanical properties of the composite films due to the incompatibility of RHA towards the PLA-silica nanocomposites (Opaprakasit, Boonpa, Jaikaew, Petchsuk, & Tangboriboonrat, 2015). According to Xiong et al. (2014), the addition of epoxidized soybean oil (ESO) nanosilica considerably improved the impact strength of the PLA-silica-ESO nanocomposites compared to neat PLA or PLA-nanosilica composite. Moreover, nanosilica introduced into sol-gel PLA matrix displayed better tensile strength compared to microsilica-PLA composites (Huang, Hung, Wen, Kang, & Yeh, 2009).

1.5.5.4 Mechanical properties of clay nanocomposites

Quaternized aromatic amine-based hybrid PVA membranes improved the mechanical properties with tensile strength up to 10%. The addition of clay at 1 vol% into PVA nanocomposites showed significant improvement in tensile strength (Allison et al., 2015). Based on the study of Sapalidis, Katsaros, Steriotis, and Kanellopoulos (2011), the addition of clay up to 10 wt% into PVA matrix improved the mechanical properties of the PVA nanocomposites.

On the other hand, the mechanical properties of the nanoclay-modified silica phenolic composites improved significantly at 3 wt% due to the reinforcing effect of nanoclay (Robert et al., 2015). Higher quantities of glass fiber and nanoclay improved the impact and compression strength of the phenolic-glass fiber-nanoclay composite foam (Hu, Cheng, Nie, & Wang, 2015).

Clay introduced into the poly(GMA) nanocomposites significantly improved the mechanical properties (Jlassi et al., 2015). The tensile strength and tensile modulus of ST-acrylonitrile copolymer impressively increased with the incorporation of nanoclay at 2 wt% (Devi & Maji, 2012). ST-butadiene rubber hybrid showed better tensile strength due to the well intercalation of clay within the composite system (Meneghetti, Shaikh, Qutubuddin, & Nazarenko, 2008).

The compatibilization effect of OMMT strengthened the interface between PLA and cellulose acetate butyrate (CAB) and thus, the tensile strength of the PLA-CAB nanocomposites gradually increased (Nam, Min, & Son, 2015). The presence of small amounts of clay marginally improved the mechanical strength of PLA-clay nanocomposite (Liao et al., 2012). Jollands and Gupta (2010) showed that PLA-MMT nanocomposites improved the tensile modulus up to 40% higher than pure PLA at the clay loading of 4 wt%.

1.5.5.5 Thermal properties of silica/fumed silica nanocomposites

PVA-silicon dioxide-titanium oxide (PVA-SiO_2-TiO_2) fibers that undergo sol-gel process showed better heat resistance than pure PVA fibers (Ma et al., 2014). Pan-milled PVA-OMMT nanocomposites exhibited significant improvement in thermal stability due to the well dispersion of OMMT on PVA matrix (Li, Wu, Wu, Xu, & Li, 2010). Additionally, the thermal stability of PVA-silica nanocomposites was significantly improved, due to the strong interactions formed by covalent bonds between both silica and polymer matrix (Jia et al., 2007).

According to Natali, Monti, Kenny, and Torre (2011), the TGA analysis showed that phenolic resin-silica nanocomposites displayed better thermal stability than pure phenolic matrix. The phenolic resol resin intercalated with 3 wt% of silica exhibited maximum thermal stability compared to pure phenolic resol resin (Fang & Suo, 2011).

From the TGA results of Grama et al. (2014), the poly(GMA) microspheres showed high thermal stability due to the incorporation of silica particles. The thermal stability of melt blended ST-butadiene rubber-silica composites was improved as the composites decomposed at higher temperature (Zhang & Tanaka, 2001).

The TGA analysis by Basilissi, Silvestro, Farina, and Ortenzi (2013) showed that the addition of nanosilica into PLA nanocomposites improved the thermal stability compared to pure PLA. In addition, modified PLA with 2-methacryloloxyethyl isocyanate and silica improved the thermal stability of the nanocomposites compared to the pure PLA (Chen, 2004). According to Li et al. (2010), the silica with 5 wt% had enhanced the thermal stability of melt-compounded silica-filled PLA nanocomposites.

1.5.5.6 Thermal properties of clay nanocomposites

The introduction of bentonite clay into PVA matrix improved the thermal stability as well as the activation energy of the nanocomposites compared to pure PVA (Turhan, Alp, Alkan, & Dogan, 2013). The addition of nanoclay into PVA improved the thermal stability due to the dispersion quality of clay (Kaboorani & Riedl, 2011).

The addition of nanoclay improved the thermal stability of the phenolic-glass fiber-nanoclay composite foam by reducing the total release rate, and thus reducing its flammability (Hu et al., 2015). According to Kato, Tsukigase, Usuki, Shimo, and Yazawa (2006), melt-compounded resole-type phenolic resin-clay nanocomposites exhibited better heat resistance compared to raw phenolic resin.

When clay as filler was introduced into poly(GMA) nanocomposites, it significantly improved their thermal stability (Jlassi et al., 2015). According to Devi and Maji (2012), the introduction of GMA-clay as a crosslinking agent improved the thermal stability of ST-acrylonitrile copolymer. The incorporation of clay improved the thermal stability of poly(ST-*b*-butadiene-ST) nanocomposites due to the strong interfacial bonding between the matrix and filler (Chen & Feng, 2008).

Smectite-rich clay as filler dispersed into the PLA matrix improves the thermal stability of the nanocomposites compared to the neat PLA (Mansa, Huang, Quintela, Rocha, & Detellier, 2015). The degradation rate of PLA-CAB-clay nanocomposite interface was lowered as the intercalated or partially exfoliated silicate platelets reduced the segmental motions of PLA and CAB (Nam et al., 2015). The nonactivated smectite clay mineral filler shortened the degradation time of the PLA-clay nanocomposites, which increased the potential for use in biomedical applications and disposable thermoplastic products (Rapacz-Kmita, Stodolak-Zych, Szaraniec, Gajek, & Dudek, 2015).

1.5.5.7 Water barrier properties of silica/fumed silica nanocomposites

The PVA-silica membrane composites increased the semipermeable water barrier effect, which enhanced the steam separation process (Bolto, Hoang, & Xie, 2012). According to Xie et al. (2011), the water flux decreased from 10 to 25 wt% when the silica content was added into PVA. This occurred due to the high silica content, which increased the hydrophobicity and thus reduced the water diffusion process.

According to Grama et al. (2014), the water absorption of poly(GMA)-silica composites decreased due to the interfacial bonding formation in the composites. Silica formed crosslinking with poly(ST-butyl acrylate-acrylic acid) that reduced the water absorption of the nanocomposites (Zhu, Cai, Yu, & Zhou, 2008).

The incorporation of silica powder extracted from rice husks into PLA biocomposites improved the water barrier properties (Battegazzore, Bocchini, Alongi, & Frache, 2014). Moreover, the addition of silica into melt-compounded PLA nanocomposites improved the water barrier effect of the nanocomposites due to the stabilization of PVA-silica network (Wen et al., 2011).

1.5.5.8 Water barrier properties of clay nanocomposites

The introduction of nanoclay into PVA-polyvinyl pyrrolidone-sodium MMT (PVA-PVP-Na^+-MMT) nanocomposites increased the tortuous paths, reducing the water permeability for diffusion (Mondal et al., 2015). In addition, solvent-casted PVA-bentonite clay nanocomposites improved the water barrier properties by reducing the water being absorbed into the nanocomposites (Sapalidis et al., 2011).

According to Anh, Hoa, Ngo, and Thai (2010), clay incorporated into ST contained vinyl ester resin increased the water resistance towards the nanocomposites. The clay platelets were fully intercalated in the PLA nanocomposites, increasing the tortuous paths, which decreased the water vapor permeability (Duan, Thomas, & Huang, 2013). The polymer barrier properties of PLA improved as the tortuosity increased, which prevented more water vapor from permeating (Delpouvet, Stoclet, Saitert, Dargent, & Marais, 2012). According to Lei and Chow (2011), the water barrier properties of the PLA nanocomposites could be significantly improved by the incorporation of OMMT and poly(ethylene glycol).

1.5.6 Techniques used to prepare polymer nanocomposites

There are many techniques for fabrication of polymer nanocomposites. The techniques that are commonly used for silica nanocomposites and clay nanocomposites include solution intercalation, condensation polymerization, free radical polymerization, solution-intercalation film-casting, in situ polymerization, sol-gel process, and melt blending (Lim et al., 2014; Zhang et al., 2015; Ejaz et al., 2014; Keshtkar et al., 2013).

1.5.6.1 Solution intercalation

Solution-intercalated PVA-MLDH nanocomposites showed improvement in thermal, physicomechanical and morphological properties compared to unmodified LDH (Mallakpour & Dinari, 2014). The addition of Na^+-MMT onto PVA prepared via solution intercalation method improved the water absorption of the nanocomposites (Lim et al., 2014).

1.5.6.2 Condensation polymerization

Aniline-formaldehyde copolymers synthesized by a condensation polymerization process in an acidic medium can be applied as corrosion inhibitors for mild steels in hydrochloric acid (Zhang et al., 2015). Condensation polymerization was applicable to phenolic resin composites, which improved the dispersion in the composites (Zhang, Lou, Ilias, Krishnamachari, & Yan, 2008).

1.5.6.3 Free radical polymerization

ST and GMA were synthesized by free radical polymerization with the introduction of benzoyl bromide as initiator on the boron nitride nanotubes (Ejaz et al., 2014). ST was polymerized with GMA through free radical polymerization with the introduction of an initiator that improved the grafting yield on other polymers (Scorah, Dhib, & Penlidis, 2006).

1.5.6.4 Solution-intercalation film-casting

PVA-silica organic-inorganic membrane was prepared using functionalized mercapto and amine groups by solution casting method for adsorption of Cu(II) ions from aqueous solutions (Keshtkar et al., 2013). According to Sapalidis et al. (2011), PVA-bentonite nanocomposites was well dispersed by a solvent casting technique, which improved the mechanical, thermal and water barrier properties. Solution-intercalation film-casting method was applied to fabricate biodegradable poly(D,L-lactic acid)-layered silicate nanocomposites that improved the thermal stability of the nanocomposites (Stathokopoulou & Tarantili, 2014).

It would be attractive to prepare polymer nanocomposites via different techniques using silica/fumed silica and nanoclay as fillers to improve the properties of the conventional polymer nanocomposites. This study aims to investigate the feasibility of preparing polymer nanocomposites with different types of polymer matrices and fillers with various technique approaches.

1.6 Conclusion

Polymer nanocomposites covered in this chapter include: PVA/fumed silica/clay (PVA-fsi-clay) nanocomposites, PVA/silica/clay (PVA-si-clay) nanocomposites, PF/fumed silica/clay (PF-fsi-clay) nanocomposites, ST-*co*-GM/fumed silica/clay (ST-*co*-GMA-fsi-clay) nanocomposites, and PLA/fumed silica/clay (PLA-fsi-clay) nanocomposites. All these nanocomposites were prepared via solution-intercalation film-casting, solution intercalation, condensation polymerization, and free radical polymerization. The physical, mechanical, thermal, and morphological analysis could be characterized by FTIR, XRF, SEM, adsorption isotherm (BET), tensile testing, TGA, and moisture absorption. With the introduction of fillers such as silica/fumed silica and nanoclay, the polymer nanocomposites showed better properties compared to conventional polymer composites.

Acknowledgments

The authors are grateful for the support of the Ministry of Higher Education Malaysia, Grant No. ERGS/02 (08)/860/2012 (12).

References

Achary, P. S., & Ramaswamy, R. (1998). Reactive compatibilization of nitrile rubber/phenolic resin blend: effect on adhesive and composite properties. *Journal of Applied Polymer Science*, *69*(6), 1187–1201.

Ahmad, A., Rafatullah, M., Suliman, O., Hakimi Ibrahim, M., Yap, Y. C., & Siddique Bazlul, M. (2009). Removal of Cu(II) and Pb(II) ions from aqueous solutions by adsorptions on sawdust of meranti wood. *Desalination*, *247*(1–3), 636–646.

Alexandre, M., & Dubois, P. (2000). Polymer-layered silicate nanocomposites: preparation, properties and uses of a new class of materials. *Materials Science & Engineering R: Reports*, *28*(1–2), 1–63.

Allison, P. G., Moser, R. D., Chandler, M. Q., Caminero-Rodriguez, J. A., Torres-Cancel, K., Rivera, O. G., et al. (2015). Mechanical, thermal, and microstructural analysis of polyvinyl alcohol/montmorillonite nanocomposites. *Journal of Nanomaterials*, *2015*, 1–9.

Anh, T. K. B., Hoa, S. V., Ngo, T. D., & Thai, M. T. T. (2010). *Effect of addition of nanoclays on properties of vinyl ester, design, manufacturing and applications of composites.* Canada: DEStech Publications.

Asghari, S., Hatami, M., & Ahmadipour, M. (2015). Preparation and investigation of novel PVA/silica nanocomposites with potential application in NLO. *Polymer-Plastics Technology and Engineering*, *54*(2), 192–201.

Ash, M. (2004). *Handbook of green chemicals.* Endicott, NY: Synapse Info Resources.

Auras, R., Lim, L. T., Selke, S. E. M., & Tsuji, H. (2010). *Poly(lactic acid): Synthesis, structures, properties, processing, and applications.* New York: John Wiley.

Ayana, B., Suin, S., & Khatua, B. B. (2014). Highly exfoliated eco-friendly thermoplastic starch (TPS)/poly(lactic acid)(PLA)/clay nanocomposites using unmodified nanoclay. *Carbohydrate Polymers*, *110*, 430–439.

Bai, Y., Huang, Z. H., & Kang, F. (2014). Electrospun preparation of microporous carbon ultrafine fibers with tuned diameter, pore structure and hydrophobicity from phenolic resin. *Carbon*, *66*, 705–712.

Bao, Z., Tao, J., & Flanigan, C. (2015). The combination of montmorillonite and silica in styrene-butadiene rubber/polybutadiene rubber tread compounds. *Polymer Composites.* https://doi.org/10.1002/pc.23653.

Basilissi, L., Silvestro, G. D., Farina, H., & Ortenzi, A. (2013). Synthesis and characterization of PLA nanocomposites containing nanosilica modified with different organosilanes II: effect of the organosilanes on the properties of nanocomposites: thermal characterization. *Journal of Applied Polymer Science, 128*(5), 3057–3063.

Battegazzore, D., Bocchini, S., Alongi, J., & Frache, A. (2014). Rice husk as bio-source of silica: preparation and characterization of PLA-silica biocomposites. *RSC Advances, 4*(97), 54703–54712.

Bayramoglu, G., Senkali, B. F., & Arica, M. Y. (2013). Preparation of clay-poly(glycidyl methacrylate) composite support for immobilization of cellulose. *Applied Clay Science, 85*, 88–95.

Bhattacharya, M., & Chaudhry, S. (2013). High-performance silica nanoparticle reinforced poly(vinyl alcohol) as templates for bioactive nanocomposites. *Materials Science & Engineering C: Materials for Biological Applications, 33*(5), 2601–2610.

Bilotti, E. (2009). *Polymer/sepiolite clay nanocomposites*. London: University of London.

Bitinis, N., Verdejo, R., Maya, E. M., Espuche, E., Cassagnau, P., & Lopez-Manchado, M. A. (2012). Physicochemical properties of organoclay filled polylactic acid/natural rubber blend bionanocomposites. *Composites Science and Technology, 72*(2), 305–313.

Bolto, B., Hoang, M., & Xie, Z. (2012). A review of water recovery by vapour permeation through membrane. *Water Research, 46*(2), 259–266.

Camargo, P. H. C., Satyanarayana, K. G., & Wypych, F. (2009). Nanocomposites: synthesis, structure, properties and new application opportunities. *Materials Research Bulletin, 12* (1), 1–39.

Chauhan, S. R., Patnaik, A., Kaith, B. S., Satapathy, A., & Dwivedy, M. (2009). Analaysis of mechanical behavior of phenol formaldehyde matrix composites using flax-g-poly (MMA) as reinforcing materials. *Journal of Reinforced Plastics and Composites, 28*, 1933–1944.

Chen, B. (2004). Polymer–clay nanocomposites: an overview with emphasis on interaction mechanisms. *British Ceramic Transactions, 103*(6), 241–249.

Chen, Z., & Feng, R. (2008). Preparation and characterization of poly(styrene-*b*-butadiene-styrene)/montmorillonite nanocomposites. *Polymer Composites, 30*(3), 281–287.

Chow, W. S. (2008). Tensile and thermal properties of poly(butylene terephtalate)/organo-montmorillonite nanocomposites. *Malaysian Polymer Journal, 3*(1), 1–13.

Chu, C. (2010). *Development of polymer nanocomposites for automotive applications*. Atlanta, GA: Georgia Institute of Technology.

Conradi, M. (2013). Nanosilica reinforced polymer composites. *Materials Technology, 47*(3), 285–293.

Das, K., Ray, D., Banerjee, I., Bandyopadhyay, N. R., Sengupta, S., Mohanty, A. K., et al. (2010). Crystalline morphology of PLA/clay nanocomposite films and its correlation with other properties. *Journal of Applied Polymer Science, 118*(1), 143–151.

Delpouvet, N., Stoclet, G., Saitert, A., Dargent, E., & Marais, S. (2012). Water barrier properties in biaxially drawn poly(lactic acid) films. *The Journal of Physical Chemistry B, 116*(15), 4615–4625.

Devi, R. R., & Maji, T. K. (2012). Chemical modification of simul wood with styrene-acrylonitrile copolymer and organically modified nanoclay. *Wood Science and Technology, 46*(1), 299–315.

Donia, A. M., Atia, A. A., & Rashad, R. T. (2011). Fast removal of Cu(II) and Hg(II) from aqueous solutions using kaolinite containing glycidyl methacrylate resin. *Desalination and Water Treatment, 30*(1–3), 254–265.

Dorfner, K. (1972). *Ion exchangers: Properties and applications*. Ann Arbor, MI: Ann Arbor Science.

Duan, Z., Thomas, N. L., & Huang, W. (2013). Water vapour permeability of polylactic acid nanocomposites. *Journal of Membrane Science*, *445*, 112–118.

Edoga, M. O., & Kovo, A. S. (2006). Development and characterization of phenol-formaldehyde molding powder. *Leonardo Electronic Journal of Practices and Technologies*, *5*(8), 41–48.

Ejaz, M., Rai, S. C., Wang, K., Zhang, K., Zhou, W., & Grayson, S. M. (2014). Surface-initiated atom transfer radical polymerization of glycidyl methacrylate and styrene from boron nitride nanotubes. *Journal of Materials Chemistry C*, *2*(20), 4073–4079.

El-Demerdash, A. M., Sadik, W. A., & El-Maghraby, A. H. (2008). Preparation, characterization, and lithographic applications of glycidyl methacrylate/methacrylic acid/t-butyl-4-vinylphenyl carbonate terpolymers. *Journal of Applied Polymer Science*, *108*(4), 2467–2471.

Fang, Z., & Suo, J. (2011). Synthesis and characterization of phenolic resol resin blended with silica sol and PVA. *Journal of Applied Polymer Science*, *119*(2), 744–751.

Feng, M., & Deng, J. (2004). Quantification of UF and PF resins in MDF fiber with an X-ray fluorescence spectrometer. *Wood and Fiber Science*, *36*(3), 337–343.

Fink, J. K. (2013). *Reactive polymers fundamentals and applications: A concise guide to industrial polymers* (2nd ed.). Norwich, NY: William Andrew Publishing.

Fonseca, C. A., Strawhecker, K. E., Manias, E., & Harrison, I. R. (2000). *Poly(vinyl alcohol)/sodium montmorillonite nanocomposites*. Boca Raton, FL: CRC Press.

Gao, J., Jiang, C., & Su, X. (2010). Synthesis and thermal properties of boron-nitrogen containing phenol-formaldehyde resin/MMT nanocomposites. *International Journal of Polymeric Materials and Polymeric Biomaterials*, *59*(8), 544–552.

Garg, M. S., & Srivastaba, D. (2014). Effect of glycidyl methacrylate (GMA) content on thermal and mechanical properties of ternary blend systems based on cardanol-based vinyl ester resinm styrene and glycidyl methacrylate. *Progress in Organic Coatings*, *77*(7), 1208–1220.

Garlotta, D. (2002). A literature review of poly(lactic acid). *Journal of Polymers and the Environment*, *9*(2), 63–84.

Grama, S., Boiko, N., Bilyy, R., Klyuchivska, O., Antonyuk, V., Stoika, R., et al. (2014). Novel fluorescent poly(glycidyl methacrylate)-silica microspheres. *European Polymer Journal*, *56*, 92–104.

Greesh, N. (2011). *Preparation of polymer-clay nanocomposites via dispersion polymerization using tailor-made polymeric surface modifiers*. South Africa: University of Stellenbosch.

He, H., Tao, Q., Zhu, J., Yuan, P., Shen, W., & Yang, S. (2013). Silylation of clay mineral surfaces. *Applied Clay Science*, *71*, 15–20.

Hedaco, R. K., Mahulikar, P. P., & Gite, V. V. (2013). Synthesis and characterization of resorcinol-based cross linked phenol formaldehyde microcapsules for encapsulation of pendimethalin. *Polymer-Plastics Technology and Engineering*, *52*(3), 243–249.

Herrera, N. N., Putaux, J. L., David, L., & Bourgeat-Lami, E. (2006). Polymer/laponite composite colloids through emulsion polymerization: influence of the clay modification level on particle morphology. *Macromolecules*, *39*(26), 9177–9184.

Hoidy, W. H., Ahmad, M. B., Al-Mulia, E. A. J., & Ibrahim, N. A. (2010). Preparation and characterization of polylactic acid/polycaprolactone clay nanocomposites. *Journal of Applied Sciences*, *10*(2), 97–106.

Hu, X., Cheng, W., Nie, W., & Wang, D. (2015). Flame retardant, thermal, and mechanical properties of glass fiber/nanoclay reinforced phenol-urea-formaldehyde foam. *Polymer Composites*. https://doi.org/10.1002/pc.23411.

Huang, J. W., Hung, Y. C., Wen, Y. L., Kang, C. C., & Yeh, M. Y. (2009). Polylactide/nano and microscale silica composite films. I. Preparation and characterization. *Journal of Applied Polymer Science*, *112*(3), 1688–1694.

Inamdar, S. S., & Rathod, K. K. (2013). Study of characteristics of partial replacement of phenol by coconut shell flour in a phenol formaldehyde resin. *International Journal of Research in Engineering and Technology*, *2*(9), 314–320.

Javier, G. R., Francisco, J. M. R., Carlos, A. O., Alma, P. E., & Saul, S. V. (2013). Structure/property relationships of poly(L-lactic acid)/mesoporous silica nanocomposites. *Journal of Polymers*, *2013*, 1–10.

Ji, G., Zhai, W., Lin, D., Ren, Q., Zheng, W., & Jung, D. W. (2013). Microcellular foaming of poly(lactic acid)/silica nanocomposites in compressed CO_2: critical influence of crystallite size on cell morphology and foam expansion. *Industrial and Engineering Chemistry Research*, *52*(19), 6390–6398.

Jia, X., Li, Y., Cheng, Q., Zhang, S., & Zhang, B. (2007). Preparation and properties of poly(vinyl alcohol)/silica nanocomposites derived from copolymerization of vinyl silica nanoparticles and vinyl acetate. *European Polymer Journal*, *43*(4), 1123–1131.

Jiang, L., Zhang, J., & Wolcott, M. P. (2007). Comparison of polylactide/nano-sized calcium carbonate and polylactide/montmorillonite composites: reinforcing effects and toughening mechanisms. *Polymer*, *48*(26), 7632–7644.

Jlassi, K., Chandran, S., Micusik, M., Benna-Zayani, M., Yagci, Y., Thomas, S., et al. (2015). Poly(glycidyl methacrylate)-grafted clay nanofiller for highly transparent and mechanically robust epoxy composites. *European Polymer Journal*, *72*, 89–101.

Jollands, M., & Gupta, R. K. (2010). Effect of mixing conditions on mechanical properties of polylactide/montmorillonite clay nanocomposites. *Journal of Applied Polymer Science*, *118*(3), 1489–1493.

Kaboorani, A., & Riedl, B. (2011). Effects of adding nano-clay on performance of polyvinyl acetate (PVA) as wood adhesive. *Composites Part A: Applied Science and Manufacturing*, *42*(8), 1031–1039.

Kang, S. G., Hong, J. H., & Kim, C. K. (2010). Morphology and mechanical properties of nanocomposites fabricated from organoclays and a novolac phenolic resin *via* melt mixing. *Industrial and Engineering Chemistry Research*, *49*(23), 11954–11960.

Karimi, A., & Wan Daud, W. M. A. (2015). Harmless hydrogels based on PVA/NA+–MMT nanocomposites for biomedical applications: fabrication and characterization. *Polymer Composites*. https://doi.org/10.1002/pc.23676.

Kato, M., Tsukigase, A., Usuki, A., Shimo, T., & Yazawa, H. (2006). Preparation and thermal properties of resole-type phenol resin-clay nanocomposites. *Journal of Applied Polymer Science*, *99*(6), 3236–3240.

Keshtkar, A. R., Irani, M., & Moosavian, M. A. (2013). Comparative study on PVA/silica membrane functionalized with mercapto and amine groups for adsorption of Cu(II) from aqueous solutions. *Journal of the Taiwan Institute of Chemical Engineers*, *44*(2), 279–286.

Kim, K., Seo, B., Lee, J. Y., Choi, B. J., Kwag, G. H., Paik, H. J., et al. (2015). Reduced filler flocculation in the silica-filled styrene-butadiene-glycidyl methacrylate terpolymer. *Composite Interfaces*, *22*(2), 137–149.

Kokabi, M., Sirousazar, M., & Hassan, Z. M. (2007). PVA-clay nanocomposite hydrogels for wound dressing. *European Polymer Journal*, *43*(10), 773–781.

Kong, Y., Zhong, Y., Shen, X., Gu, L., Cui, S., & Yang, M. (2013). Synthesis of monolithic mesoporous silicon carbide from resorcinol-formaldehyde/silica composites. *Materials Letters*, *99*, 108–110.

Ku, H., Yap, Y. S., Lee, T. S., & Trada, M. (2012). Tensile properties of phenol-formaldehyde nanoclay reinforced composites: a pilot study. *Journal of Applied Polymer Science, 126*(2), 142–150.

Kurian, M., & Sugunan, S. (2006). Wet peroxide oxidation of phenol over mixed pillared montmorillonites. *Chemical Engineering Journal, 115*(3), 139–146.

Lai, S. M., Chen, J. R., Han, J. L., Yu, Y. F., & Lai, H. Y. (2013). Preparation and properties of melt-blended polylactic acid/polyethylene glycol-modified silica nanocomposites. *Journal of Applied Polymer Science, 130*(1), 496–503.

Lai, S. M., Wu, S. H., Lin, G. G., & Don, T. M. (2014). Unusual mechanical properties of melt-blended poly(lactic acid) (PLA)/clay nanocomposites. *European Polymer Journal, 52*, 193–206.

LeBaron, P. C., Wang, Z., & Pinnavaia, T. J. (1999). Polymer-layered silicate nanocomposites: an overview. *Applied Clay Science, 15*(1–2), 11–29.

Lei, Y. Y., & Chow, W. S. (2011). Kinetics of water absorption and thermal properties of poly(lactic acid)/organomontmorillonite/poly(ethylene glycol) nanocomposites. *Journal of Vinyl and Additive Technology, 17*(1), 40–47.

Li, F., Wu, S., Wu, Y., Xu, R., & Li, G. (2010). Preparation of novel poly(vinyl alcohol)/SiO_2 composite nanofiber membranes with mesostructure and their application for removal of Cu^{2+} from waste water. *Chemical Communications, 46*(10), 1694–1696.

Li, Q., Chen, L., Zhang, J., Zheng, K., Zhang, X., Fang, F., et al. (2015). Enhanced mechanical properties, thermal stability of phenolic formaldehyde foam/silcia nanocomposites *via in situ* polymerization. *Polymer Engineering and Science, 55*(12), 2783–2793.

Liao, X., Nawaby, A. V., & Naguib, E. (2012). Porous poly(lactic acid) and PLA-nanocomposite structure. *Journal of Applied Polymer Science, 124*(1), 585–594.

Lightsey, J. W., Kneiling, D. J., Long, J. M., & Kolbert, A. C. (1999). Compatibilized silica and polymer silica-reinforced masterbatch containing same. United States Patent, 5985953.

Lim, M., Kim, D., Han, H., Khan, S. B., & Seo, J. (2014). Water sorption and water-resistance properties of poly(vinyl alcohol)/clay nanocomposite films: effects of chemical structure and morphology. *Polymer Composites, 36*(4), 660–667.

Lin, H. P., Chien, C. Y., Tang, C. Y., & Lin, C. Y. (2006). Synthesis of p6mm hexagonal mesoporous carbons and silicas using Pluronic F127-PF resin polymer blends. *Microporous and Mesoporous Materials, 93*(1–3), 344–348.

Lin, Y., Liu, L., Xu, G., Zhang, D., Guan, A., & Wu, G. (2015). Interfacial interactions and segmental dynamics of poly(vinyl acetate)/silica nanocomposites. *The Journal of Physical Chemistry C, 119*(3), 12956–12966.

Lu, Q., Chen, F., Shen, Q., Qin, Y., Huang, Z. X., & Zhang, L. M. (2014). Pyrolysis behaviour of boron-containing phenol-formaldehyde resin (BPFR) modified by B_2O_3. *Key Engineering Materials, 616*, 315–318.

Ma, H., Shi, T., & Song, Q. (2014). Synthesis and characterization of novel PVA/SiO_2-TiO_2 hybrid fibers. *Fibers, 2*(4), 275–284.

Maio, L. D., Garofalo, E., Scarfato, P., & Incarnato, L. (2015). Effect of polymer/organoclay composition on morphology and rheological and properties of polylactide nanocomposites. *Polymer Composites, 36*(6), 1135–1144.

Mallakpour, S., & Dinari, M. (2014). Novel bionanocomposites of poly(vinyl alcohol) and modified chiral layered double hydroxides: synthesis, properties and a morphological study. *Progress in Organic Coatings, 77*(3), 583–589.

Mallikarjuna, S. R. (2009). *Synthesis and properties of polymer/layered silicate clay nanocomposites*. India: University of Pune.

Mamantha, B. S., & Jagadish, R. L. (2013). Effect of nanoclay on water absorption and thickness swelling of wood based composite. *International Journal of Advanced Research in Science and Technology*, *2*(3), 160–164.

Mansa, R., Huang, C. T., Quintela, A., Rocha, F., & Detellier, C. (2015). Preparation and characterization of novel clay/PLA nanocomposites. *Applied Clay Science*, *115*, 87–96.

Martinez-Tong, D. E., Najar, A. S., Soccio, M., Nogales, A., Bitinis, N., Lopez-Manchado, M. A., et al. (2014). Quantitative mapping of mechanical properties in polylactic acid/natural rubber/organoclay bionanocomposites as revealed by nanoindentation with atomic force microscopy. *Composites Science and Technology*, *104*, 34–39.

Meneghetti, P., Shaikh, S., Qutubuddin, S., & Nazarenko, S. (2008). Synthesis and characterization of styrene-butadiene rubber-clay nanocomposites with enhanced mechanical and gas barrier properties. *Rubber Chemistry and Technology*, *81*(5), 821–841.

Mirzapour, A., Asadollahi, M. H., Baghshaei, S., & Akbari, M. (2014). Effect of nanosilica on the microstructure, thermal properties and bending strength of nanosilica modified carbon fiber/phenolic nanocomposite. *Composites Part A: Applied Science and Manufacturing*, *63*, 159–167.

Mondal, A. N., Cheng, C., Yao, Z., Pan, J., Hossain, M. M., Khan, M. I., et al. (2015). Novel quaternized aromatic amine based hybrid PVA membranes for acid recovery. *Journal of Membrane Science*, *490*, 29–37.

Myoung, S. H., Im, S. S., & Kim, S. H. (2015). Non-isothermal crystallation behavior of PLA/acetylated cellulose nanocrystal/silica nanocomposites. *Polymer International*, *65*(1), 115–124.

Nam, B. U., Min, K. D., & Son, Y. (2015). Investigation of the nanostructure, thermal stability, and mechanical properties of polylactic acid/cellulose acetate butyrate/clay nanocomposites. *Materials Letters*, *150*, 118–121.

Natali, M., Monti, M., Kenny, J., & Torre, L. (2011). Synthesis and thermal characterization of phenolic resin/silica nanocomposites prepared with high shear rate-mixing technique. *Journal of Applied Polymer Science*, *120*(5), 2632–2640.

Nekhamanurak, B., Patanathabutr, P., & Hongsriphan, N. (2012). Mechanical properties of hydrophilicity modified $CaCO_3$-poly(lactic acid) nanocomposites. *International Journal of Applied Physics and Mathematics*, *2*(2), 98–103.

OECD SIDS. (2002). *Screening information dataSet (SIDS) high production volume chemicals*. UNEP Chemicals: Switzerland.

Ogawa, M., Okumoto, S., & Kuroda, K. (1998). Control of interlayer microstructures of a layered silicate by surface modification with organochlorosilanes. *Journal of American Chemical Society*, *120*(29), 7361–7362.

Oksman, K., Skrifvars, M., & Selin, J. F. (2003). Natural fibres as reinforcement in polylactic acid (PLA) composites. *Composites Science and Technology*, *63*(9), 1317–1324.

Olad, A. (2011). *Polymer/clay nanocomposites, advances in diverse industrial applications of nanocomposites*. Iran: InTech.

Opaprakasit, P., Boonpa, S., Jaikaew, N., Petchsuk, A., & Tangboriboonrat, P. (2015). Preparation of surface-modified silica particles from rice husk ash and its composites with degradable polylactic acid. *Macromolecular Symposia*, *354*(1), 48–54.

Park, K. W. (2012). *Ion exchange and application of layered silicate*. Department of Chemistry and Research Institute of Natural Science: Republic of Korea.

Pirzada, T., Arvidson, S. A., Saquing, C. D., Shah, S. S., & Khan, S. A. (2012). Hybrid silica-PVA nanofibers *via* sol-gel electrospinning. *Langmuir*, *28*(13), 5834–5844.

Pukánszky, B. (2005). Interfaces and interphases in multicomponent materials: past, present, future. *European Polymer Journal*, *41*(4), 645–662.

Rangari, V. R., Hassan, T. A., Zhou, Y., Mahfuz, H., Jeelani, S., & Prorok, B. C. (2006). Cloisite clay-infused phenolic foam nanocomposites. *Journal of Applied Polymer Science, 103*(1), 308–314.

Rapacz-Kmita, A., Stodolak-Zych, E., Szaraniec, B., Gajek, M., & Dudek, P. (2015). Effect of clay mineral on the accelerated hydrolytic degradation of polylactide in the polymer/clay nanocomposites. *Materials Letters, 146*, 73–76.

Ray, S. S. (2012). Polylactide-based bionanocomposites: a promising class of hybrid materials. *Accounts of Chemical Research, 45*(10), 1710–1720.

Ray, S. S., & Okamoto, S. M. (2003). Polymer/layered silicate nanocomposites: a review from preparation to processing. *Progress in Polymer Science, 28*(11), 1539–1641.

Rescignano, N., Fortunati, E., Montesano, S., Emiliani, C., Kenny, J. M., Martino, S., et al. (2014). PVA bio-nanocomposites: a new take-off using cellulose nanocrystals and PLGA nanoparticles. *Carbohydrate Polymers, 99*, 47–58.

Robert, T. M., Chandran, M. S., Jishnu, S., Sunitha, K., Rajeev, R. S., Mathew, D., et al. (2015). Nanoclay modified silica phenolic composites: mechanical properties and thermal response under stimulated atmospheric re-entry conditions. *Polymers Advanced Technologies, 26*(1), 104–109.

Robertson, C. G., Lin, C. J., Bogoslovov, R. B., Rackaitis, M., Sadhukhan, P., Quinn, J. D., et al. (2011). Flocculation, reinforcement, and glass transition effects in silica-filled styrene-butadiene rubber. *Rubber Chemistry and Technology, 84*(4), 507–519.

Roelofs, J. C. A., & Berben, P. H. (2006). Preparation and performance of synthetic organoclays. *Applied Clay Science, 33*(1), 13–20.

Rosi, M., Iskandar, F., Abdullah, M., & Khairurrijal. (2014). Hydrogel-polymer electrolytes based on polyvinyl alcohol hydroxyethylcellulose for supercapacitor applications. *International Journal of Electrochemical Science, 9*, 4251–4256.

Sapalidis, A. A., Katsaros, F. K., Steriotis, T. A., & Kanellopoulos, N. K. (2011). Properties of poly(vinyl alcohol)-bentonite clay nanocomposite films in relation to polymer-clay interactions. *Journal of Applied Polymer Science, 123*(3), 1812–1821.

Scorah, M. J., Dhib, R., & Penlidis, A. (2006). Modelling of free radical polymerization of styrene and methyl methacrylate by tetrafunctional initiator. *Chemical Engineering Science, 61*(15), 4827–4859.

Sengwa, R. J., Choudhary, S., & Sankhla, S. (2010). Dielectric properties of montmorillonite clay filled poly(vinyl alcohol)/poly(ethylene oxide) blend nanocomposites. *Composites Science and Technology, 70*(11), 1621–1627.

Singh, P., & Ghosh, A. K. (2014). Torsional, tensile and structural properties of acrylonitrile-butadiene-styrene clay nanocomposites. *Materials and Design, 55*, 137–145.

Singla, P., Mehta, R., & Upadhyay, S. N. (2012). Clay modification by the use of organic cations. *Green and Sustainable Chemistry, 2*(1), 21–25.

Sirousazar, M., Kokabi, M., & Hassan, Z. M. (2011). Swelling behaviour and structural characteristics of polyvinyl alcohol/montmorillonite nanocomposite hydrogels. *Journal of Applied Polymer Science, 123*(1), 50–58.

Son, H. A., Yoon, K. Y., Lee, G. J., Cho, J. W., Choi, S. K., Kim, J. W., et al. (2015). The potential applications in oil recovery with silica nanoparticle and polyvinyl alcohol stabilized emulsion. *Journal of Petroleum Science and Engineering, 126*, 152–161.

Stathokopoulou, C., & Tarantili, P. A. (2014). Preparation, characterization and drug studies from poly(D,L-lactic acid)/organoclay nanocomposite films. *Journal of Macromolecular Science, Part A: Pure and Applied Chemistry, 52*(2), 117–124.

Streckova, M., Sopcak, T., Medvecky, L., Bures, R., Faberova, M., Batko, I., et al. (2012). Preparation, chemical and mechanical properties of microcomposites materials based on Fe powder and phenol-formaldehyde resin. *Chemical Engineering Journal, 180*, 343–353.

Studzinski, M., Jeziorska, R., Szadkowska, A., & Zielecka, M. (2014). Modified nano-silica-filled polypropylene composites with glycidyl methacrylate grafted ethylene/n-octene copolymer as compatibilizer. *Polimery, 59*(9), 625–635.

Sykorova, M., Janosova, V., Stroffekova, O., Kostalova, D., Havranek, E., & Rackova, L. (2009). Determination of selected elements by XRF and total phenolics in leaves and crude methanol extract of leaves of Arctostaphylos UVA-URSI. *Journal of Analytical Chemistry, 65*(1), 56–63.

Tanigami, T., Yano, K., Yamaura, K., & Matsuzawa, S. (1995). Anomalous swelling of poly(vinyl alcohol) film in mixed solvents of dimethylsulfoxide and water. *Polymer, 36* (15), 2941–2946.

Teng, M., Qiao, J., Li, F., & Bera, P. B. (2012). Electrospun mesoporous carbon nanofibers produced from phenolic resin and their use in the adsorption of large dye molecules. *Carbon, 50*(8), 2877–2886.

Thomas, S., Kuruvilla, J., Malhotra, S. K., Goda, K., & Sreekela, M. S. (2012). Advances in polymer composites: macro-and microcomposites-state of art, new challenges, and opportunities. *Polymer Composites: vol. 1* (pp. 1–16).

Turhan, Y., Alp, Z. G., Alkan, M., & Dogan, M. (2013). Preparation and characterization of poly(vinyl alcohol)/modified bentonite nanocomposites. *Microporous and Mesoporous Materials, 174*, 144–153.

Vidya, F. (2012). *Modification of linear low density polyethylene for improved photo and biodegradation*. Kochi, India: Cochin University of Science and Technology.

Wang, C., Wei, J., Xia, B., Chen, X., & He, B. (2012). Effect of nano-silica on the mechanical, thermal and crystalline properties of poly(vinyl alcohol)/nano-silica films. *Journal of Applied Polymer Science, 128*(3), 1652–1658.

Wei, C., Zeng, M., Xiong, X., Liu, H., Luo, K., & Liu, T. (2015). Friction properties of sisal fiber/nano-silica reinforced phenol formaldehyde composites. *Polymer Composites, 36* (3), 433–438.

Wen, X., Zhang, K., Wang, Y., Han, L., Han, C., Zhang, H., et al. (2011). Study of the thermal stabilization mechanism of biodegradable poly(L-lactide)/silica nanocomposites. *Polymer International, 60*(2), 202–210.

Wu, F., Zhang, B., Yang, W., Liu, Z., & Yang, M. (2014). Inorganic silica functionalized with PLLA chains *via* grafting methods to enhance the melt strength of PLLA/silica nanocomposites. *Polymer, 55*(22), 5760–5772.

Xie, Z., Hoang, M., Duong, T., Ng, D., Dao, B., & Gray, S. (2011). Sol-gel derived poly(vinyl alcohol)/maleic acid/silica hybrid membrane for desalination by pervaporation. *Journal of Membrane Science, 383*(1–2), 96–103.

Xiong, Z., Dai, X., Na, H., Tang, Z., Zhang, R., & Zhu, J. (2014). A toughened PLA/nanosilica composite obtained in the presence of epoxidized soybean oil. *Journal of Applied Polymer Science, 132*(1). https://doi.org/10.1002/app.41220.

Yaris, S., & Cross, H. (2002). *Organo-clay complexes and interactions*. New York/Basel: Marcel Dekker.

Yoshida, E. (2012). Selective controlled/living photoradical polymerization of glycidyl methacrylate, using an nitroxide mediator in the presence of a photosensitive triarylsulfonium salt. *Polymer, 4*(3), 1580–1589.

Yu, C. B., Wei, C., Lv, J., Liu, H. X., & Meng, L. T. (2012). Preparation and thermal properties of mesoporous silica/phenolic resin nanocomposites *via in situ* polymerization. *Express Polymer Letters*, *6*(10), 783–793.

Zaharri, N. D., Othamn, N., & Mohd Ishak, Z. A. (2013). Effect of zeolite modification *via* cationic exchange method on mechanical, thermal, and morphological properties of ethylene vinyl acetate/zeolite composites. *Advances in Materials Science and Engineering*, *2013*, 1–9.

Zhang, J., Lou, J., Ilias, S., Krishnamachari, P., & Yan, J. (2008). Thermal properties of poly(lactic acid) fumed silica nanocomposites: experiments and molecular dynamics simulations. *Polymer*, *49*(9), 2381–2386.

Zhang, X., & Tanaka, H. (2001). Copolymerization of glycidyl methacrylate with styrene and applications of the copolymer as paper-strength additive. *Journal of Applied Polymer Science*, *80*(3), 334–339.

Zhang, Y., & Le, Y. (2014). Structure and property of polyvinyl alcohol/precipitated silica composite hydrogels for microorganism immobilization. *Composites Part B: Engineering*, *56*, 749–755.

Zhang, Y., Nie, M., Wang, X., Zhu, Y., Shi, F., Yu, J., et al. (2015). Effect of molecular structure of aniline-formaldehyde copolymers on corrosion inhibition of mild steel in hydrochloric acid solution. *Journal of Hazardous Materials*, *289*, 130–139.

Zhao, Z., Zhao, X., Gong, G., Zheng, J., Liang, T., Yin, C., et al. (2012). Influence of particle type and silane coupling agent on properties of particle-reinforced styrene-butadiene rubber. *Polymer-Plastics Technology and Engineering*, *51*(3), 268–272.

Zhu, A., Cai, A., Yu, Z., & Zhou, W. (2008). Film characterization of poly(styrene-butylacrylate-acrylic acid)-silica nanocomposites. *Journal of Colloid and Interface Science*, *322*(1), 51–58.

Zuo, X. H., Li, W., Yu, Z. L., Ma, F. M., & Ruan, M. (2013). Tribological behavior of phenolic resin composites modified using tetraethyl orthosilicate. *Tribology Transactions*, *56*(1), 115–120.

Preparation and characterizations of silica and various clay dispersed nanocomposites

2

Md. Rezaur Rahman, Josephine Lai Chang Hui, Sinin bin Hamdan
Department of Chemical Engineering and Energy Sustainability, Faculty of Engineering, University Malaysia Sarawak (UNIMAS), Kota Samarahan, Sarawak, Malaysia

2.1 Materials

2.1.1 Polymers

In this research, the polymers such as polyvinyl alcohol (PVA), phenol formaldehyde (PF), styrene-*co*-glycidyl methacrylate (ST-*co*-GMA), and polylactic acid (PLA) were used as polymer matrix. The molecular weights of PVA, phenol, formaldehyde, ST, GMA, and PLA were 89,000–98,000 g/mol, 94 g/mol, 30 g/mol, 104 g/mol, 142 g/mol, and 60,000 g/mol respectively. All of the polymers were supplied by Sigma Aldrich (USA), except ST, which was supplied by J.T. Baker (USA).

2.1.2 Fillers

Glass sheet was obtained from Samajaya Industrial Park and was crushed into small pieces. The pieces were ground into silica powders. The silica powders were sieved to obtain a mesh size of 38 μm. Silicon dioxide powder (fumed silica) was also used in this study. The particle size of fumed silica was 8 μm. Four types of nanoclay, namely Nanomer 1.28E, 1.30E, 1.31PS, and 1.34TCN, were used throughout this research. The bulk density of all nanoclay was 200–500 kg/m^3 and the average particle size was around 20 μm. Nanomer 1.28E was montmorillonite clay modified with 25–30 wt% trimethylstearyl ammonium. Nanomer 1.30E was montmorillonite clay modified with 25–30 wt% octadecylamine. Nanomer 1.31PS was montmorillonite clay modified with 15–35 wt% octadecylamine and 0.5–5 wt% aminopropyl triethoxysilane. Nanomer 1.34TCN was montmorillonite clay modified with 25–30 wt% methyl dihydroxyethyl hydrogenated tallow ammonium. The aforementioned clays, silica and fumed silica were supplied by Sigma Aldrich (USA).

2.1.3 Chemicals

Sodium hydroxide and dichloromethane were used as solvent to prepare polymer nanocomposites. All the chemicals were supplied by Sigma Aldrich (USA). Benzoyl peroxide supplied by Merck, Germany, was used as initiator in ST-MMA-fsi-clay nanocomposites.

Silica and Clay Dispersed Polymer Nanocomposites. https://doi.org/10.1016/B978-0-08-102129-3.00002-6

2.2 Preparations of nanocomposites

2.2.1 Preparation of PVA-fsi-clay and PVA-si-clay nanocomposites by solution intercalation technique

The monomer system was prepared using PVA without any initiator. To produce PVA nanocomposites, 27.0 g of PVA, 0.5 g of nanoclay, and 2.5 g of silica/fumed silica were mixed as shown in Table 2.1.

PVA and clay were dried for 24 h in a vacuum oven at 50°C and 60°C respectively. The samples were prepared in a deionized water and methanol (3:1, *v*/*v*) solvent mixture. The mixture was mixed together with silica/fumed silica and stirred at 70°C for 24 h. Then, the homogeneous solutions were ultrasonicated in an ultrasonic bath at room temperature for 1 h. The solid films were produced by casting onto a glass mold. The untreated solvent was evaporated within 3 days. The films with a thickness of ~150 μm were placed in a vacuum oven at 50°C for 24 h to get rid of the extra solvent. The dried films were stored in a desiccator prior to any characterization.

2.2.2 Preparation of PF-fsi-clay nanocomposites by condensation polymerization technique

The composition of polymer was prepared using phenol and formaldehyde without any initiator. The ratio of various chemicals was shown in Table 2.2.

Nanocomposites were prepared through condensation polymerization technique. In a 500-mL beaker, 26.00 mL of 37% formaldehyde solution was placed and mixed thoroughly. Then, 20.00 mL of phenol was added into the mixture. After homogenous mixing, 1.00 g of silicon dioxide powder and 0.25 g of nanoclay were added simultaneously into the mixture. After the chemicals were mixed, 3.00 mL of concentrated sodium hydroxide was added into the mixture. The mixture was heated and the hot mixture was poured into a petri dish. The petri dish was then placed into an oven at 60°C. After 30 min, the mixture solidified and pink nanocomposite was formed. The nanocomposites were kept in a desiccator for further characterization and analysis.

Table 2.1 Preparation of PVA-fsi and PVA-si composites system with different types of clay loading

Amount of PVA (g)	Amount of fumed silica or silica (fsi or si) (g)	Amount and type of clay (g)
30	0	0
27	2.5	0.5 (1.28E)
27	2.5	0.5 (1.30E)
27	2.5	0.5 (1.31PS)
27	2.5	0.5 (1.34TCN)

Table 2.2 **Preparation of PF-fsi composites system with different types of clay loading**

Amount of phenol (mL)	Amount of formaldehyde (mL)	Amount of fumed silica (g)	Amount and type of clay (g)	Amount of sodium hydroxide (mL)
20.00	26.00	–	–	3.00
20.00	26.00	1.00	–	3.00
20.00	26.00	1.00	0.25 (1.28E)	3.00
20.00	26.00	1.00	0.25 (1.30E)	3.00
20.00	26.00	1.00	0.25 (1.31PS)	3.00
20.00	26.00	1.00	0.25 (1.34TCN)	3.00

Table 2.3 **Preparation of ST-*co*-GMA-fsi composites system with different types of clay loading**

Amount of styrene (mL)	Amount of glycidyl methacrylate (mL)	Amount of benzoyl peroxide (g)	Amount of fumed silica (g)	Amount of clay (g)
50	50	2	–	–
50	50	2	0.2	–
50	50	2	0.2	1.0 (1 wt%)
50	50	2	0.2	2.0 (2 wt%)
50	50	2	0.2	3.0 (3 wt%)
50	50	2	0.2	4.0 (4 wt%)

2.2.3 Preparation of ST-co-GMA-fsi-clay nanocomposites by free radical polymerization technique

The polymer nanocomposite was prepared using styrene, glycidyl methacrylate, fumed silica, and clay (1.30E) in four weight percentages in the presence of benzoyl peroxide. The composition of nanocomposites was prepared as shown in Table 2.3.

Benzoyl peroxide was added with 50 mL of styrene, 50 mL of glycidyl methacrylate along with 0.2 g of fumed silica power at different clay loadings (1.0, 2.0, 3.0, and 4.0 g). The mixture was heated for 10 min at 80°C. The mixture was then cast on a glass surface and kept in vacuum oven for a day for controlled evaporation of the solvent. The clear nanocomposite films with thickness ranging from 500 to 700 μm were obtained and kept in desiccator for further characterization and analysis.

2.2.4 Preparation of PLA-fsi-clay nanocomposites by solution-intercalation film-casting technique

The polymer nanocomposite was prepared using PLA, fumed silica, and four types of nanoclay in four different weight percentages. The composition of nanocomposites was prepared as shown in Table 2.4.

Quantities of 24.45, 23.70, and 22.45 g of PLA was dissolved, respectively, in 244.5 mL, 237.0 mL, and 224.5 mL of dichloromethane. Dichloromethane was added into four types of clay in four separated beakers to form clay dispersions. Both the PLA solution and clay suspension were ultrasonicated separately for 30 min at room temperature and subsequently mixed. Fumed silica was added before the final mixture was further sonicated for 30 min. The mixture was then cast on a glass surface and kept in a vacuum oven for 2 days for controlled evaporation of the solvent. The clear nanocomposite films with thickness ranging from 500 to 700 μm were obtained and subsequently dried at 80°C under vacuum for 2 days. The nanocomposites were kept in a desiccator for further characterization and analysis.

2.3 Fourier transform infrared spectroscopy analysis

The infrared spectra of the nanocomposite were recorded on a Shimadzu IRAffinity-1. The technique used was attenuated total reflection (ATR). The spectral resolution of this equipment was 12,000 spectra. The transmittance range of the scan was 4000 to 600 cm^{-1}.

2.4 X-ray fluorescence analysis

For each sample, scanning was conducted with a DP-6000 Delta Premium PXRF (Olympus, Waltham, MA, USA). The instrument features Rh X-ray tubes operated at 10–40 kV with integrated large area silicon drift detector (165 eV). The instrument was operated in Geochem Mode, which was able to detect various types of elements. Geochem Mode consists of two beams that operate sequentially. Each beam was set to

Table 2.4 **Preparation of PLA-fsi composites system with different types of clay and clay loadings**

Amount of PLA (g)	Amount of fumed silica (g)	Amount of clay (g)
24.45	0.05	0.50
23.70	0.05	1.25
22.45	0.05	2.50

scan for 30 s such that one whole scan was completed in 60 s. Prior to sample analysis, the X-ray fluorescence (XRF) was standardized with a "316" alloy clip. Each sample was scanned in triplicate to obtain an average elemental composition. The XRF was repositioned between each scan. Elemental data, along with the limit of detection was logged for each element.

2.5 Scanning Electron microscopy analysis

The interfacial bonding between the polymer matrices (PVA, PF, ST-*co*-GMA, and PLA) with fillers (fumed silica/silica and clay) were examined using a scanning electron microscope (SEM) (JSM-7610F) supplied by JEOL Company Limited, Japan. The accelerating voltage of SEM was 15 kV. The specimens were first coated with a thin layer of gold before being viewed microscopically. The micrographs were taken at magnifications ranging from 500 × to 1500 ×.

2.6 Adsorption isotherm

The nitrogen adsorption isotherms of PVA-fsi-clay, PVA-si-clay, PF-fsi-clay, ST-*co*-GMA-fsi-clay, and PLA-fsi-clay nanocomposites at 77 K were obtained using a Quantachrome, Asic-7 physicosorption analyzer. In this analysis, the nanocomposites were degassed at 250°C in a vacuum for 1 h before the nitrogen adsorption isotherm was constructed. The surface area and pore volume of nanocomposites were evaluated by the Brunauer-Emmett-Teller (BET) model. The surface area, S_{BET} was calculated using Eq. (2.1).

$$S_{BET} = \frac{V_m \times A_s \times N_A}{M_m} \tag{2.1}$$

where S_{BET} = specific surface area (m^2/g), V_m = monolayer value (cm^3/g), A_s = settlement area of a nitrogen molecule (m^2/mol), N_A = Avogadro number, and M_m = molecular weight of nitrogen (g/mol).

Pore volume, V_p was calculated using Eq. (2.2).

$$V_p = \left(\frac{V_N}{V_{av}}\right)\frac{M_m}{D_{ads}} \tag{2.2}$$

where V_p = physical volume of pores, V_N = volume of nitrogen (gaseous), V_{av} = adsorbed volume, M_m = molecular weight of nitrogen (g/mol), and D_{ads} = adsorbate density at the measurement temperature.

Pore radius, r_p was calculated using Kelvin equation as shown in Eq. (2.3).

$$r_p = \frac{2\delta V_v}{RT\ln\left(\frac{P}{P_0}\right)} \tag{2.3}$$

where r_p = pore radius, δ = surface tension, V_v = molar volume, R = universal gas constant, T = temperature (in Kelvin), and P/P_0 = relative pressure.

2.7 Tensile testing

Thin films were cut with a rectangular die. The gauge length, width, and thickness of the samples were 25, 4, and 0.15 mm, respectively. All the thin films were tested at room temperature using AG-X Plus Series Precision Universal Testers (300kN Floor Model) supplied by Shimadzu Corporation. The cross head speed used was 1 mm/min. The quoted results were averaged over four specimens. The results obtained were stored in Trapezium X testing software. The stored results could be outputted in Microsoft Excel file.

2.8 Thermogravimetric analysis

Thermogravimetric analysis (TGA) measurements were carried out on 5–10 mg PVA-fsi-clay, PVA-si-clay, PF-fsi-clay, ST-*co*-GMA-fsi-clay, and PLA-fsi-clay nanocomposites at a heating rate of 10°C/min in a nitrogen atmosphere using a Thermogravimetric Analyzer (TA Instrument SDT Q600). All the nanocomposites were subjected to TGA in high purity nitrogen under a constant flow rate of 5 mL/min. Thermal decomposition of each sample occurred in a programmed temperature range of 0–700°C. The continuous weight loss and temperature were recorded and analyzed. The activation energy was calculated from TGA graphs based on the Arrhenius equation as shown in Eq. (2.4).

$$k = Ae^{\frac{-E_a}{RT}} \tag{2.4}$$

where k = reaction rate constant, A = frequency factor for the reaction, E_a = activation energy, R = universal gas constant, and T = temperature (in Kelvin).

2.9 Moisture absorption test

The moisture absorption was carried out at 110°C for 3 h using electronic moisture balance (MOC-120H) supplied by Shimadzu Corporation, Kyoto, Japan. Dry nanocomposites were immersed in distilled water at 25°C. The nanocomposites were removed after immersion. All the samples were placed as flat as possible on the pan in order to ensure the heat was applied evenly to the nanocomposites during the measurement. After that, the nanocomposites were weighted while the heater lid was closed firmly. The measuring time was displayed in second (s) and changes in weight were displayed in percentage (%). The measurement data was outputted as Microsoft Excel file by transferring through a RS232C cable.

The moisture water absorbed, W_{ab} was calculated as shown in Eq. (2.5).

$$\text{Moisture absorbed percentage}, W_{ab}\,(\%) = \frac{\text{Weight of wet nanocomposites}\,(W_i) - \text{weight of dry nanocomposites}\,(W_f)}{\text{Weight of wet nanocomposites}\,(W_i)} \times 100 \tag{2.5}$$

2.10 Conclusion

The physical, mechanical, thermal, and morphological properties of polymer nanocomposites could be well fabricated through suitable techniques such as solution intercalation, condensation polymerization, free radical polymerization, and solution-intercalation film-casting and analyzed by using characterizations such as Fourier transform infrared spectroscopy, XRF, scanning electron microscopy, adsorption isotherm (BET), tensile testing, thermogravimetric analysis (TGA), and moisture absorption.

Acknowledgments

The authors are grateful for the support of the Ministry of Higher Education Malaysia, Grant No. ERGS/02 (08)/860/2012 (12).

Impact of nanoclay on physicomechanical, morphological, optical, BET isotherm, and thermal analysis of polyvinyl alcohol/fumed silica/ clay nanocomposites

3

Md. Rezaur Rahman, Josephine Lai Chang Hui, Sinin bin Hamdan
Department of Chemical Engineering and Energy Sustainability, Faculty of Engineering, University Malaysia Sarawak (UNIMAS), Kota Samarahan, Sarawak, Malaysia

3.1 Introduction

Nowadays, polyvinyl alcohol (PVA) is used as a binder in the preparation of ceramics. It can also be used as a copolymer for pyroelectric films for medical purposes (Szafran, Wisniewski, & Rokicki, 2004). The molecular structure of PVA is suitable to be used in membrane separation, drug delivery systems, artificial biomedical devices and fuel cell electrolytes. However, PVA is not applicable to the aqueous environment as it has the hydrophilic nature. The addition of filler materials into PVA matrix become polymer nanocomposite. This polymer nanocomposite will enhance the physicomechanical and thermal strength (Park, Peng, Gidley, Xue, & Pinnavaia, 2006).

Silicon dioxide, also known as fumed silica, has been largely used as filler to improve the properties of polymers. It can be defined as finely divided amorphous silicon dioxide particles produced by high temperature in an oxygen-hydrogen flame. Silicon dioxide has a large number of silanol groups which are not hydrogen-bonded. Silanol groups are isolated and distributed over the surface (Gun'ko, Zarko, Leboda, & Chibowski, 2001).

Clay is another type of filler that plays an important role in polymer composite production. Clays are defined as a group of materials with different crystal structures and mineral contents such as iron, magnesium, alkali metals, and cations (Garcia-Chavez et al., 2003). The most commonly used clay, namely montmorillonite, has been improving the physical and mechanical properties of nanocomposites (Yang, Tharappiwattananon, & Long, 1998). The modification of clay by ammonium ions with trihydroxyl groups provides more tethering points with polyurethane molecules, which enhances the mechanical properties of polymer composites.

Silica and Clay Dispersed Polymer Nanocomposites. https://doi.org/10.1016/B978-0-08-102129-3.00003-8

Additionally, nanoclay increases the tensile strength, tensile modulus, and elongation at break of the composites with the loading of nanoclay up to 3 wt%. The impact strength decreased with higher loading of nanoclay (Kord, 2012).

Polymer nanocomposites represent a very attractive route to upgrading and diversifying properties of the polymers. In modern research, polymer nanocomposites exhibit significant improvement on mechanical properties and thermal properties, which could be used for biomedical applications such as drug delivery systems. The addition of clay into polymers produced polymer-clay nanocomposites (PCNs). The advantages of PCNs include improved tensile modulus and strength, good thermal stability, and reduced gas permeability at low filler concentrations (Lu & Mai, 2005). Furthermore, the highly dispersed fumed silica enhances the thermal properties in polymer-silica composite systems (Torro-Palau, Fernandez-Garcia, Orgiles-Barcelo, & Martin-Martinez, 2001). The fumed silica nanopowder enhances the tensile strength, impact strength, and thermal stability (Zheng, Zheng, & Ning, 2003).

The present study investigated the effect of fumed silica and different types of clays on the properties of PVA-fsi-clay nanocomposites. The physicomechanical, morphological, and thermal properties of pure PVA and PVA-fsi-clay nanocomposites were characterized. The compatibility of different type of clays with PVA-fsi was also reported.

3.2 Results and discussion

3.2.1 Spectral analysis

The IR spectra of pure PVA and PVA-fsi-clay nanocomposites are shown in Fig. 3.1. The functional groups of ν(Si-O-Si), ν(Si-OH), and δ(Si-O-Si) represented the characteristic peaks at 1100–1000 cm^{-1}, 950–900 cm^{-1}, and 800–700 cm^{-1} respectively (Mansur, Orefice, & Mansur, 2004). The characteristic band 1650 and 1420 cm^{-1} were attributed to ν(C=C) and $\delta(CH_2)$ respectively. For the $\nu(CH_2)$ group, the IR spectrum was between 3100 and 2900 cm^{-1} (Perrira, Vasconcelis, & Orefice, 2000). According to the finding, partial loss of silanol groups bonded with silica particles and immobilized vinyl groups on the surface of functionalized silica nanoparticles were detected (Jia, Li, Cheng, Zhang, & Zhang, 2007). The spectrum 950–900 cm^{-1} represented the silanol group's bonds of silica nanoparticles. Due to the combination of clay and PVA coupled with silica nanoparticles, the amplitudes of Si-O-Si were increased compared with original silica nanoparticles. Besides, the hydroxyl group's peak at 3290 cm^{-1} was significantly reduced when stronger covalent bonds formed between matrix and fillers that reduced the pores, which prevented water molecules from penetrating into the polymer nanocomposites.

The characteristic bands of 3296 and 2958 cm^{-1} proved the ν(OH) and $\nu(CH_2)$ in PVA as shown in Fig. 3.1a. IR spectrum for PVA-fsi-clay (1.30E) nanocomposite was much broader compared to pure PVA and other PVA-fsi-clay nanocomposites. This was due to the fact that silica nanoparticles appended at the side chain hampered the

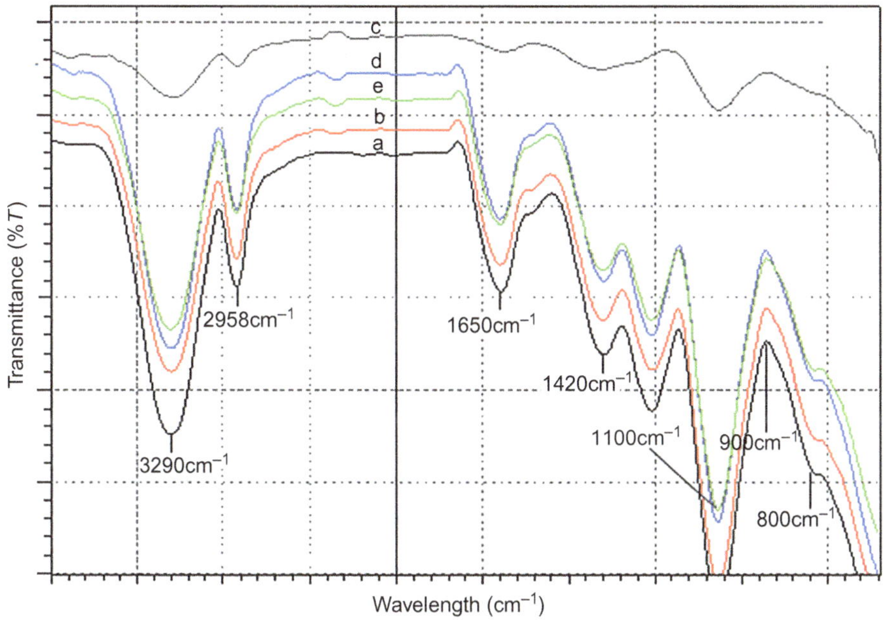

Fig. 3.1 FTIR spectra of (a) pure PVA, (b) PVA-fsi-clay (1.28E) nanocomposite, (c) PVA-fsi-clay (1.30E) nanocomposite, (d) PVA-fsi-clay (1.31PS) nanocomposite, and (e) PVA-fsi-clay (1.34TCN) nanocomposite.

formation of hydrogen bonds between PVA chains. The Fourier transform infrared spectroscopy (FTIR) results concluded that PVA and clay were compatible with fumed silica and had altered tri-dimensional structure (Jia et al., 2007).

3.2.2 Scanning electron microscopy (SEM) analysis

The scanning electron microscopy (SEM) micrographs of pure PVA and aggregation of silica and clay nanocomposites are shown in Fig. 3.2A–E. The flat surface with some cracking, as shown in Fig. 3.2A, indicated weak interfacial bonding of vinyl groups in pure PVA (Bhattacharya & Chaudhry, 2013). From Fig. 3.2B–E, some agglomeration between PVA, fumed silica, and clay occurred, which proved aggregation of fumed silica in the interspherulitic region. The uniform surface of PVA-fsi-clay (1.30E) and PVA-fsi-clay (1.31PS) nanocomposites proved the compatibility among the PVA-fsi with clay, which was shown in Fig. 3.2C and D. From Fig. 3.2B and E, it showed that PVA-fsi-clay (1.28E) and PVA-fsi-clay (1.34TCN) nanocomposites had agglomeration due to high surface energy and poor adhesion between clay and PVA-fsi. The poor compatibility of PVA-fsi with clay occurred due to the nonuniform surface of PVA-fsi, which could not react well with the clay particles (Luo, Zhao, Cai, & Du, 2012). Among all the nanocomposites, PVA-fsi-clay

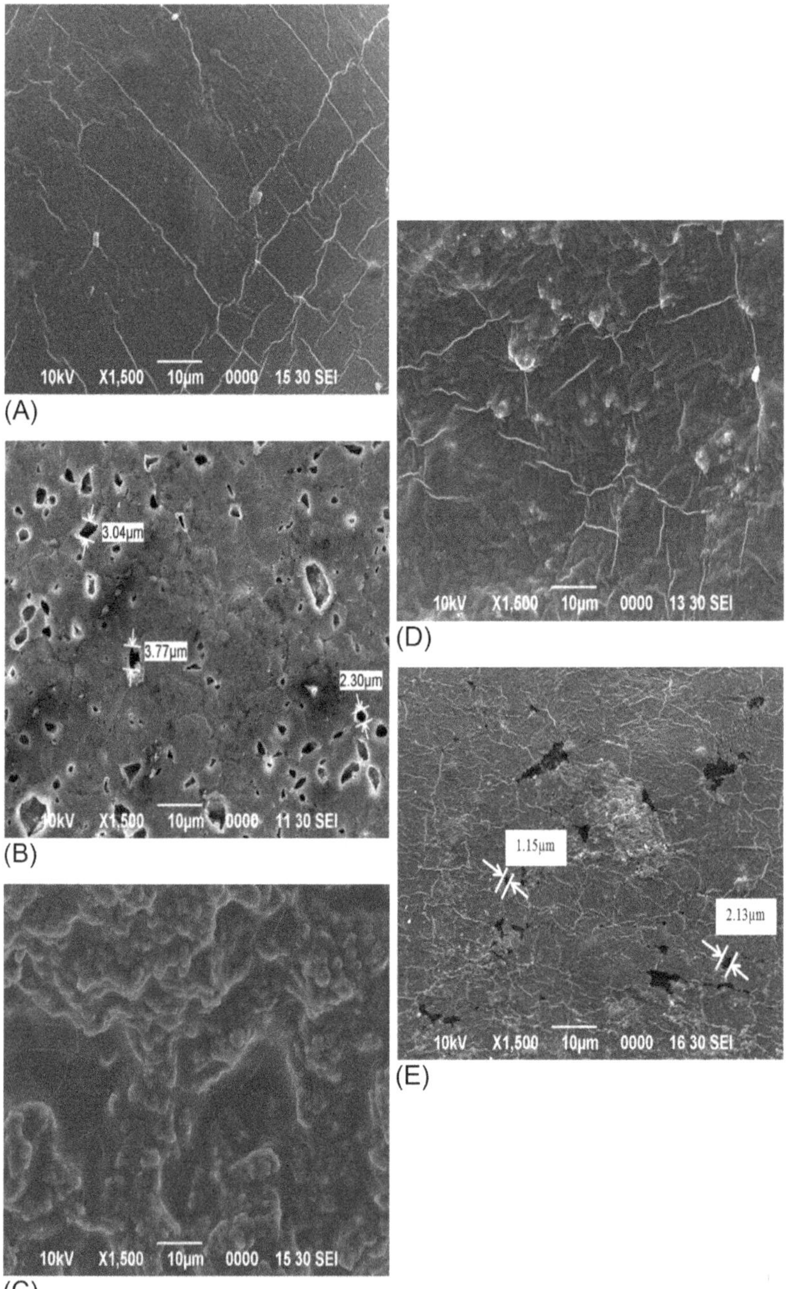

Fig. 3.2 SEM micrographs of (A) pure PVA, (B) PVA-fsi-clay (1.28E) nanocomposite, (C) PVA-fsi-clay (1.30E) nanocomposite, (D) PVA-fsi-clay (1.31PS) nanocomposite, and (E) PVA-fsi-clay (1.34TCN) nanocomposite.

(1.30E) nanocomposite had better interfacial bonding and strong compatibility, which was reflected in the improvement of the mechanical properties and thermal stability.

3.2.3 Adsorption isotherm analysis

The N_2 adsorption isotherms for pure PVA and PVA-fsi-clay nanocomposites are shown in Fig. 3.3. Specific surface area (S_{BET}) was calculated by the BET equation (Brunauer, Emmett, & Teller, 1938). The surface areas for the pure PVA and PVA-fsi-clay (1.28E), PVA-fsi-clay (1.30E), PVA-fsi-clay (1.31PS), and PVA-fsi-clay (1.34TCN) nanocomposites were found to be 1.93, 3.81, 468.4, 167.96, and 23.54 m^2/g, respectively, which indicated an increase in surface area of nanocomposites compared to pure PVA. Due to the good dispersion of clay within the PVA-fsi, the pores were decreased and thus enhanced their accessibility for nitrogen adsorption. The isotherm patterns showed the presence of a hysteresis loop, which was a characteristic feature of the type IV isotherms according to the original International Union of Pure and Applied Chemistry (IUPAC) classification (Sing et al., 1985). In the case of pure PVA, the isotherm initially showed an initial ascending section up to $P/P_0 = 0.107$. Afterwards it showed a rather straight section, which was extended up to $P/P_0 = 0.97$. Finally, the isotherm exhibited an upward sweep near saturation pressure. Similar isotherm patterns were observed in the PVA-fsi-clay (1.28E), PVA-fsi-clay (1.30E), PVA-fsi-clay (1.31PS), and PVA-fsi-clay (1.34TCN) nanocomposites, with the initial ascending section extended up to $P/P_0 = 0.107$, 0.103, 0.107, and 0.107, respectively. The enhancement in the N_2 adsorption at high P/P_o values was observed for PVA-fsi-clay (1.30E) nanocomposite. After the incorporation of clay into PVA-fsi, BET surface area and pore volume of PVA-fsi-clay (1.30E) and PVA-fsi-clay (1.31PS) nanocomposite greatly increased, while the pore size decreased as shown in Table 3.1

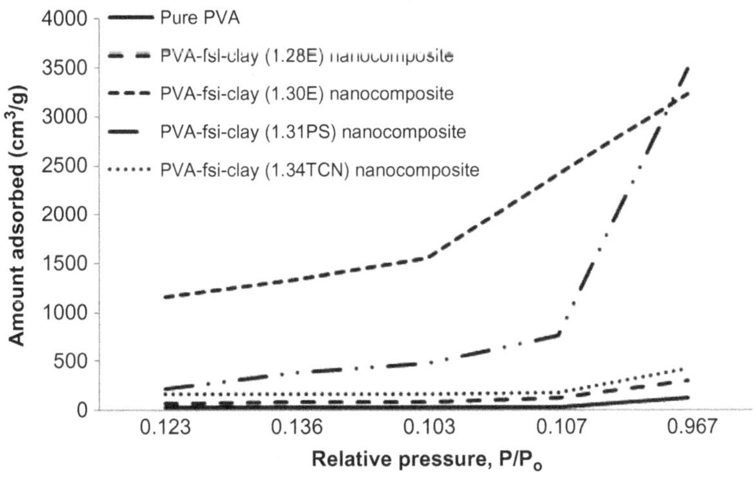

Fig. 3.3 N_2 adsorption isotherms of pure PVA and PVA-fsi-clay nanocomposites.

Table 3.1 Physical properties detected from N_2 adsorption at 77 K on pure PVA and PVA-fsi-clay nanocomposites

Samples	Specific surface area, S_{BET} (m^2/g)	Average pore volume, V_m (10^{-4} cm^3/g)	d_{BET} (nm)	Type of isotherms
Pure PVA	1.93	0.00008	2.00	IV
PVA-fsi-clay (1.28E) nanocomposite	3.81	0.00010	1.58	IV
PVA-fsi-clay (1.30E) nanocomposite	468.40	19.90000	1.57	IV
PVA-fsi-clay (1.31PS) nanocomposite	167.96	8.36000	1.57	IV
PVA-fsi-clay (1.34TCN) nanocomposite	23.54	1.99000	1.57	IV

(Hasan, Banerjee, & Lee, 2014). The adsorption isotherms indicated that the pores were mesoporous. It proved that PVA-fsi-clay (1.30E) nanocomposite had the highest surface area and average pore volume with less pore size, which was reflected in the surface morphology and thermal analysis.

3.2.4 Tensile properties

The tensile strength and tensile modulus of pure PVA and different types of PVA-fsi-clay nanocomposites are shown in Figs. 3.4 and 3.5, respectively. From Fig. 3.4, it could be observed that all the PVA-fsi-clay nanocomposites had a higher tensile strength compared to pure PVA. It showed that the tensile strength increased significantly when the clay was added to the PVA-fsi (Yang, Hu, Wang, Qin, & Guo, 2004).

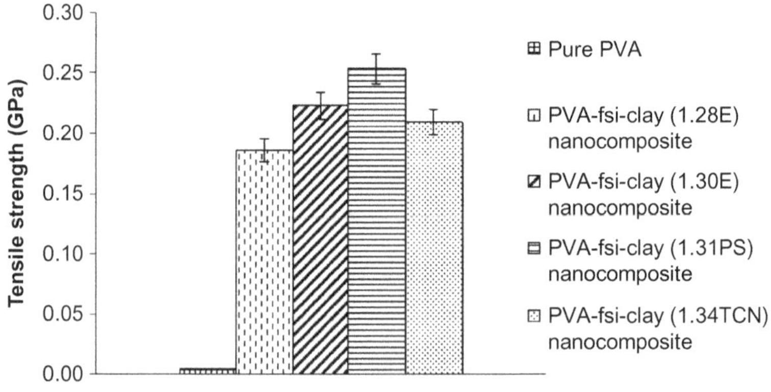

Fig. 3.4 Tensile strength of pure PVA and PVA-fsi-clay nanocomposites.

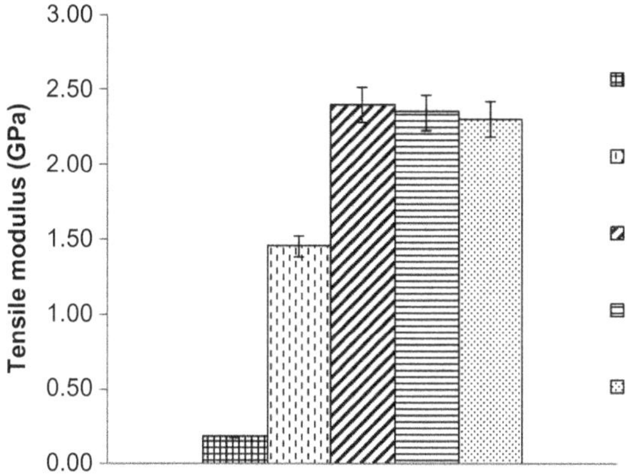

Fig. 3.5 Tensile modulus of pure PVA and PVA-fsi-clay nanocomposites.

In addition, silica was left unbound to PVA particles as the nanocomposites aggregate reduced the interparticle distance (Nakane, Yamashita, Iwakura, & Suzuki, 1999). Among the entire nanocomposites, PVA-fsi-clay (1.30E) nanocomposite showed the highest tensile strength compared with PVA-fsi-clay (1.28E), PVA-fsi-clay (1.31PS), and PVA-fsi-clay (1.34TCN) nanocomposites.

Fig. 3.5 shows that most of the PVA-fsi-clay nanocomposites, especially PVA-fsi-clay (1.30E) nanocomposite, had higher tensile modulus as the clay (1.30E) contained octadecylamine, which improved the compatibility of clay and PVA-fsi. The improved interfacial bonding could improve the load transfer from matrix to clay, which was attributed to the alkyl chains on the surface of fsi-clay (Yang et al., 2004). This proved the clay (1.30E) was compatible when added to the PVA-fsi, which enhanced the tensile strength and tensile modulus. The outcome of the better compatibility of PVA-fsi-clay (1.30E) nanocomposite was reflected in the thermal stability, as well as surface morphology.

3.2.5 Thermogravimetric analysis

The thermal stability of PVA-fsi-clay nanocomposites are shown in Fig. 3.6. The thermal stability of different clay loaded nanocomposites significantly increased compared to the pure PVA. Three steps in the thermal decomposition of pure PVA and PVA-fsi-clay nanocomposites could be observed. Initial temperature of every step was defined as a critical point of weight loss for the sample in the thermogravimetric analysis (TGA) curve (Liu, Zhang, & Zhou, 2013). This step was associated with the moisture loss or evaporation of trapped solvent.

According to the TGA thermograph, the weight loss was about 8.67%, 0.56%, 0.48%, 1.61%, and 1.49% for pure PVA, PVA-fsi-clay (1.28E), PVA-fsi-clay (1.30E), PVA-fsi-clay (1.31PS), and PVA-fsi-clay (1.34TCN) nanocomposites, respectively. Different weight losses were visible due to the removal of moisture. For second step degradation, the weight loss was found at 23.4% for pure PVA

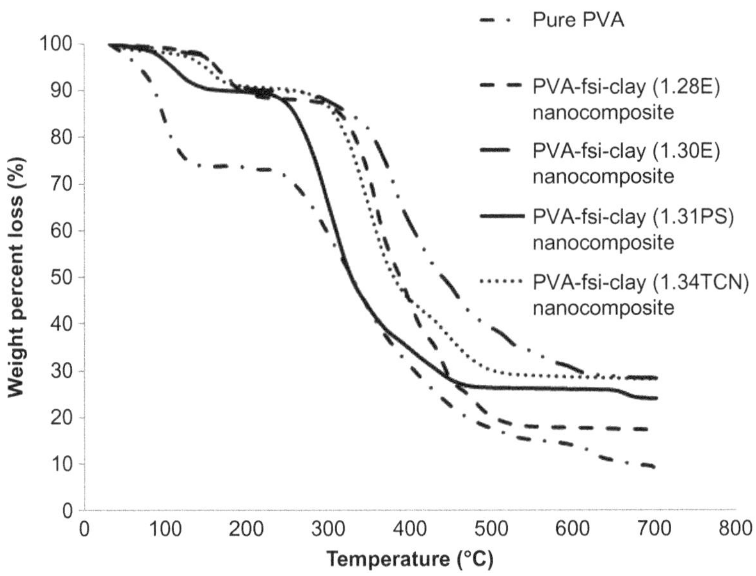

Fig. 3.6 TGA curves of pure PVA and PVA-fsi-clay nanocomposites.

and 10.0% for PVA-fsi-clay nanocomposites (Jia et al., 2007). Due to the restriction of mobility of polymer chains, different types of clay led to different weight loss during the decomposition. It showed that clay and silica were well intercalated with polymer matrix by covalent bonds (Jia et al., 2007). Well-distributed silicate layers of clay could prevent the passage of volatile decomposed product throughout the nanocomposites. The addition of clay and silica improved the thermal stability of nanocomposites as shown in Fig. 3.6 (Nourbakhsh, Karegarfard, Ashori, & Nourbakhsh, 2010).

At the final degradation which started at 250°C and ends at 430°C, the weight loss found for pure PVA, PVA-fsi-clay (1.28E), PVA-fsi-clay (1.30E), PVA-fsi-clay (1.31PS), and PVA-fsi-clay (1.34TCN) nanocomposites were 63.6%, 60.3%, 42.1%, 64.4%, and 54.5% respectively. The final weight of nanocomposites was significantly less than pure PVA due to the degradation temperature of silica itself, which was 360°C. The bonding compatibility of PVA-fsi and clay (1.30E) was higher than PVA-fsi-clay (1.28E), PVA-fsi-clay (1.30E), PVA-fsi-clay (1.31PS), and PVA-fsi-clay (1.34TCN) nanocomposites, as proven by FTIR result.

Activation energy is used to investigate the thermal stability of PVA-fsi-clay nanocomposites. Arrhenius equation was used to determine the activation energy (Chanmal & Jog, 2008). Table 3.2 summarizes the activation energy of pure PVA and PVA-fsi-clay nanocomposites determined by Arrhenius equation. It was found that the activation energy of PVA-fsi-clay nanocomposites was significantly higher than the pure PVA. The higher activation energy implied greater thermal stability. PVA-fsi-clay (1.30E) nanocomposite had the highest thermal stability among other PVA-fsi-clay nanocomposites.

Table 3.2 Activation energy of pure PVA and PVA-fsi-clay nanocomposites determined by Arrhenius equation

Samples	T_i (°C)[a]	T_m (°C)[b]	T_f (°C)[c]	W_{Ti} (%)[d]	W_{Tm} (%)[e]	W_{Tf} (%)[f]	Activation energy, E_a (kJ/mol)
Pure PVA	236.0	326.0	347.0	72.8	50.7	43.8	20.4
PVA-fsi-clay (1.28E) nanocomposite	280.0	347.0	410.0	89.1	66.0	43.5	25.6
PVA-fsi-clay (1.30E) nanocomposite	303.0	431.0	511.0	86.0	52.0	38.0	31.7
PVA-fsi-clay (1.31PS) nanocomposite	280.0	368.0	441.0	87.8	54.1	38.6	30.1
PVA-fsi-clay (1.34TCN) nanocomposite	258.0	303.0	368.0	85.3	61.8	39.0	28.6

[a] Temperature corresponding to the beginning of decomposition.
[b] Temperature corresponding to the maximum rate of mass loss.
[c] Temperature corresponding to the end of decomposition.
[d] Mass loss at temperature corresponding to the beginning of decomposition.
[e] Mass loss at temperature corresponding to the maximum rate of mass loss.
[f] Mass loss at temperature corresponding to the end of decomposition.

3.3 Conclusion

PVA-fsi-clay nanocomposites were prepared via solution intercalation. The reduction of hydroxyl groups and the dispersion of clay into PVA-fsi were confirmed by FTIR spectroscopy. From the surface morphology, it was found that the dispersion of clay improved the antiwetting property, which enhanced the compatibility of the nanocomposites. Surface analysis proved that the PVA-fsi-clay (1.30E) nanocomposites had higher surface area and pore volume with less pore size compared to other nanocomposites and pure PVA. The tensile strength and tensile modulus of clay (1.30E) loaded nanocomposites was the highest and confirmed the best compatibility of clay (1.30E) into PVA-fsi. Clay's dispersion in PVA-fsi nanocomposites increased the thermal stability significantly compared to pure PVA. Among the clay loaded nanocomposites, the thermal stability of PVA-fsi-clay (1.30E) nanocomposite was the highest with higher activation energy. It was concluded that the best compatibility of PVA-fsi-clay (1.30E) nanocomposite enhanced the thermal stability and surface morphology.

Acknowledgments

The authors are grateful for the support of the Ministry of Higher Education Malaysia, Grant No. ERGS/02 (08)/860/2012 (12).

References

Bhattacharya, M., & Chaudhry, S. (2013). High-performance silica nanoparticle reinforced poly(vinyl alcohol) as templates for bioactive nanocomposites. *Materials Science & Engineering C: Materials for Biological Applications*, *33*(5), 2601–2610.

Brunauer, B., Emmett, P. H., & Teller, E. (1938). Adsorption of gases in multimolecular layers. *Journal of the American Chemical Society*, *60*(2), 309–319.

Chanmal, C. V., & Jog, J. P. (2008). Dielectric relaxations in PVDF/$BaTiO_3$ nanocomposites. *eXPRESS Polymer Letter*, *2*(4), 294–301.

Garcia-Chavez, K. I., Hernandez-Escobar, C. A., Flores-Gallardo, S. G., Soriano-Corral, S., Saucedo-Salazar, E., & Zaragoza-Contreras, E. A. (2003). Morphology and thermal properties of clay/PMMA nanocomposites obtained by mini emulsion polymerization. *Micron*, *49*, 21–27.

Gun'ko, V. M., Zarko, V. I., Leboda, R., & Chibowski, E. (2001). Aqueous suspension of fumed oxides: particle size distribution and zeta potential. *Advances in Colloid and Interface Science*, *91*(1), 1–112.

Hasan, M., Banerjee, A. N., & Lee, M. (2014). Enhanced thermo-optical performance and high BET surface area of graphene@PVA nanocomposite fibers prepared by simple facile deposition technique: N_2 adsorption study. *Journal of Industrial and Engineering Chemistry*, *21*, 828–834.

Jia, X., Li, Y., Cheng, Q., Zhang, S., & Zhang, B. (2007). Preparation and properties of poly(vinyl alcohol)/silica nanocomposites derived from copolymerization of vinyl silica nanoparticles and vinyl acetate. *European Polymer Journal*, *43*(4), 1123–1131.

Kord, B. (2012). Effect of nanoparticles loading on properties of polymeric composite based on hemp fiber/polypropylene. *Journal of Thermoplastic Composite Materials*, *25*(7), 793–806.

Liu, M., Zhang, Y., & Zhou, C. (2013). Nanocomposites of halloysite and polylactide. *Applied Clay Science*, *75–76*, 52–59.

Lu, C., & Mai, Y. W. (2005). Influence of aspect ratio on barrier properties of polymer-clay nanocomposites. *Physical Review Letters*, *95*(8), 88–303.

Luo, Y., Zhao, J., Cai, Y., & Du, S. (2012). Effect of amino-functionalization on the interfacial adhesion of multi-walled carbon nanotubes/epoxy nanocomposites. *Materials Design*, *33*, 405–412.

Mansur, H. S., Orefice, R. L., & Mansur, A. A. P. (2004). Characterization of poly(vinyl alcohol)/poly(ethylene glycol) hydrogels and PVA-derived hybrids by small-angle X-ray scattering and FTIR spectroscopy. *Polymer*, *45*(21), 7193–7202.

Nakane, K., Yamashita, T., Iwakura, K., & Suzuki, F. (1999). Properties and structure of poly(vinyl alcohol)/silica composites. *Journal of Applied Polymer Science*, *74*(1), 133–138.

Nourbakhsh, A., Karegarfard, A., Ashori, A., & Nourbakhsh, A. (2010). Effects of particle size and coupling agent concentration on mechanical properties of particulate-filled polymer composites. *Journal of Thermoplastic Composite Materials*, *23*(2), 169–174.

Park, I., Peng, H. G., Gidley, D. W., Xue, S., & Pinnavaia, T. J. (2006). Epoxy-silica mesocomposites with enhanced tensile properties and oxygen permeability. *Chemistry of Materials*, *18*(3), 650–656.

Perrira, A. P. V., Vasconcelis, W. L., & Orefice, R. L. (2000). Novel multicomponent silicate-poly(vinyl alcohol) hybrids with controlled reactivity. *Journal of Non-Crystalline Solids*, *273*(1–3), 180–185.

Sing, K. S. W., Everett, D. H., Haul, R. A. W., Moscou, L., Pierotti, R. A., Rouquerol, J., et al. (1985). Reporting physisorption data for gas/solid systems with special reference to the determination of surface area and porosity (Recommendations 1984). *International Union of Pure and Applied Chemistry*, *57*(4), 603–619.

Szafran, M., Wisniewski, P., & Rokicki, G. (2004). Effect of glass transition temperature of polymeric binders on properties ceramic materials. *Journal of Thermal Analysis and Calorimetry*, *77*(1), 319–327.

Torro Palau, A. M., Fernandez-Garcia, J. C., Orgiles-Barcelo, A. C., & Martin-Martinez, J. M. (2001). Characterization of polyurethanes containing different silicas. *International Journal of Adhesion and Adhesives*, *21*(1), 1–9.

Yang, J., Hu, J., Wang, C., Qin, Y., & Guo, Z. (2004). Fabrication and characterization of soluble multi-walled carbon nanotubes reinforced P(MMA-*co*-EMA) composites. *Macromolecular Materials and Engineering*, *289*(9), 828–832.

Yang, R. T., Tharappiwattananon, N., & Long, R. Q. (1998). Ion-exchanged pillared clays for selective catalytic reduction of NO by ethylene in the presence of oxygen. *Journal of Applied Catalysis B: Environmental*, *19*(3–4), 289–304.

Zheng, Y., Zheng, Y., & Ning, R. (2003). Effects of nanoparticles SiO_2 on the performance of nanocomposites. *Materials Letters*, *57*(19), 2940–2944.

Polyvinyl alcohol/silica/clay nanocomposites: effect of clay on surface morphology, electrical and thermo-mechanical properties

4

Md. Rezaur Rahman, Josephine Lai Chang Hui, Sinin bin Hamdan
Department of Chemical Engineering and Energy Sustainability, Faculty of Engineering, University Malaysia Sarawak (UNIMAS), Kota Samarahan, Sarawak, Malaysia

4.1 Introduction

PVA is a widely used polymer due to its unique properties. PVA is a polyhydroxyl compound that has been applied in paper coating, articular cartilage, and thickener materials in paints. It has excellent compatibility and attractive properties (Singh & Singh, 2013). PVA has good mechanical and thermal properties that can be used in flexible water-soluble packaging films. However, the hydroxyl (—OH) groups attached in the PVA structure make it hydrophilic and soluble in water. Due to its hydrophilic nature, it can be degraded through hydrolysis reaction (Stevens, 2002). Nowadays, PVA nanocomposites have become one of the main research interests due to their potential applications in biomedical devices, drug delivery systems, carrier for cells immobilization, carrier for signaling molecules, and bioseparation membranes (Mansur, Orefice, & Mansur, 2004).

Silica is widely used as a pozzolanic admixture, which is effective in enhancing mechanical properties to a great extent. Silica is produced in the electric arc furnaces and treated as a byproduct from production of elemental silicon. Silica particles play an important role when embedded in a polymer matrix for their large surface area, as well as their smooth and nonporous surface (Conradi, 2013).

Clay is defined as a class of materials which make up layered silicates. The addition of MMT improved the physicomechanical properties of polymer nanocomposites (Yang, Tharappiwattananon, & Long, 1998). Organic alkyl exchanged clay is commonly used to enhance the hydrophobicity. It can help to expand the interlamellar spaces of silicate platelets (Hossain, Kim, Hwang, & Lim, 2009). The four types of nanoclay that are used to prepare the PVA nanocomposites include Nanoclay 1.28E, 1.30E, 1.31PS, and 1.34TCN. Nanoclay 1.28E helped to enhance tensile strength and thermal properties of the nanocomposites. Nanoclay 1.30E was designed to be easily dispersed in resins to form nanocomposites, and helps to improve

Silica and Clay Dispersed Polymer Nanocomposites. https://doi.org/10.1016/B978-0-08-102129-3.00004-X

performance at minimal added weight. Nanoclay 1.31PS contributed better dispersion, heat stability, and mechanical properties. Nanoclay 1.34TCN enhanced mechanical properties at low loadings and minimized the surface imperfections compared to the conventional fillers (Sigma Aldrich, 2000). However, the drawback was the incompatibility between hydrophilic clay and hydrophobic polymer, which often caused agglomeration of clay mineral in the polymer matrix (Zeng, Yu, Lu, & Paul, 2005).

PVA-clay composite hydrogels could be applied on wound dressing, as it meets the essential requirements of good mechanical properties (Kokabi, Sirousazar, & Hassan, 2007). Clay acting as a filler was added into PVA-starch, providing the maximum strength of the nanocomposites. However, the transparency of the nanocomposites decreased as the clay is added to the matrix. This proved that 1 wt% of clay is the optimum clay loading towards PVA-starch (Sadhu, Soni, Varmani, & Garg, 2014). The dispersion of MMT into pure PVA to form nanocomposites showed that the thermal stability decreased as the MMT loading increased. This proved that thermal stability of PVA-MMT nanocomposites increased as MMT acted as reinforcing particles. The nanocomposites showed significant improvement in tensile strength at lower clay loading compared to the pure PVA. However, the tensile strength decreased when the clay loading increased. This showed that there was limitation of the clay loading into PVA matrix (Allison et al., 2015).

Polymer-silica nanocomposite is another area that gained a lot of attraction from both academician and industrial researchers due to their excellent electrochemical performance, proton conductivity, higher tensile strength, and thermal stability. The scanning electron microscopy (SEM) micrographs of PVA-silica nanocomposites showed that silica dispersed in the PVA matrix without the large aggregation of particles. With the addition of silica into PVA matrix, PVA-silica nanocomposites were stronger and more useful than pure PVA. Furthermore, PVA-silica nanocomposites were thermally stable with high tensile strength by sol-gel method (Humayun, 2010).

In this study, micron sized silica was used and various clay were dispersed in the PVA matrix by solution casting method to produce PVA-si-clay nanocomposites. The physicomechanical, morphological, and thermal properties of PVA-si-clay nanocomposites were characterized and investigated. The compatibility among the clay, silica, and PVA were also determined.

4.2 Results and discussion

4.2.1 *Spectral analysis*

The Fourier transform infrared (FTIR) spectra of pure PVA and PVA-si-clay nanocomposites are shown in Fig. 4.1. The characteristic peaks between 1100–1000 cm^{-1} and 950–900 cm^{-1} were attributed to ν(Si-O-Si) and ν(Si-OH), respectively. The characteristic band 1650 and 1420 cm^{-1} were for ν(C=C) and $\delta(CH_2)$, respectively. The IR spectrum of 3100 to 2900 cm^{-1} was representing

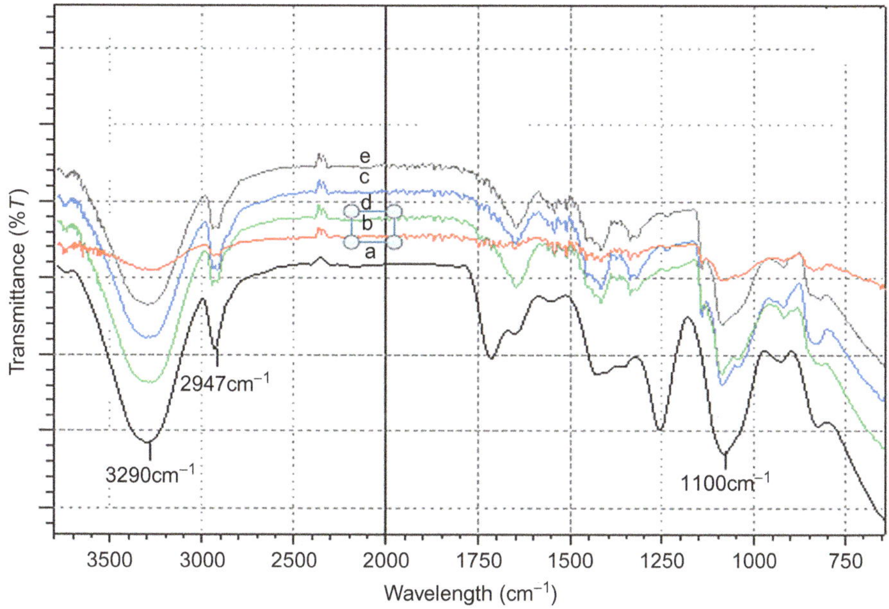

Fig. 4.1 FTIR spectra of (a) pure PVA, (b) PVA-si-clay (1.28E) nanocomposite, (c) PVA-si-clay (1.30E) nanocomposite, (d) PVA-si-clay (1.31PS) nanocomposite, and (e) PVA-si-clay (1.34TCN) nanocomposite.

$\nu(CH_2)$ (Perrira, Vasconcelis, & Orefice, 2000). The silanol groups and immobilized of vinyl groups were detected on the surface of the modified silica particles (Jia, Li, Cheng, Zhang, & Zhang, 2007).

From Fig. 4.1, the amplitudes for silanol group bonds of silica particles decreased on vinyl silica particles (950–900 cm^{-1}). However, the amplitudes of Si-O-Si increased compared to the original silica particles, as the clay and PVA had coupled with silica particles. All characteristic peaks proved that PVA and clay had reacted with silica.

The characteristic bands of pure PVA were ν(O—H) at 3290 cm^{-1} and $\nu(C—H_2)$ at 2947 cm^{-1}. The peaks between 3500 and 3000 cm^{-1} were mainly related to hydroxyl groups. It was noted that IR spectrum for PVA-si-clay (1.28E) nanocomposite had the least peak intensity percentage compared to the pure PVA, with differences of 4% at 3290 cm^{-1} and 2% at 2947 cm^{-1}. This confirmed that the incorporation of clay (1.28E) into PVA-si system created strong covalent bonds between polymer matrix and fillers, and led to the significant reduction of the hydroxyl groups. The peak intensity percentage of pure PVA, PVA-si-clay (1.30), PVA-si-clay (1.31PS), and PVA-si-clay (1.34TCN) nanocomposites at 3290 cm^{-1} were 32%, 28%, 30%, and 16%. At 2947 cm^{-1}, the peak intensity percentage for pure PVA, PVA-si-clay (1.30), PVA-si-clay (1.31PS), and PVA-si-clay (1.34TCN) nanocomposites were 6%, 6%, 6%,

Table 4.1 Oxide contents of pure PVA and PVA-si-clay nanocomposites.

Samples	Si	Ca	Fe	S	Cr	K	Cu
Pure PVA	–	30.3	36.6	–	–	–	32.9
PVA-si-clay (1.28E) nanocomposite	40.9	33.0	15.0	1.7	1.7	2.5	–
PVA-si-clay (1.30E) nanocomposite	32.7	32.3	26.3	4.4	2.2	1.7	–
PVA-si-clay (1.31PS) nanocomposite	31.4	34.0	22.9	1.7	2.2	2.1	–
PVA-si-clay (1.34TCN) nanocomposite	33.3	35.2	21.5	6.5	2.1	1.2	–

and 4%. This was due to the weakening of hydrogen bonds between PVA chains as the silica particles were attached at the side chain. Thus, the structure of PVA chains had changed (Jia et al., 2007). According to the findings, it proved that there was strong compatibility between PVA, silica, and clay in the nanocomposites.

4.2.2 X-ray fluorescence analysis

The compositions of the pure PVA and PVA-si-clay nanocomposites are presented in Table 4.1. The major quantities in the nanocomposites included silicon, calcium, and iron (Nayak & Singh, 2007). From the infrared and compositional analysis, the results indicated that the clay particles were compatible by fitting well into the PVA-si. Table 4.1 shows that pure PVA did not contain any silica particles but PVA-si-clay nanocomposites contained higher percentage of silica group. This confirmed the presence of a silanol group that was assigned to Si-O-Si stretching vibration. The clay mineral content in the nanocomposites was presented by the high percentage of K (Karunakan et al., 2013). Therefore, the appearance of silica groups improved the thermal and mechanical properties of the nanocomposites compared to the pure PVA.

4.2.3 Scanning electron microscopy analysis

The SEM micrographs of pure PVA and PVA-si-clay nanocomposites are shown in Fig. 4.2A–E. From Fig. 4.2A, the pure PVA displayed a surface with cracks. There was no intercalation on the pure PVA surface due to the absence of monomer intercalation (Perrira et al., 2000). From Fig. 4.2B–E, some agglomeration was spotted, which proved the silica aggregation in the interspherulitic region. Fig. 4.2B presents the uniform surface of PVA-si-clay (1.28E) nanocomposite due to the well intercalation between PVA-si with clay. Clay (1.30E) and clay (1.31PS) were dispersed

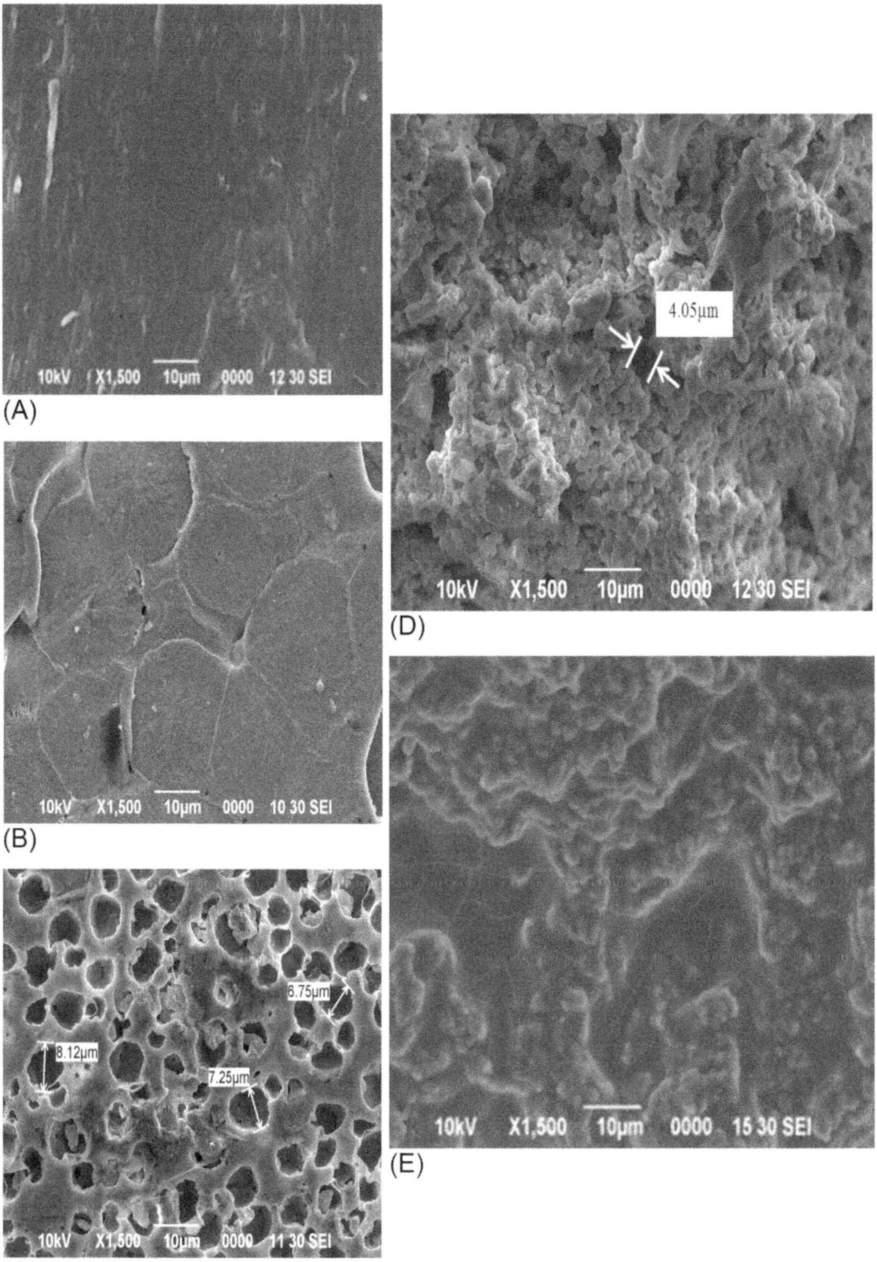

Fig. 4.2 SEM micrographs of (A) pure PVA, (B) PVA-si-clay (1.28E) nanocomposite, (C) PVA-si-clay (1.30E) nanocomposite, (D) PVA-si-clay (1.31PS) nanocomposite, and (E) PVA-si-clay (1.34TCN) nanocomposite.

unevenly on the PVA-si as shown in Figs. 4.2C and D, respectively. The high surface energy, inert surface of clay particles, and the poor adhesion between clay and PVA-si caused agglomerates on the nanocomposites' surfaces (Luo, Zhao, Cai, & Du, 2012). Fig. 4.2E shows the compatibility of the PVA-si with clay in nanocomposites. However, the dispersion of silica or clay was nonuniform in the nanocomposite film. The strong interfacial bonding between PVA-si and clay (1.28E) showed higher mechanical and thermal properties compared to PVA-si-clay (1.30E), PVA-si-clay (1.31PS) and PVA-si-clay (1.34TCN) nanocomposites.

4.2.4 Adsorption isotherm analysis

The N_2 adsorption isotherm of pure PVA and PVA-si-clay nanocomposites are shown in Fig. 4.3. Specific surface area (S_{BET}) was calculated by BET equation (Brunauer, Emmett, & Teller, 1938). The surface areas of PVA-si-clay nanocomposites increased compared to pure PVA. This was due to the addition of nanofiller, which contributed to greater interaction between polymer matrix and fillers (Ching, Rahman, Ching, Sukiman, & Chuah, 2015). From Fig. 4.3, PVA-si-clay nanocomposites showed type IV isotherms based on the original IUPAC classification (Sing et al., 1985). All the PVA samples initially showed an initial ascending section, continued with a rather straight section, and extended up to an upward sweep near saturation pressure. Fig. 4.3 showed that all type of nanocomposites were mesoporous. From Table 4.2, PVA-si-clay (1.28E) nanocomposite showed the highest increment in the N_2 adsorption at $P/P_0 = 0.996$. It exhibited the highest surface area and average pore volume with the reduction of pore size (Hasan, Banerjee, & Lee, 2014), which was reflected in the surface morphology.

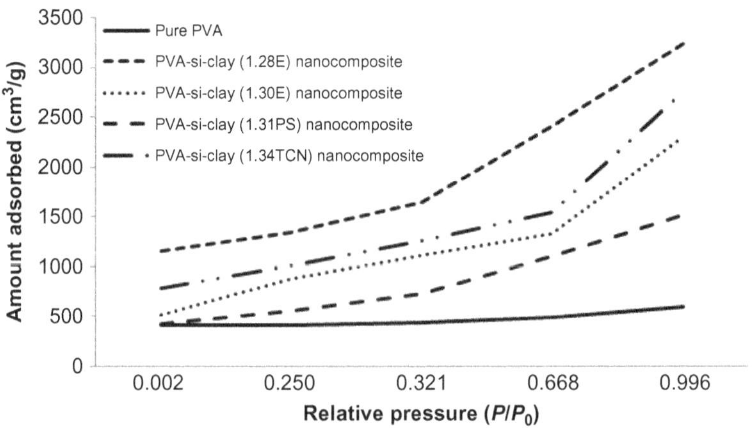

Fig. 4.3 N_2 adsorption isotherms of pure PVA and PVA-si-clay nanocomposites.

Table 4.2 Physical properties detected from N_2 adsorption at 77 K on pure PVA and PVA-si-clay nanocomposites

Samples	Specific surface area, S_{BET} (m^2/g)	Average pore volume, V_m (10^{-4} cm^3/g)	d_{BET} (nm)	Type of isotherms
Pure PVA	1.93	0.00008	2.00	IV
PVA-si-clay (1.28E) nanocomposite	11.16	9.12200	1.57	IV
PVA-si-clay (1.30E) nanocomposite	8.26	5.20900	1.58	IV
PVA-si-clay (1.31PS) nanocomposite	6.05	4.06000	1.58	IV
PVA-si-clay (1.34TCN) nanocomposite	8.86	8.36100	1.57	IV

4.2.5 Tensile properties

Figs. 4.4 and 4.5 represent the tensile strength and tensile modulus of pure PVA and different types of PVA-si-clay nanocomposites respectively. From Fig. 4.4, PVA-si-clay (1.28E) and PVA-si-clay (1.30E) nanocomposites had higher tensile strength compared to both pure PVA and other PVA-si-clay nanocomposites. The addition

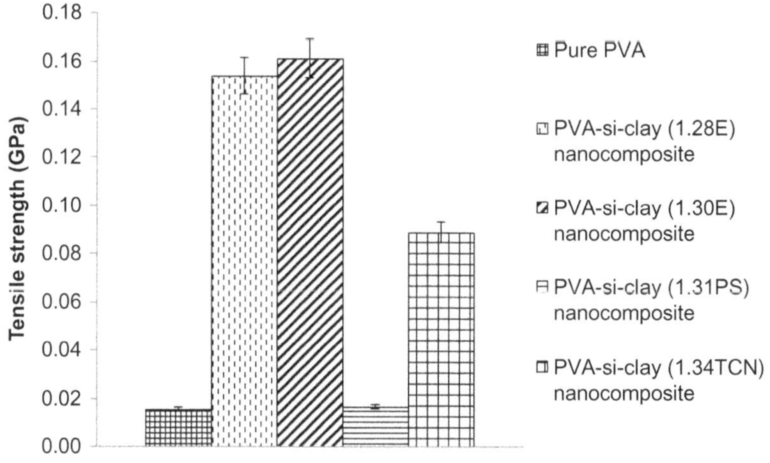

Fig. 4.4 Tensile strength of pure PVA and PVA-si-clay nanocomposites.

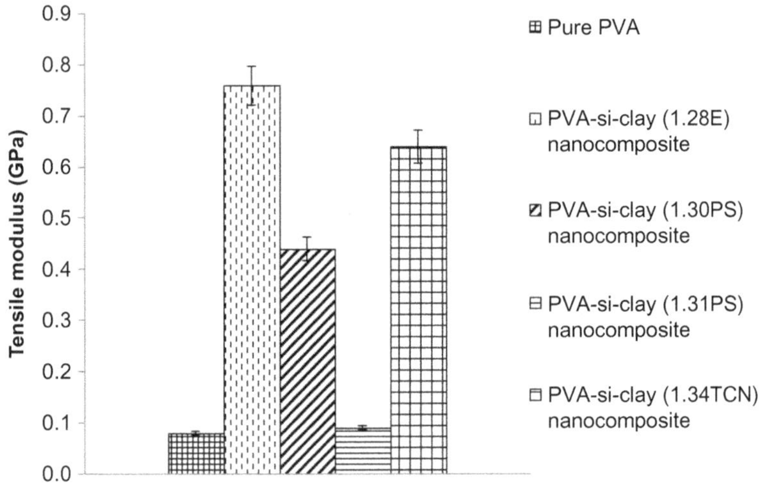

Fig. 4.5 Tensile modulus of pure PVA and PVA-si-clay nanocomposites.

of clay significantly improved the tensile strength of PVA-si-clay nanocomposites (Nakane, Yamashita, Iwakura, & Suzuki, 1999).

According to Fig. 4.5, PVA-si-clay (1.28E) nanocomposite had the highest tensile modulus compared to pure PVA and other PVA-si-clay nanocomposites. This was due to the clay (1.28E) content of trimethyl stearyl ammonium, which imparted anti-wetting properties and caused the surface of PVA-si-clay (1.28E) nanocomposite become hydrophobic (Seeni Meera, Sankar, Murali, Janiskar, & Mandal, 2012). Due to the hydrophobicity, the strong intercalation bonding between PVA-si and clay (1.28E) was formed. The addition of clay into PVA-si (1.28E) enhanced the compatibility between PVA-si and clay, which improved the tensile strength and tensile modulus. The strong compatibility of PVA-si-clay (1.28E) nanocomposite was also determined by thermal properties and surface morphology results.

4.2.6 Thermogravimetric analysis

Fig. 4.6 shows the thermal stability of pure PVA and PVA-si-clay nanocomposites. PVA-si-clay nanocomposites showed better thermal stability compared to pure PVA. Different types of clay led to different weight loss, as there was a restriction in polymer chain mobility and suppression of decomposition. The clay and silica were well intercalated with polymer matrix by covalent bonds. The presence of silica particles, along with clay in the polymer chain, prevented complete decomposition of PVA-si-clay nanocomposites' main chain (Jia et al., 2007).

According to the TGA (thermogravimetric analysis) thermographs, the first step degradation occurred at 85°C for pure PVA and PVA-si-clay nanocomposites. The weight loss for pure PVA, PVA-si-clay (1.28E), PVA-si-clay (1.30E), PVA-si-clay

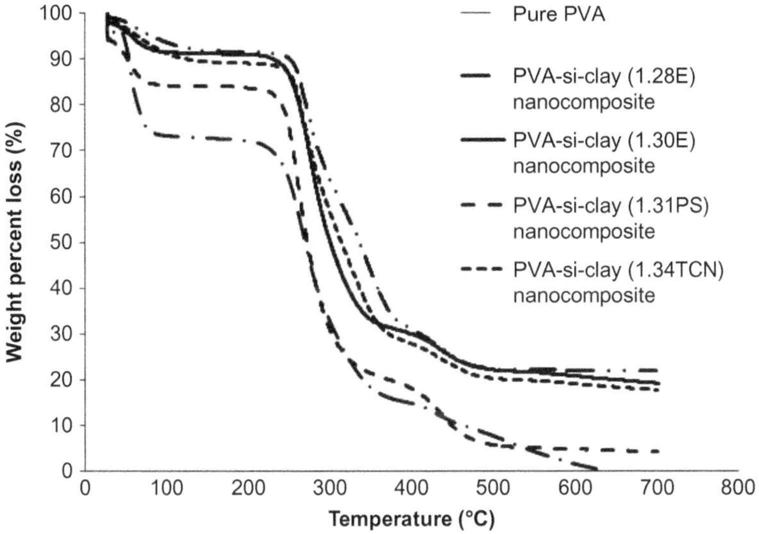

Fig. 4.6 TGA curves of pure PVA and PVA-si-clay nanocomposites.

(1.31PS), and PVA-si-clay (1.34TCN) nanocomposites were about 16%, 5%, 5%, and 8%, respectively, due to the moisture removal. The second step degradation occurred at 85–250°C. The weight loss was found to be 10% for PVA and 7% for PVA-si-clay nanocomposites. This indicated that the nanocomposites were more thermally stable compared to pure PVA due to the incorporation of silica and clay. The final degradation temperature was from 250°C to 430°C and weight loss was 8%, 1.5%, 1.3%, 1.3%, and 1.3% for pure PVA, PVA-si-clay (1.28E), PVA-si-clay (1.30E), PVA-si-clay (1.31PS), and PVA-si-clay (1.34TCN) nanocomposites, respectively. The final degradation weight of nanocomposites was drastically less than pure PVA, as the degradation temperature of silica itself was 360°C. The clays acted as barriers to maximize heat insulation (Chang, Jang, Ihn, & Sur, 2003). Clay helped to reduce the permeability of volatile products through the nanocomposites during degradation, and thus the weight loss of PVA nanocomposites were greatly reduced. The bonding compatibility of PVA-si and clay (1.28E), which was proven by tensile strength and tensile modulus, was higher than PVA-si-clay (1.30E), PVA-si-clay (1.31PS), and PVA-si-clay (1.34TCN) nanocomposites.

The activation energy of nanocomposites calculated by Arrhenius equation is summarized in Table 4.3 (Chanmal & Jog, 2008). The activation energy of PVA-si-clay nanocomposites, especially PVA-si-clay (1.28E) nancomposite, was significantly higher than pure PVA. Clay acted as excellent insulator, which prevented the heat transfer from an external source towards the PVA-si. Therefore, more heat was applied to burn the PVA-si-clay nanocomposites than pure PVA (Taghizadeh & Sabouri, 2013). This implied that clay loaded nancomposites had higher thermal stability.

Table 4.3 **Activation energy of pure PVA and PVA-si-clay nanocomposites determined by Arrhenius equation**

Samples	T_i (°C)[a]	T_m (°C)[b]	T_f (°C)[c]	W_{Ti} (%)[d]	W_{Tm} (°C)[e]	W_{Tf} (°C)[f]	**Activation energy, E_a (kJ/mol)**
Pure PVA	217.3	261.0	283.6	71.4	57.7	41.6	18.1
PVA-si-clay (1.28E) nanocomposite	267.5	300.8	350.0	79.7	57.5	35.2	40.9
PVA-si-clay (1.30E) nanocomposite	266.5	289.0	319.3	85.6	70.4	57.6	29.2
PVA-si-clay (1.31PS) nanocomposite	242.3	269.5	291.3	93.2	75.6	56.2	31.9
PVA-si-clay (1.34TCN) nanocomposite	240.0	264.7	285.8	81.2	62.2	41.3	40.6

[a] Temperature corresponding to the beginning of decomposition.
[b] Temperature corresponding to the maximum rate of mass loss.
[c] Temperature corresponding to the end of decomposition.
[d] Mass loss at temperature corresponding to the beginning of decomposition.
[e] Mass loss at temperature corresponding to the maximum rate of mass loss.
[f] Mass loss at temperature corresponding to the end of decomposition.

4.3 Conclusion

In this study, PVA-si-clay nanocomposites were prepared via solution intercalation method. The reduction of hydroxyl group and the dispersion of clay were confirmed by FTIR spectroscopy. The FTIR result proved that PVA and clay were compatible with silica and had altered PVA chains tri-dimensional structure. PVA-si-clay nanocomposites contained high percentage of S and K groups, which proved the presence of Si-O-Si stretching vibration and clay mineral content. SEM showed that strong interfacial bonding was formed between PVA-si and clay (1.28E). BET results proved that PVA-si-clay (1.28E) nanocomposite had the highest surface area and average pore volume with less pore size. PVA-si-clay (1.28E) nanocomposite showed higher tensile strength and tensile modulus compared to pure PVA, PVA-si-clay (1.30E), PVA-si-clay (1.31PS), and PVA-si-clay (1.34TCN) nanocomposites. The bonding compatibility of PVA-si and clay (1.28E) was higher than PVA-si-clay (1.30E), PVA-si-clay (1.31PS), and PVA-si-clay (1.34TCN) nanocomposites, respectively. PVA-si-clay nanocomposites showed higher thermal stability and activation energy compared to the pure PVA. According to the findings and the literature, PVA nanocomposites were more compatible towards silica and clays. The nanocomposites enhanced the physicomechanical and thermal properties compared to the pure PVA. All the nanocomposites fabricated were environmentally friendly with minimum production cost. These nanocomposites could be applied in biomedical devices, such as coating surface on small artery replacements, as well as functioning as an absorption membrane.

Acknowledgments

The authors are grateful for the support of the Ministry of Higher Education Malaysia, Grant No. ERGS/02 (08)/860/2012 (12).

References

Allison, P. G., Moser, R. D., Chandler, M. Q., Caminero-Rodriguez, J. A., Torres-Cancel, K., Rivera, O. G., et al. (2015). Mechanical, thermal, and microstructural analysis of polyvinyl alcohol/montmorillonite nanocomposites. *Journal of Nanomaterials*, *2015*, 1–9.

Brunauer, B., Emmett, P. H., & Teller, E. (1938). Adsorption of gases in multimolecular layers. *Journal of the American Chemical Society*, *60*(2), 309–319.

Chang, J. H., Jang, T. G., Ihn, K. J., & Sur, G. S. (2003). Poly(vinyl alcohol) nanocomposites with different clays: pristine clays and organoclays. *Journal of Applied Polymer Science*, *90*(12), 3204–3214.

Chanmal, C. V., & Jog, J. P. (2008). Dielectric relaxations in PVDF/$BaTiO_3$ nanocomposites. *eXPRESS Polymer Letter*, *2*(4), 294–301.

Ching, Y. C., Rahman, A., Ching, K. Y., Sukiman, N. L., & Chuah, C. H. (2015). Preparation and characterization of polyvinyl alcohol-based composite reinforced with nanocellulose and nanosilica. *BioResources*, *10*(2), 3364–3377.

Conradi, M. (2013). Nanosilica reinforced polymer composites. *Materials and Technology*, *47*(3), 285–293.

Hasan, M., Banerjee, A. N., & Lee, M. (2014). Enhanced thermo-optical performance and high BET surface area of graphene@PVA nanocomposite fibers prepared by simple facile deposition technique: N_2 adsorption study. *Journal of Industrial and Engineering Chemistry*, *21*, 828–834.

Hossain, M. D., Kim, W. S., Hwang, H. S., & Lim, K. T. (2009). Role of water on PMMA/clay nanocomposites synthesized by in situ polymerization in ethanol and supercritical carbon dioxide. *Journal of Colloid and Interface Science*, *336*(2), 443–448.

Humayun, M. (2010). *Synthesis and study of thin film composites of poly(vinyl alcohol)-silica and poly(vinyl alcohol)-alumina*. Pakistan: University of Peshawar.

Jia, X., Li, Y., Cheng, Q., Zhang, S., & Zhang, B. (2007). Preparation and properties of poly(vinyl alcohol)/silica nanocomposites derived from copolymerization of vinyl silica nanoparticles and vinyl acetate. *European Polymer Journal*, *43*(4), 1123–1131.

Karunakan, G., Suriyaprabha, R., Manivasakan, P., Yuvakkumar, R., Rajendran, V., Prabu, P., et al. (2013). Effect of nanosilica and silicon sources on plant growth promoting rhizobacteria, soil nutrients and maize seed germination. *IET Nanobiotechnology*, *7*(3), 70–77.

Kokabi, M., Sirousazar, M., & Hassan, Z. M. (2007). PVA-clay nanocomposite hydrogels for wound dressing. *European Polymer Journal*, *43*(10), 773–781.

Luo, Y., Zhao, J., Cai, Y., & Du, S. (2012). Effect of amino-functionalization on the interfacial adhesion of multi-walled carbon nanotubes/epoxy nanocomposites. *Materials Design*, *33*, 405–412.

Mansur, H. S., Orefice, R. L., & Mansur, A. A. P. (2004). Characterization of poly(vinyl alcohol)/poly(ethylene glycol) hydrogels and PVA-derived hybrids by small-angle X-ray scattering and FTIR spectroscopy. *Polymer*, *45*(21), 7193–7202.

Nakane, K., Yamashita, T., Iwakura, K., & Suzuki, F. (1999). Properties and structure of poly(vinyl alcohol)/silica composites. *Journal of Applied Polymer Science*, *74*(1), 133–138.

Nayak, P. S., & Singh, B. K. (2007). Instrumental characterization of clay by XRF, XRD and FTIR. *Bulletin of Materials Science*, *30*(3), 235–238.

Perrira, A. P. V., Vasconcelis, W. L., & Orefice, R. L. (2000). Novel multicomponent silicate-poly(vinyl alcohol) hybrids with controlled reactivity. *Journal of Non-Crystalline Solids*, *273*(1-3), 180–185.

Sadhu, S. D., Soni, A., Varmani, S. G., & Garg, M. (2014). Preparation of starch-polyvinyl alcohol (PVA) blend using potato and study of its mechanical properties. *International Journal of Pharmaceutical Science Invention*, *3*(3), 33–37.

Seeni Meera, K. M., Sankar, R. M., Murali, A., Janiskar, S. N., & Mandal, A. B. (2012). Sol–gel network silica/modified montmorillonite clay hybrid nanocomposites for hydrophobic surface coatings. *Colloids and Surfaces B: Biointerfaces*, *90*(1), 204–210.

Sigma Aldrich (2000). *Application data for nanoclay, surface modified*. Retrieved http://www.sigmaaldrich.com/catalog/search?term=nanoclay&interface=All&N=0&mode=match%20partialmax&lang=en&region=MY&focus=product. Accessed 15. 12. 2014.

Sing, K. S. W., Everett, D. H., Haul, R. A. W., Moscou, L., Pierotti, R. A., Rouquerol, J., et al. (1985). Reporting physisorption data for gas/solid systems with special reference to the determination of surface area and porosity (Recommendations 1984). *International Union of Pure and Applied Chemistry*, *57*(4), 603–619.

Singh, V., & Singh, D. (2013). Polyvinyl alcohol–silica nanohybrids: an efficient carrier matrix for amylase immobilization. *Process Biochemistry*, *48*(1), 96–102.

Stevens, E. S. (2002). *Green plastics: An introduction to the new science of biodegradable plastics*. New Jersey: Princeton University Press.

Taghizadeh, M., & Sabouri, N. (2013). Biodegradation behaviors and water adsorption of poly(vinyl alcohol)/starch/carboxymethyl cellulose/clay nanocomposites. *International Nano Letters*, *3*(1), 1–8.

Yang, R. T., Tharappiwattananon, N., & Long, R. Q. (1998). Ion-exchanged pillared clays for selective catalytic reduction of NO by ethylene in the presence of oxygen. *Journal of Applied Catalysis B: Environmental*, *19*(3-4), 289–304.

Zeng, Q. H., Yu, A. B., Lu, G. Q., & Paul, D. R. (2005). Clay-based polymer nanocomposites: research and commercial development. *Journal of Nanoscience and Nanotechnology*, *5*(10), 1574–1592.

Nanoclay dispersed phenol formaldehyde/fumed silica nanocomposites: effect of diverse clays on physicomechanical and thermal properties

5

Md. Rezaur Rahman, Josephine Lai Chang Hui, Sinin bin Hamdan
Department of Chemical Engineering and Energy Sustainability, Faculty of Engineering, University Malaysia Sarawak (UNIMAS), Kota Samarahan, Sarawak, Malaysia

5.1 Introduction

In recent years, there has been high demand for innumerable new products that have been produced through the development of new and improved polymers. The changes in the properties of individual polymer can be observed when there are changes in arrangement of monomeric units or in the reaction conditions. The method usually used to discover the possibilities of new polymers includes the modification of the existing polymers (Ranade, Wunder, & Baran, 2006). Addition of filler or reinforcement into polymer matrix reduces the disadvantages of pure polymers.

Nowadays, the common fillers that are widely used consist of fumed silica and clay. Fumed silica can be defined as finely divided amorphous silicon dioxide particles produced by high temperature in an oxygen-hydrogen flame (Gun'ko, Zarko, Leboda, & Chibowski, 2001). Addition of fumed silica enhances the mechanical properties as well as restricts the thermal degradation of phenolic resin. Incorporation of MMT (montmorillonite nanoclay) improves the physicomechanical and thermal properties of nanocomposites (Yang, Tharappiwattananon, & Long, 1998).

PF resins are a type of thermosets polymer matrix. These resins are used widely due to their low cost, ease of processing, thermal resistance, chemical resistance, and good mechanical properties (Yan et al., 2014). When 2 wt% of silica was added into phenolic resin via in situ polymerization, it enhanced the thermal stability of the nanocomposites compared to pure PF (Yu, Wei, Lv, Liu, & Meng, 2012). The tensile strength and thermal stability of phenolic resin were greatly enhanced when 2 wt% of silica was added into the nanocomposites (Li et al., 2015). The addition of 2 wt% of silica into PF matrix showed that the thermal stability of PF nanocomposites significantly increased. The mechanical properties of PF-silica nanocomposites were two times higher than pure PF (Yu et al., 2012). Addition of fillers such as boron carbide and fumed silica into PF resin improved the bonding performance and enhanced mechanical properties of the carbon materials. However, the addition of silica into

Silica and Clay Dispersed Polymer Nanocomposites. https://doi.org/10.1016/B978-0-08-102129-3.00005-1

PF beyond 15 wt% inversely affected the tensile strength, compressive strength, and impact resistance of the composite (Zhuang, Qian, Lv, & Wan, 2010).

PF with a linear structure enters the MMT layers and forms the network structure by curing process. PF-MMT nanocomposites improved the modulus of elasticity (Liu, Zhang, & Zhou, 2013). The addition of clay into PF composites enhanced the tensile properties that could be applied in structural application (Ku, Yap, Lee, & Trada, 2012). Clay was completely exfoliated in resole PF resins, which improved the impact strength of the resultant. It was also used to improve the impact strength and fracture toughness of cellulose-PF resin-based surface laminates up to 16% and 66% (Ma et al., 2005). Clay accelerated the resole PF curing process to improve the thermal stability and tensile strength of the nanocomposites. Incorporation of clay led to good interfacial interaction with phenolic resin matrix and improved the mechanical properties of the composites (Eesaee & Shojaei, 2014).

Although phenolic resins have been widely used as coatings and adhesives composites, the applications of phenolic resins were limited due to its inherent brittleness and poor resistance to crack initiation and growth (Johnsen, Kinloch, Mohammed, Taylor, & Sprenger, 2007). Because fumed silica and nanoclay proved their potential effect in improving many properties of nanocomposites, mixing of fumed silica and clay at a molecular level to phenolic resins could overcome the aforementioned problem.

Many researchers had carried out research on PF and clay-based nanocomposites. However, there was no research work that has been reported on PF-fsi-clay nanocomposites, especially using resole-type resin. From the previous study, the addition of fumed silica at 2 wt% and clay at 0.5 wt% showed the best mechanical and thermal properties on the PF matrix (Yu et al., 2012). Therefore, the aim of this study is to fabricate PF-fsi-clay nanocomposite using PF, FS, and clay via condensation polymerization technique, and to investigate the effect of different clays on physicomechanical, morphological, and thermal properties of nanocomposites.

5.2 Results and discussion

5.2.1 Spectral analysis

The infrared (IR) spectra of pure PF, PF-fsi, and various PF-fsi-clay nanocomposites are shown in Fig. 5.1. The spectra were separated into two regions where the —OH stretching vibration in 4000–3000 cm^{-1} region and fingerprint in 1800–400 cm^{-1} region (Zhuang et al., 2010). The characteristic band at 1620 cm^{-1} was attributed to ν(C=C) groups, while the spectrum 4000–3300 cm^{-1} represented the existence of a surface hydroxyl groups due to the chemisorbed water. According to the findings, PF-fsi-clay (1.34TCN) nanocomposite showed flat spectrum in the range of 4000–3000 cm^{-1}, which was different from pure PF, PF-fsi, PF-fsi-clay (1.28E), PF-fsi-clay (1.30E), and PF-fsi-clay (1.31PS) nanocomposites. This was due to the silica nanoparticles attached at side chain, which hindered the formation of hydrogen bonds between phenol formaldehyde chains (Lai et al., 2015). For pure PF, PF-fsi, PF-fsi-clay (1.28E), PF-fsi-clay (1.30E), and PF-fsi-clay (1.31PS) nanocomposites, a peak at around 1100 cm^{-1} region appeared, representing the absorptions due to

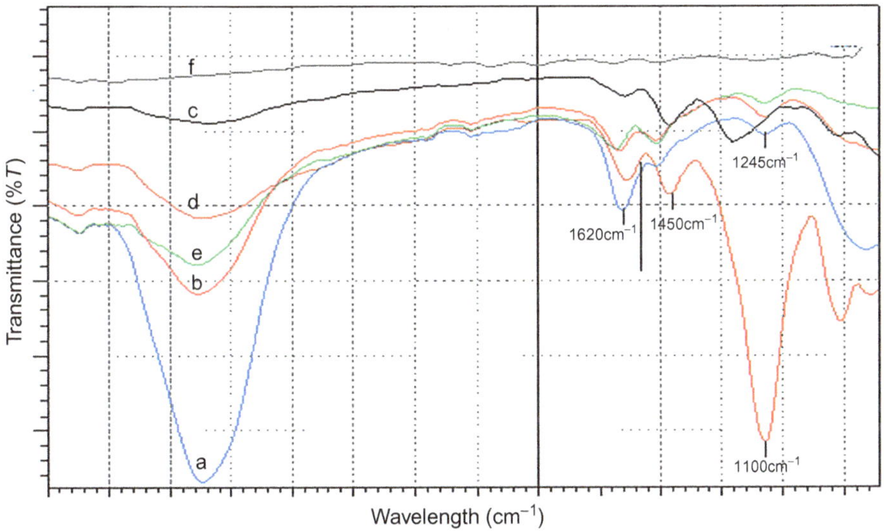

Fig. 5.1 FTIR spectra of (a) pure PF, (b) PF-fsi nanocomposite, (c) PF-fsi-clay (1.28E) nanocomposite, (d) PF-fsi-clay (1.30E) nanocomposite, (e) PF-fsi-clay (1.31PS) nanocomposite, and (f) PF-fsi-clay (1.34TCN) nanocomposite.

Si-O-Si (asymmetric stretch). Additional absorption bands like asymmetric Si-O-C stretch (cross-linking of silica and PVA) and asymmetric C-O stretch (due to ether linkage) could inhibit in this region. The peaks at 1450, 1550, and 1650 cm^{-1} represented the presence of benzene rings of phenol. In addition, the interfacial bonding of phenol-O-C structure can be detected at 1245 cm^{-1}, which proved the well intercalation of clay onto PF-fsi (Qu et al., 2012). However, FTIR spectrum for PF-fsi-clay (1.34TCN) nanocomposites was slightly broader at these areas compared to pure PF, PF-fsi, and other nanocomposites.

This result indicated that clay (1.34TCN) strongly increased the compatibility between PF and FS, which enhanced the overall properties of nanocomposites as compared to other clays. The result of better compatibility was also reflected in the thermal stability as well as surface morphology in PF-fsi-clay (1.34TCN) nanocomposite.

5.2.2 X-ray fluorescence analysis

The X-ray fluorescence (XRF) elemental analysis of pure PF, PF-fsi, PF-fsi-clay (1.28E), PF-fsi-clay (1.30E), PF-fsi-clay (1.31PS), and PF-fsi-clay (1.34TCN) nanocomposites are presented in Table 5.1. Overview (wide scan) photoelectron spectra of these films indicated the presence of silicon, nitrogen, oxygen, and carbon (Besling, Goossens, Meester, & Schoonman, 1998). Table 5.1 shows an overlapping of symmetric Si—C and Si—N bonds for all the samples. This happened due to C and N atoms bonding with Si only. The numbers of Si—C and Si—N bonds were nearly identical, as proven in Table 5.1 (Luo, Zhao, Cai, & Du, 2012). PF-fsi-clay (1.34TCN) nanocomposite showed the highest amount of Si compound, which confirmed the

Table 5.1 Oxide contents of pure PF, PF-fsi, and PF-fsi-clay nanocomposites

Samples	Si	C	S	N	K	Co	Ni
Pure PF	47.8	23.1	5.8	17.3	4.6	0.7	0.6
PF-fsi nanocomposite	70.1	17.6	–	13.2	–	–	–
PF-fsi-clay (1.28E) nanocomposite	60.8	18.3	–	17.1	–	2.8	–
PF-fsi-clay (1.30E) nanocomposite	57.8	13.3	11.3	12.1	–	2.5	2.6
PF-fsi-clay (1.31PS) nanocomposite	68.5	16.3	–	14.0	–	1.2	–
PF-fsi-clay (1.34TCN) nanocomposite	75.6	8.8	13.3	2.7	–	–	–

presence of the silanol group that was assigned to Si-O-Si stretching vibration. Thus, this proved that the PF-fsi and clay (1.34TCN) were well-linked with each other. From the findings, PF-fsi-clay (1.34TCN) nanocomposite displayed better linkage between PF-fsi and clay as reflected by the FTIR results.

5.2.3 Scanning electron microscopy analysis

An scanning electron microscopy (SEM) test was conducted, and aggregations of silica and clay nanocomposites are shown in Fig. 5.2A–F. Fig. 5.2A indicates that the fracture surface of PF (without the addition of fumed silica and clay) was relatively smooth. This proves that pure PF was a typically a brittle material and no monomer intercalation occurred (Azlin & Zuraifah, 2013). The addition of FS into PF shows that the agglomeration between PF and FS was poor, which indicates weak adhesion and compatibility between PF and FS as shown in Fig. 5.2B. Fig. 5.2C–F shows the aggregations between PF, fumed silica, and clay. It was found from Fig. 5.2C that the interfacial bonding between PF-fsi and clay (1.28E) was stronger compared to pure PF, PF-fsi, PF-fsi-clay (1.30E), and PF-fsi-clay (1.31PS) nanocomposites, respectively. This proved that PF-fsi was well mixed with clay (1.28E). This clay provides better compatibility toward the PF-fsi. Thus, clay (1.28E) led to better interface bonding in the nanocomposite system. According to Fig. 5.2D and E, it was observed that clay (1.30E) and clay (1.31PS) were dispersed unevenly in the PF-fsi with the presence of agglomeration. This was due to the poor adhesion between clay and PF-fsi. Fig. 5.2E shows that the inert surface of PF-fsi could not react well with clay particles and indicates the poor compatibility of PF-fsi with clay particles (Luo et al., 2012). The uniform surface of PF-fsi-clay (1.34TCN) nanocomposite proved better compatibility among the nanocomposites, as shown in Fig. 5.2F. There was no clear boundary between PF-fsi and clay, which clearly indicated the excellent adhesion and compatibility between PF-fsi and clay (Du, 2015). The outcome of better interfacial bonding and strong compatibility of PF-fsi-clay (1.34TCN) nanocomposite enhanced mechanical and thermal stability of the nanocomposite.

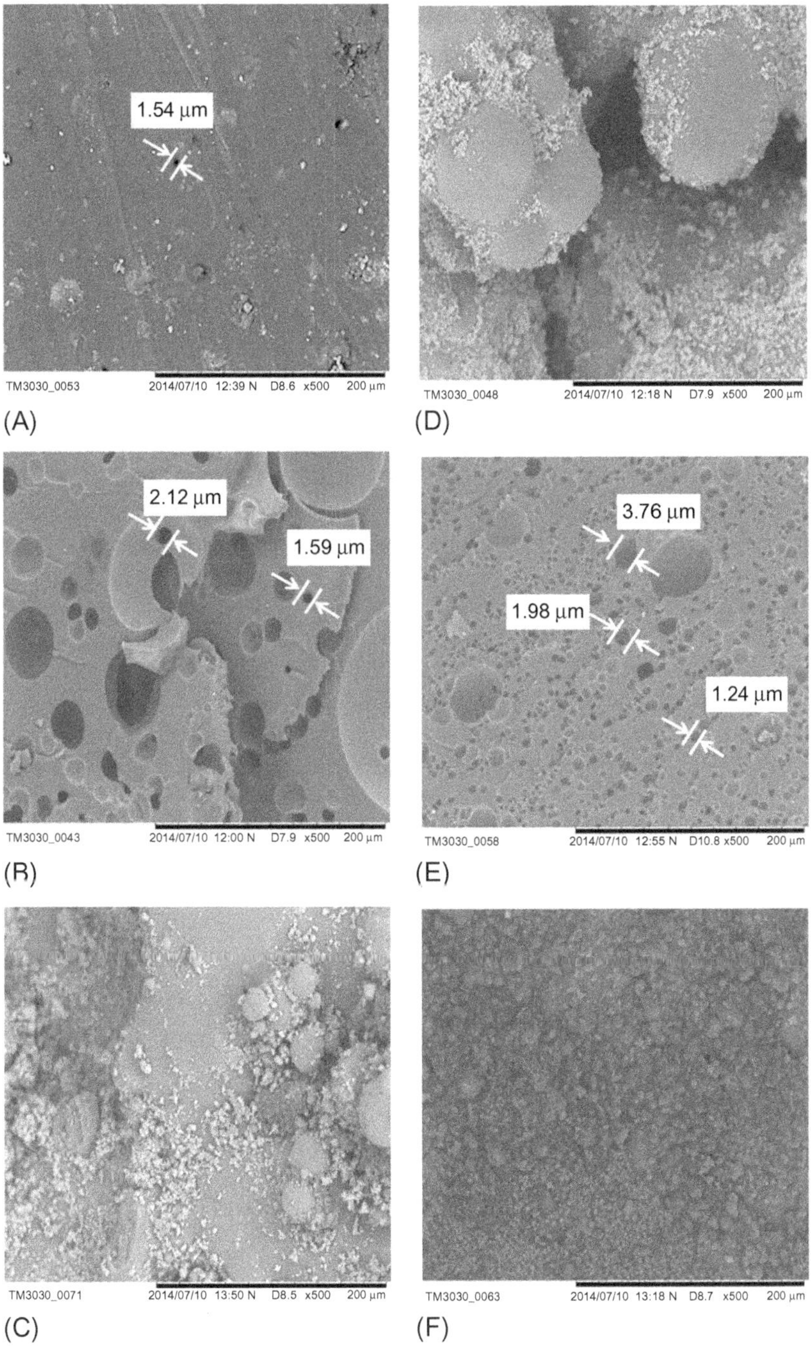

Fig. 5.2 SEM micrographs of (A) pure PF, (B) PF-fsi nanocomposite, (C) PF-fsi-clay (1.28E) nanocomposite, (D) PF-fsi-clay (1.30E) nanocomposite, (E) PF-fsi-clay (1.31PS) nanocomposite, and (F) PF-fsi-clay (1.34TCN) nanocomposite.

5.2.4 Adsorption isotherm analysis

The N_2 adsorption isotherms of pure PF, PF-fsi, PF-fsi-clay (1.28E), PF-fsi-clay (1.30E), PF-fsi-clay (1.31PS), and PF-fsi-clay (1.34TCN) nanocomposites are shown in Fig. 5.3. Specific surface area, S_{BET}, was calculated using the BET equation and tabulated in Table 5.2 (Brunauer, Emmett, & Teller, 1938). It showed that pure PF had the lowest specific surface area at $20.29\,m^2/g$, while PF-fsi-clay (1.34TCN) nanocomposite showed the highest specific surface area at $80.75\,m^2/g$. This proves that the addition of clay into PF-fsi improved the bonding within the nanoparticles in the nanocomposites. From the adsorption isotherm analysis, clay (1.34TCN) had higher surface area compared to clay (1.28E), clay (1.30E), and clay (1.31PS) respectively. The small pores detected increased the accessibility for nitrogen adsorption into the nanocomposites due to the well dispersed clay within PF-fsi. Type IV isotherms was detected for pure PF, PF-fsi, and PF-fsi-clay nanocomposites (Sing et al., 1985).

In the case of pure PF, the initial isotherm ascended up to $P/P_0 = 0.04$ and continuously extended up to relative pressure at 0.94. PF-fsi, PF-fsi-clay (1.28E), PF-fsi-clay (1.30E), PF-fsi-clay (1.31PS), and PF-fsi-clay (1.34TCN) nanocomposites exhibited similar patterns of isotherms with the initial ascending section and the extended region started at the relative pressure of 0.04, 0.23, 0.23, 0.23, and 0.40 respectively. It was observed that at saturation pressure, PF-fsi-clay (1.34TCN) nanocomposite had the highest amount adsorbed. From Table 5.2, it showed that BET surface area and pore volume of PF-fsi-clay (1.28E), PF-fsi-clay (1.30E), PF-fsi-clay (1.31PS), and PF-fsi-clay (1.34TCN) nanocomposites greatly increased with smaller pore size compared to pure PF and PF-fsi nanocomposite (Knop & Pilato, 1985). Overall, PF-fsi-clay (1.34TCN) nanocomposite had the highest surface area and average pore volume with smaller pore size, as well as smooth surface morphology.

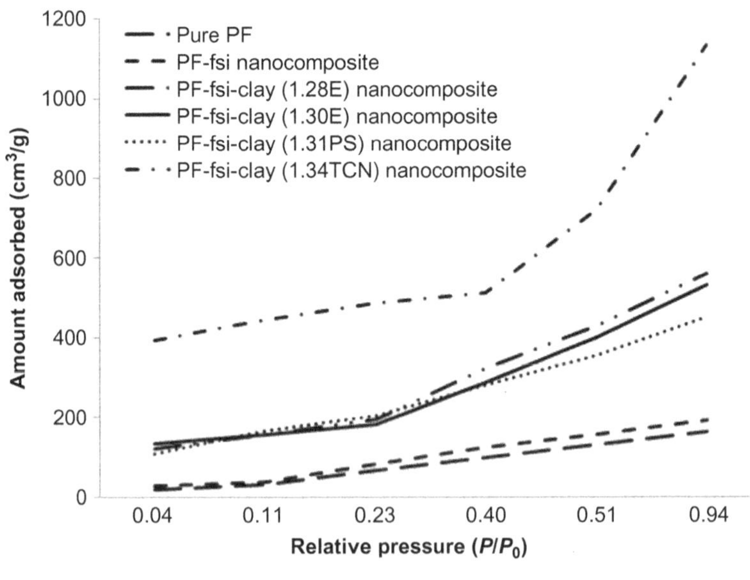

Fig. 5.3 N_2 adsorption isotherms of pure PF, PF-fsi, and PF-fsi-clay nanocomposites.

Table 5.2 Physical properties detected from N_2 adsorption at 77 K on pure PF, PF-fsi, and PF-fsi-clay nanocomposites

Samples	Specific surface area, S_{BET} (m^2/g)	Average pore volume, V_m (10^{-4} cm^3/g)	d_{BET} (nm)	Type of isotherms
Pure PF	20.29	0.98	5.10	IV
PF-fsi nanocomposite	28.56	1.99	2.26	IV
PF-fsi-clay (1.28E) nanocomposite	74.31	8.75	1.80	IV
PF-fsi-clay (1.30E) nanocomposite	38.23	8.28	1.80	IV
PF-fsi-clay (1.31PS) nanocomposite	35.02	8.08	2.00	IV
PF-fsi-clay (1.34TCN) nanocomposite	80.75	29.20	1.79	IV

5.2.5 Tensile properties

The tensile strength and tensile modulus of pure PF, PF-fsi, four types of clay loaded PF-fsi-clay nanocomposites are shown in Figs. 5.4 and 5.5 respectively. From Fig. 5.4, PF-fsi-clay (1.34TCN) nanocomposite showed the highest tensile strength. Fig. 5.4 also indicates that the addition of clay to PF-fsi increased the tensile strength of the nanocomposite up to 56%. This was due to the aggregation of clay that shortened the inter-particle distance within the nanocomposites (Lai et al., 2015).

PF-fsi-clay (1.34TCN) nanocomposite showed the highest tensile modulus as shown in Fig. 5.5. This was due to the clay (1.34TCN), which contained methyl dihydroxyethyl hydrogenated tallow ammonium that enhanced anti-wetting properties and prevented the surface of PF-fsi-clay (1.34TCN) nanocomposite from becoming hydrophilic (Balaji, Sasikumar, & Elayaperumal, 2014). It was also observed that the addition of clay (1.34TCN) provided the highest compatibility with PF-fsi compared to other clay toward the nanocomposites. The intercalation structure and the nanoscale dispersion of clay provided large interfacial interaction. This improved a significant restriction on the polymer chain movement at the polymer-clay interface region (Eesaee & Shojaei, 2014). The addition of both fumed silica and clay had positive effects on the bonding between matrix and fillers, and thus improved tensile strength and modulus. The higher compatibility of PF-fsi-clay (1.34TCN) nanocomposite was also proven in the thermal stability and surface morphology.

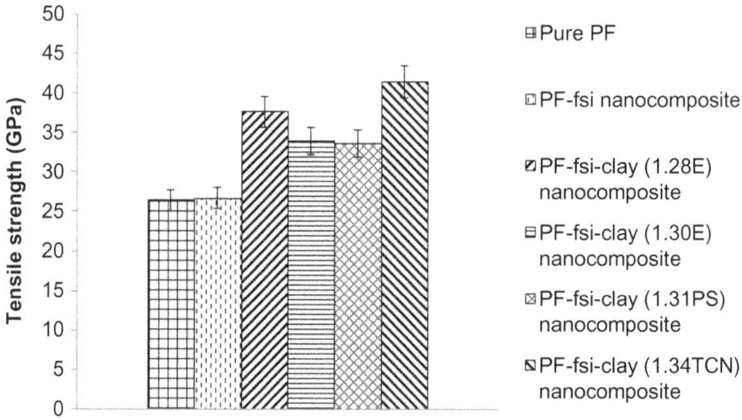

Fig. 5.4 Tensile strength of pure PF, PF-fsi, and PF-fsi-clay nanocomposites.

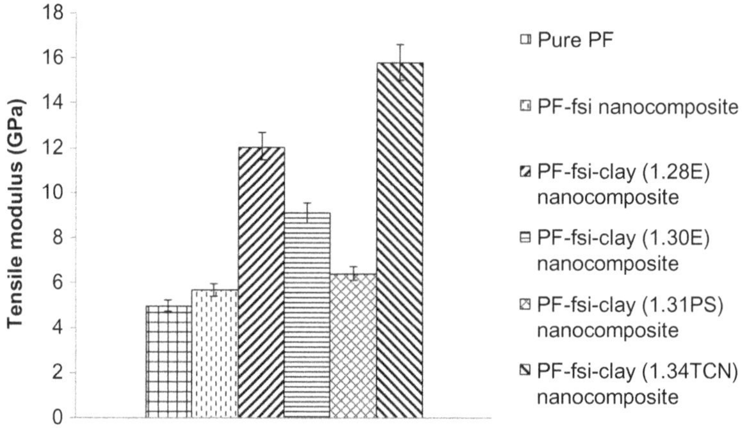

Fig. 5.5 Tensile modulus of pure PF, PF-fsi, and PF-fsi-clay nanocomposites.

5.2.6 Thermogravimetric analysis

The improved thermal properties of polymer nanocomposites explain the uniform dispersion of fillers and intensive interactions between fillers and polymer. Thermogravimetric analysis (TGA) was commonly used to measure mass loss when subjected to heating (Balaji et al., 2014). TGA curves of the various nanocomposites are shown in Fig. 5.6. According to the findings, the thermal degradation of all nanocomposites occurred in three stages. The first stage was obtained at low temperature (below 200°C). This indicated that all nanocomposites were relatively stable in a nitrogen atmosphere. During the curing process, there was a weight loss of about 1%–2% due to evolution of gases trapped in the polymer matrix (Balaji et al., 2014). The intermediate temperature stage started at 200°C and ended at around 500°C. At this stage, the changes in PF-fsi-clay nanocomposites were significant compared to pure PF and PF-fsi due to the release of carbon dioxide and the loss of water vapor (Seeni Meera, Sankar, Murali, Janiskar, & Mandal, 2012). It was observed that the weight loss curve shifted due to the addition of fumed silica, and burnt above 500°C for last thermal degradation stage (Lai et al., 2015).

At the first stage, the weight loss was about 23.2%, 22.9%, 21.7%, 18.7%, 21.3%, and 15.6% for pure PF, PF-fsi, PF-fsi-clay (1.28E), PF-fsi-clay (1.30E), PF-fsi-clay (1.31PS), and PF-fsi-clay (1.34TCN) nanocomposites respectively. The weight loss at second stage (210°C and 250°C) were 13.0%, 11.3%, 16.3%, 9.2%, 11.5%, and 6.2% while the weight loss at the final degradation temperature (410°C and 580°C) were 16.5%, 15.8%, 9.7%, 11.7%, 13.3%, and 8.3% for PF, PF-fsi, PF-fsi-clay (1.28E), PF-fsi-clay (1.30E), PF-fsi-clay (1.31PS), and PF-fsi-clay (1.34TCN) nanocomposites respectively. Dispersion of silica and clay improved the thermal stability of the nanocomposites. The result also indicates that thermal stability of different clay loaded nanocomposites significantly increased compared to the pure PF and PF-fsi nanocomposite. The bonding of PF-fsi along with clay, especially clay (1.34TCN), enhanced the compatibility of the nanocomposites, which showed lower weight percentage loss at the final degradation compared it to pure PF, PF-fsi, PF-fsi-clay (1.28E), PF-fsi-clay (1.30E), and PF-fsi-clay (1.31PS) nanocomposites respectively.

Table 5.3 summarizes the activation energy of all nanocomposites by applying Arrhenius equation (Chanmal & Jog, 2008). It was found that the activation energy of PF-fsi-clay (1.34TCN) nanocomposite was significantly higher than pure PF, PF-fsi, PF-fsi-clay (1.28E), PF-fsi-clay (1.30E), and PF-fsi-clay (1.31PS) nanocomposites. The overall result confirmed that PF-fsi-clay nanocomposites were thermally stable compared to pure PF and PF-fsi nanocomposites.

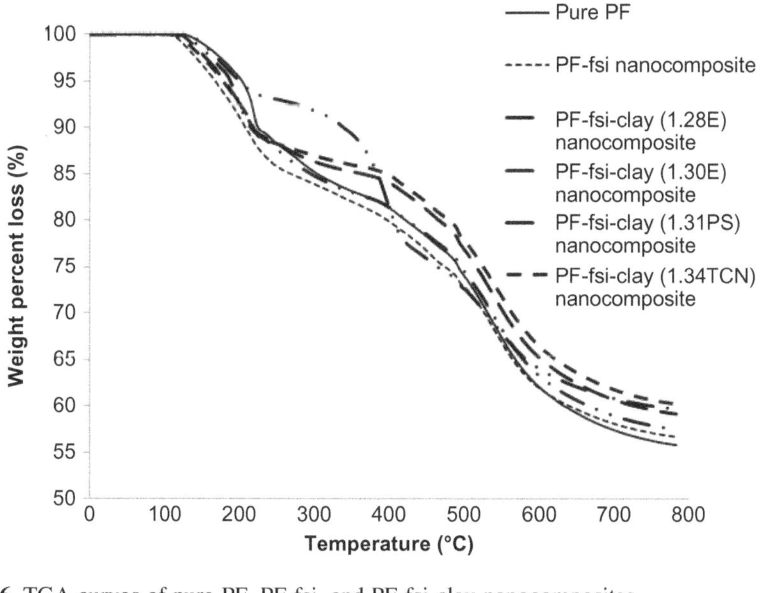

Fig. 5.6 TGA curves of pure PF, PF-fsi, and PF-fsi-clay nanocomposites.

Table 5.3 Activation energy of pure PF, PF-fsi, and PF-fsi-clay nanocomposites determined by Arrhenius equation

Samples	T_i (°C)[a]	T_m (°C)[b]	T_f (°C)[c]	W_{Ti} (%)[d]	W_{Tm} (°C)[e]	W_{Tf} (°C)[f]	Activation energy, E_a (kJ/mol)
Pure PF	249.9	473.7	578.3	88.3	76.8	64.1	2.1
PF-fsi nanocomposite	235.3	447.2	565.4	86.9	77.1	64.9	2.9
PF-fsi-clay (1.28E) nanocomposite	232.1	417.8	523.5	93.5	78.3	70.7	13.6
PF-fsi-clay (1.30E) nanocomposite	224.6	449.6	541.5	89.5	81.3	71.8	12.6
PF-fsi-clay (1.31PS) nanocomposite	232.1	449.9	556.8	88.9	78.7	68.2	7.3
PF-fsi-clay (1.34TCN) nanocomposite	218.6	410.1	564.4	90.0	84.4	77.4	14.1

[a] Temperature corresponding to the beginning of decomposition.
[b] Temperature corresponding to the maximum rate of mass loss.
[c] Temperature corresponding to the end of decomposition.
[d] Mass loss at temperature corresponding to the beginning of decomposition.
[e] Mass loss at temperature corresponding to the maximum rate of mass loss.
[f] Mass loss at temperature corresponding to the end of decomposition.

5.3 Summary

PF-fsi-clay nanocomposites were successfully prepared via condensation polymerization method. FTIR spectra showed that the PF-fsi-clay (1.34TCN) nanocomposite had a slightly broader peak at 3370 cm^{-1} compared to other PF-fsi-clay nanocomposites. From the XRF result, PF-fsi-clay (1.34TCN) nanocomposite contained the highest percentage of silicon group, which proved the well linkage between polymer matrix and fillers. Nanocomposite with clay (1.34TCN) showed extreme changes and smooth surface morphology, which proved better compatibility with PF-fsi. The BET result showed PF-fsi-clay (1.34TCN) nanocomposite had the highest surface area and average pore volume with smaller pore size. PF-fsi-clay (1.34TCN) nanocomposite had the highest tensile strength and modulus followed by pure PF, PF-fsi, PF-fsi-clay (1.28E), PF-fsi-clay (1.30E), and PF-fsi-clay (1.31PS) nanocomposites. In addition, the thermal stability of PF-fsi-clay (1.34TCN) nanocomposite was thermally more stable compared to other clay nanocomposites. Thus, PF-fsi-clay (1.34TCN) nanocomposite exhibited higher activation energy compared to other clay nanocomposites. It was concluded that the clay loaded PF nanocomposites had better physicomechanical and thermal properties compared to pure PF and PF-fsi nanocomposites.

Acknowledgments

The authors are grateful for the support of the Ministry of Higher Education Malaysia, Grant No. ERGS/02 (08)/860/2012 (12).

References

Azlin, M. Z. N., & Zuraifah, M. Z. S. (2013). Preparation and 3D morphology observation of phenolic resin-silica nanocomposites. *International Journal of Materials, Mechanics and Manufacturing*, *1*(2), 107–109.

Balaji, R., Sasikumar, M., & Elayaperumal, A. (2014). Effect of microsilica in woven ceramic fiber/phenolic resin composites. *International Journal of Research in Engineering and Technology*, *3*(5), 609–613.

Besling, W. F. A., Goossens, A., Meester, B., & Schoonman, J. (1998). Laser-induced chemical vapor deposition of nanostructured silicon carbonitride thin films. *Journal of Applied Physics*, *83*(1), 544–553.

Brunauer, B., Emmett, P. H., & Teller, E. (1938). Adsorption of gases in multimolecular layers. *Journal of the American Chemical Society*, *60*(2), 309–319.

Chanmal, C. V., & Jog, J. P. (2008). Dielectric relaxations in PVDF/$BaTiO_3$ nanocomposites. *eXPRESS Polymer Letter*, *2*(4), 294–301.

Du, X. (2015). *Study on the mechanism of the synthesis of phenylbiguanide and the kinetics of curing epoxy*. China: Wuhan University of Technology.

Eesaee, M., & Shojaei, A. (2014). Effects of nanoclays on the mechanical properties and durability of novolac phenolic resin/woven glass fiber composite at various environments. *Composites: Part A*, *63*, 149–158.

Gun'ko, V. M., Zarko, V. I., Leboda, R., & Chibowski, E. (2001). Aqueous suspension of fumed oxides: particle size distribution and zeta potential. *Advances in Colloid and Interface Science*, *91*(1), 1–112.

Johnsen, B. B., Kinloch, A. J., Mohammed, R. D., Taylor, A. C., & Sprenger, S. (2007). Toughening mechanisms of nanoparticle-modified epoxy polymers. *Polymer*, *48*(2), 530–541.

Knop, A., & Pilato, L. A. (1985). *Phenolic Resins*. Heidelberg: Springer.

Ku, H., Yap, Y. S., Lee, T. S., & Trada, M. (2012). Tensile properties of phenol-formaldehyde nanoclay reinforced composites: A pilot study. *Journal of Applied Polymer Science*, *126*(2), 142–150.

Lai, J. C. H., Rahman, M. R., Hamdan, S., Liew, F. K., Rahman, M. M., & Hossen, M. F. (2015). Impact of nanoclay on physicomechanical and thermal analysis of polyvinyl alcohol/ fumed silica/clay nanocomposites. *Journal of Applied Polymer Science*, *132*(15).

Li, Q., Chen, L., Zhang, J., Zheng, K., Zhang, X., Fang, F., et al. (2015). Enhanced mechanical properties, thermal stability of phenolic formaldehyde foam/silcia nanocomposites via in situ polymerization. *Polymer Engineering and Science*, *55*(12), 2783–2793.

Liu, M., Zhang, Y., & Zhou, C. (2013). Nanocomposites of halloysite and polylactide. *Applied Clay Science*, *75–76*, 52–59.

Luo, Y., Zhao, J., Cai, Y., & Du, S. (2012). Effect of amino-functionalization on the interfacial adhesion of multi-walled carbon nanotubes/epoxy nanocomposites. *Materials Design*, *33*, 405–412.

Ma, H., Wei, G., Liu, Y., Zhang, X., Gao, J., Huang, F., et al. (2005). Effect of elastomeric nanoparticles on properties of phenolic resin. *Polymer*, *46*(23), 10568–10573.

Qu, J. C., Ren, C. L., Dong, Y. L., Chang, Y. P., Zhou, M., & Chen, X. G. (2012). Facile synthesis of multifunctional graphene oxide/AgNPs-Fe3O4 nanocomposite: A highly integrated catalysts. *Chemical Engineering Journal*, *211–212*, 412–420.

Ranade, R. A., Wunder, S. L., & Baran, G. R. (2006). Toughening of dimethacrylate resins by addition of ultra high molecular weight polyethylene (UHMWPE) particles. *Polymer*, *47*(12), 4318–4327.

Seeni Meera, K. M., Sankar, R. M., Murali, A., Janiskar, S. N., & Mandal, A. B. (2012). Sol–gel network silica/modified montmorillonite clay hybrid nanocomposites for hydrophobic surface coatings. *Colloids and Surfaces B: Biointerfaces*, *90*(1), 204–210.

Sing, K. S. W., Everett, D. H., Haul, R. A. W., Moscou, L., Pierotti, R. A., Rouquerol, J., et al. (1985). Reporting physisorption data for gas/solid systems with special reference to the determination of surface area and porosity (Recommendations 1984). *International Union of Pure and Applied Chemistry*, *57*(4), 603–619.

Yan, D., Li, X., Jiang, Y., Zhang, H. B., Jia, B. B., Ma, H. L., et al. (2014). Thermally conductive phenol formaldehyde composites filled with carbon fillers. *Materials Letters*, *118*, 212–216.

Yang, R. T., Tharappiwattananon, N., & Long, R. Q. (1998). Ion-exchanged pillared clays for selective catalytic reduction of NO by ethylene in the presence of oxygen. *Journal of Applied Catalysis B: Environmental*, *19*(3–4), 289–304.

Yu, C. B., Wei, C., Lv, J., Liu, H. X., & Meng, L. T. (2012). Preparation and thermal properties of mesoporous silica/phenolic resin nanocomposites via in situ polymerization. *eXPRESS Polymer Letters*, *6*(10), 783–793.

Zhuang, X., Qian, X., Lv, J., & Wan, Y. (2010). An alternative method to remove PEO–PPO–PEO template in organic–inorganic mesoporous nanocomposites by sulfuric acid extraction. *Applied Surface Science*, *256*(17), 5343–5348.

Study on physical, mechanical, morphological and thermal properties of styrene-*co*-glycidyl methacrylate/fumed silica/clay nanocomposites

6

Md. Rezaur Rahman, Sinin bin Hamdan
Department of Chemical Engineering and Energy Sustainability, Faculty of Engineering, University Malaysia Sarawak (UNIMAS), Kota Samarahan, Sarawak, Malaysia

6.1 Introduction

Polymer matrices have been widely used as a matrix for composites production. They have great properties, including excellent toughness and adhesion. These resins undergo ring opening when reacting with substances that possess methylene groups (Hanifah, Hamzah, & Heng, 2013).

Styrene (ST) is a monomer that helps to enhance the rigidity of the copolymer. ST is prepared through the dispersion polymerization, suspension polymerization, bulk polymerization, atom transfer radical polymerization (ATRP), and radical emulsion polymerization (Nita, Chiriac, Cimmino, Duraccio, & Vasile, 2008). ST-based composites can be applied in the automotive fields (Zanchet, Carli, Giocanela, Brandalise, & Crespo, 2012)

Glycidyl methacrylate (GMA) is another monomer that has flexible applications. They are in the commercially available vinyl monomer carrying epoxy group. GMA is useful for polymer matrix modification, which can be utilized as a matrix for drug delivery (Clemons, 2013). Benzoyl peroxide is used as initiator for ST-based copolymer or GMA-based copolymer (Kunita, Girotto, Muniz, & Rubira, 2006).

Due to low melting temperature, ST-*co*-GMA does not possess adequate thermal and mechanical properties. Therefore, nanofillers such as fumed silica and nanoclay are incorporated to overcome this problem. Inorganic fumed silica, which acts as nanofiller, was incorporated into polymer matrices to enhance the interfacial area as well as the mechanical properties of the nanocomposites. It was widely used due to its high surface area. Clay is a nanofiller that can be introduced into polymer nanocomposites. It is a layered structure that consists of an octahedral alumina sheet stacked between two tetrahedral silica sheets by a weak dipolar force, which leads to interlayer galleries (Grim, 1968). Clay was widely used due to its low cost, high availability, and the fact that it is environmental friendly. The most commonly used clay is

Silica and Clay Dispersed Polymer Nanocomposites. https://doi.org/10.1016/B978-0-08-102129-3.00006-3

montmorillonite nanoclay (MMT), which is used in many applications such as the construction, polymer, and medical industries (Conzatti, 2012).

Kim et al. (2015) confirmed the compatibility of fumed silica in ST-*co*-GMA by reducing the agglomerate formed. The incorporation of fumed silica in ST-*co*-GMA had improved the mechanical properties as the covalent bonds between the epoxy groups of ST-*co*-GMA and silanol groups from silica, which increased the filler-matrix interaction. Fumed silica incorporated into the ST matrix enhanced the tensile strength (Yin, Zhang, & Gong, 2013). According to Barrantes, Valentin, Rodriquez, Garrido, and Paris (2012), poly(ST)-fumed silica had undergone free radical polymerization that had enhanced the tensile strength with high dispersion and excellent compatibility between the polymer matrix and filler. The incorporation of fumed silica onto poly(GMA) matrix enhanced thermal stability of the nanocomposites (Grama et al., 2014). Fourier transform infrared (FTIR) spectroscopy of Ou et al. (2013) showed that poly(GMA) was well intercalated with fumed silica nanoparticle by free radical polymerization.

According to Nemati, Taromi, and Mirzataheri (2015), silica was well dispersed in the styrene matrix, which enhanced the compatibility within the composites. A lower percentage of clay improved the thermal stability of ST-*co*-GMA compared to a higher percentage of clay (Devi & Maji, 2012). The intercalation of clay into poly(ST) matrix improved the tensile strength through melt intercalation (Xu, Li, & Yu, 2005). Poly(GMA)-clay nanocomposites that had undergone free radical in situ polymerization had enhanced thermal and mechanical performances compared to epoxy-bentonite nanocomposites (Jlassi et al., 2015). Based on Bayramoglu, Senkali, and Arica (2013), introduction of clay into GMA was confirmed to improve the thermal stability.

Single filler onto ST- or GMA-based nanocomposites improved the mechanical and thermal properties significantly. However, the surface morphology and physical properties did not improve by single filler loaded nanocomposites. To overcome the aforementioned drawback, the present study investigated dual filler loadings on ST-*co*-GMA and their compatibility. The physical, mechanical, morphological, and thermal properties of ST-GMA-fsi-clay (1.30E) were also reported in this study.

6.2 Results and discussion

6.2.1 *Spectral analysis*

The FTIR spectra of the ST-*co*-GMA, ST-*co*-GMA-fsi, and ST-*co*-GMA-fsi-clay (1.30E) nanocomposites are shown in Fig. 6.1. The spectra confirmed the presence of the functional groups in ST-*co*-GMA-fsi-clay (1.30E) nanocomposites with the chemical reaction as shown in Fig. 6.2.

From Fig. 6.1a, the absorption peak at $3350\,cm^{-1}$ represented the O-H stretching vibration within the GMA monomer. The C-H stretching vibration was detected at $2970\,cm^{-1}$ in all the samples (Mirous, 2006). However, the intensity of the peak at 2 wt% of clay loaded ST-*co*-GMA-fsi-clay (1.30E) nanocomposite was broader due

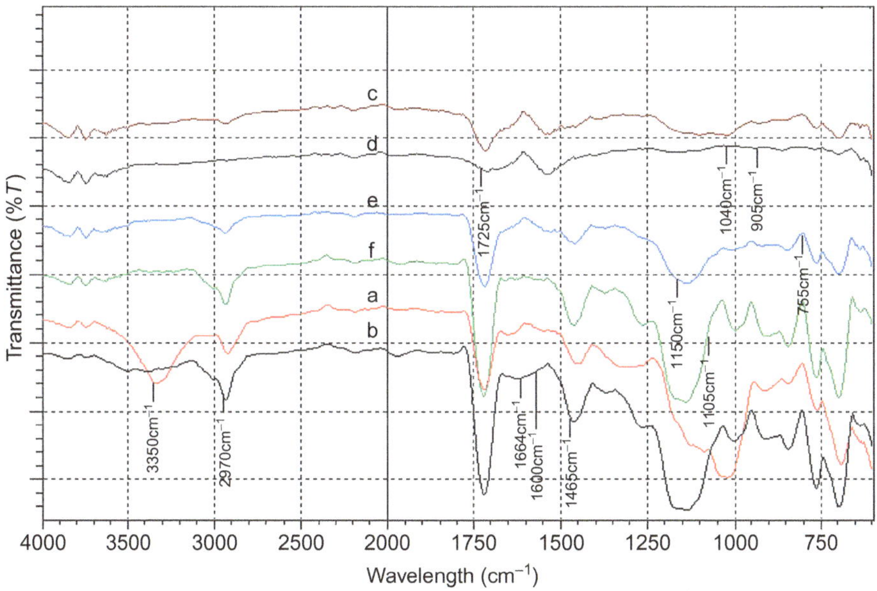

Fig. 6.1 FTIR spectra of (a) ST-*co*-GMA, (b) ST-*co*-GMA-fsi nanocomposite, (c) ST-*co*-GMA-fsi-clay (1.30E) nanocomposite, 1 wt%, (d) ST-*co*-GMA-fsi-clay (1.30E) nanocomposite, 2 wt%, (e) ST-*co*-GMA-fsi-clay (1.30E) nanocomposite, 3 wt%, and (f) ST-*co*-GMA-fsi-clay (1.30E) nanocomposite, 4 wt%.

to the optimum loading. The existence of ST-*co*-GMA was proven with the appearance of absorption peak at 1725 cm^{-1}, which corresponded to the carbonyl group vibration in all the samples (Hanifah et al., 2013). The peak at 1664 cm^{-1} was attributed to the deformation vibration in the H-O-H groups (Mukasa-Tebandeke et al., 2015). This peak was significant when 1 and 2 wt% of clay (1.30E) was added into the ST-*co*-GMA-fsi nanocomposites as shown in Fig. 6.1c and d. The C=C double bond from ST with GMA was indicated by the absorption peak at 1600 cm^{-1} (Hanifah et al., 2013).

In addition, the absorption peak at 1465 cm^{-1} was attributed to the vibration of CH_2 groups. This peak occurred when the carbon atom from GMA was attached to the carbon atom from ST, as well as forming aliphatic carbons in the main chain and side chain of the copolymer. Broader peaks were observed at 1 and 2 wt% of ST-*co*-GMA-clay (1.30E) nanocomposites, which indicated stronger bonding between ST and GMA compared to other nanocomposites as shown in Fig. 6.1c and d. The C-O stretching vibration was observed due to the carbonyl group, which showed the peak at 1150 cm^{-1} (Bicak, Bulutcu, Senkal, & Gazi, 2001). The absorption peak at 1105 cm^{-1} represented the Si-O-C asymmetric stretching mode in ring opening for ST and GMA (Bellamy, 1975). The intensity of this peak decreased when fumed silica and clay were added into the ST-*co*-GMA. This was due to the CH_3 groups attached to the Si-O-Si ring that filled the voids within ST-*co*-GMA (Jing, Lee, & Choi, 2002). The absorption peak at 1040 cm^{-1} was assigned to Si-O-Si bonding in the

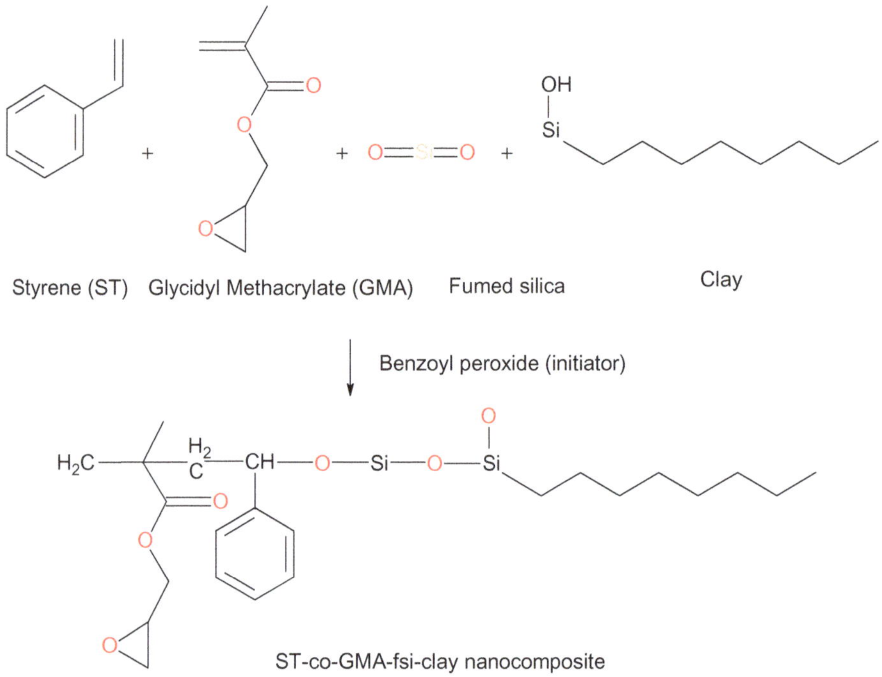

Fig. 6.2 Schematic reaction of ST-*co*-GMA-fsi-clay (1.30E) nanocomposites in the presence of benzoyl peroxide.

nanocomposites, which confirmed the bonding between the silica and clay (1.30E) (Lv, Fang, & Chen, 2009). IR spectra at 950 cm^{-1} represented the small number of unreacted Si-CH_3 bonds that existed within the nanocomposites (Jing et al., 2002). The clay (1.30E) showed less transmittance among all the nanocomposites, which confirmed strong bonding with ST-*co*-GMA-fsi. Therefore, it could be concluded that 2 wt% of clay (1.30E) was more suitable to be incorporated into ST-*co*-GMA-fsi nanocomposites among all the nanocomposites.

6.2.2 Scanning electron microscopy analysis

Scanning electron microscopy (SEM) is used to explain the morphology of the ST-*co*-GMA, ST-*co*-GMA-fsi, and different weight loadings of clay loaded ST-*co*-GMA-fsi-clay (1.30E) nanocomposites.

The fracture surface of ST-*co*-GMA is shown in Fig. 6.3A. This indicated the well interaction between ST and GMA (Yang, Hu, Wang, Qin, & Guo, 2004). From Fig. 6.3B, ST-*co*-GMA-fsi nanocomposite showed a moderately even surface, which proved that silica improved the compatibility with ST-*co*-GMA by forming strong covalent bonds (Kim et al., 2015). From Fig. 6.3C and D, the nanocomposites showed uniform surface morphology due to the 1 and 2 wt% of clay loading. This occurred as the clay aggregated into small particles after polymerization and was dispersed

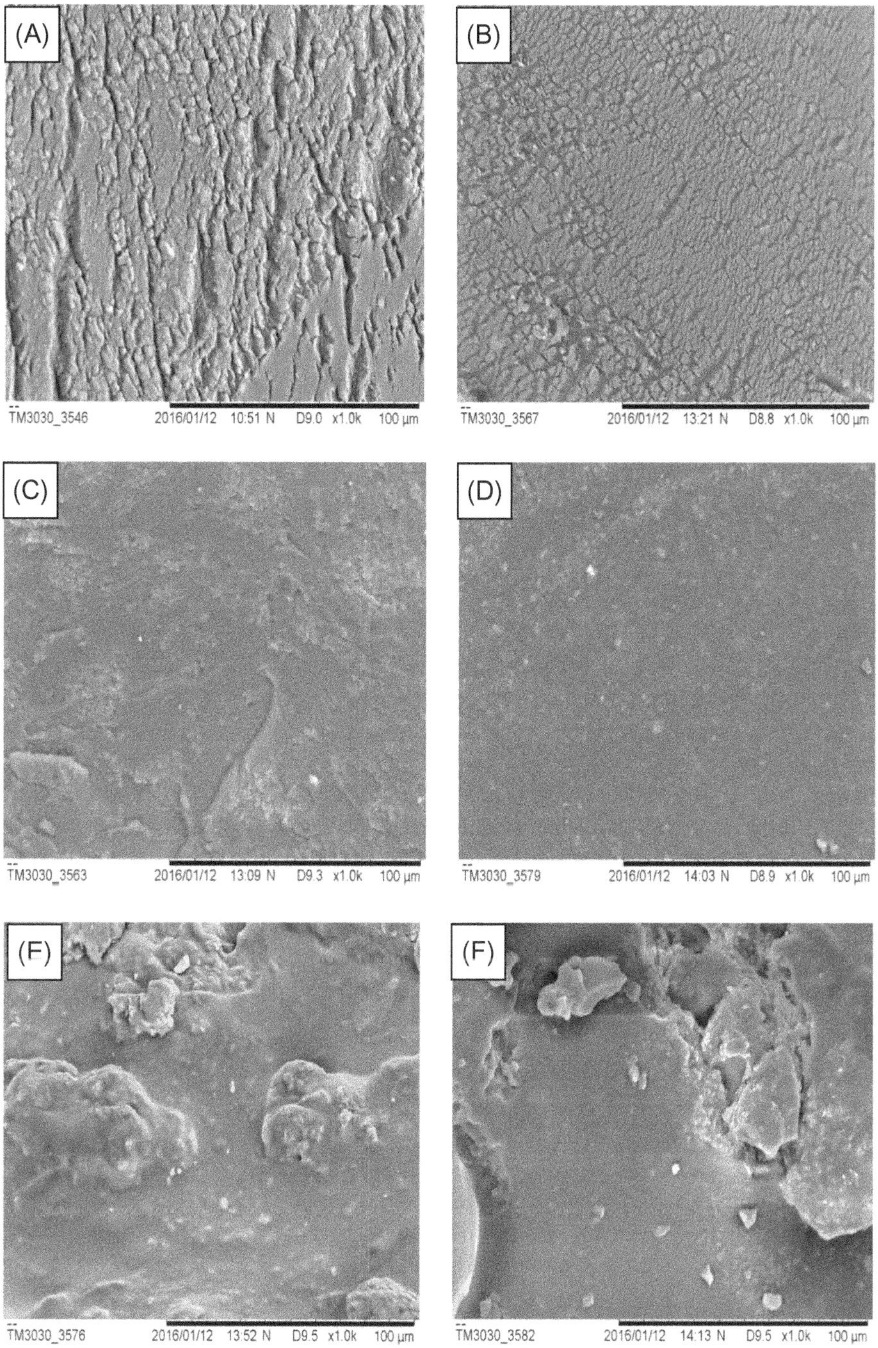

Fig. 6.3 SEM micrographs of (A) ST-*co*-GMA, (B) ST-*co*-GMA-fsi nanocomposite, (C) ST-*co*-GMA-fsi-clay (1.30E) nanocomposite, 1 wt%, (D) ST-*co*-GMA-fsi-clay (1.30E) nanocomposite, 2 wt%, (E) ST-*co*-GMA-fsi-clay (1.30E) nanocomposite, 3 wt%, and (F) ST-*co*-GMA-fsi-clay (1.30E) nanocomposite, 4 wt%.

homogeneously into the ST-*co*-GMA-fsi (Fan, Liu, Chen, & Qi, 2002). Higher clay loading (3 and 4 wt%) into ST-*co*-GMA-fsi would cause agglomeration as shown in Fig. 6.3E and F. This was due to clay loading over 2 wt%, which prevented the intercalation mechanism and led to a large agglomerate (Flitz, 2008).

From the observations, it could be concluded that there was good compatibility as well as dispersion of 2 wt% of clay (1.30E) into ST-*co*-GMA-fsi. The uniform surface of 2 wt% of ST-*co*-GMA-fsi-clay (1.30E) nanocomposites proved the better intercalation between clay and ST-*co*-GMA-fsi, as reflected in the mechanical and thermal properties.

6.2.3 Adsorption isotherm analysis

The N_2 adsorption isotherms for ST-*co*-GMA, ST-*co*-GMA-fsi, and ST-*co*-GMA-fsi-clay (1.30E) nanocomposites are shown in Fig. 6.4. Specific surface area (S_{BET}) is calculated by the BET equation (Brunauer, Emmett, & Teller, 1938).

Fig. 6.4 showed that 2 wt% of ST-*co*-GMA-fsi-clay (1.30E) nanocomposite showed a rather straight section, which extended up to $P/P_0 = 0.55$ whereas 1, 3, and 4 wt% of ST-*co*-GMA-fsi-clay (1.30E) nanocomposites showed extension up to $P/P_0 = 0.43$. ST-*co*-GMA and ST-*co*-GMA-fsi nanocomposites showed lowest extension, which was up to $P/P_0 = 0.26$. This proved that 2 wt% of ST-*co*-GMA-fsi-clay (1.30E) nanocomposite had strong interfacial bonding, which reduced the amount adsorbed on the nanocomposite.

From Table 6.1, the incorporation of silica into ST-*co*-GMA showed higher surface area and average pore volume compared to ST-*co*-GMA. This showed that silica improved the interfacial bonding between ST and GMA. The ST-*co*-GMA-fsi-clay (1.30E) nanocomposite had higher surface area and average pore volume with smaller pore size among all the nanocomposites. This was due to the uniform dispersion of

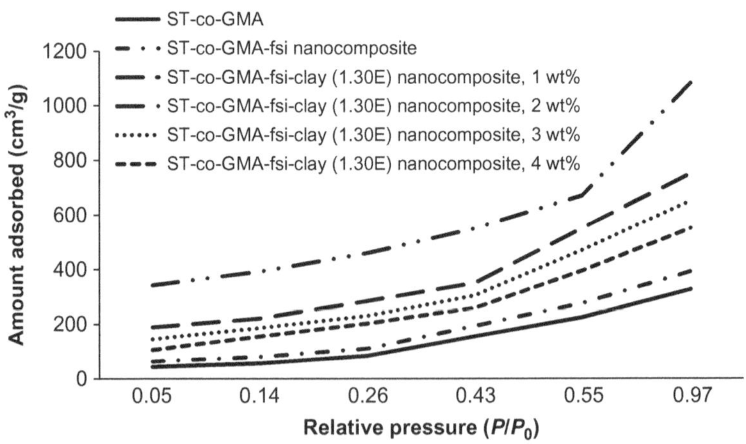

Fig. 6.4 N_2 adsorption isotherms of ST-*co*-GMA, ST-*co*-GMA-fsi, and ST-*co*-GMA-fsi-clay (1.30E) nanocomposites.

Table 6.1 Physical properties detected from N_2 adsorption at 77 K on ST-*co*-GMA, ST-*co*-GMA-fsi, and ST-*co*-GMA-fsi-clay (1.30E) nanocomposites

Samples	Specific surface area, S_{BET} (m^2/g)	Average pore volume, V_m (10^{-4} cm^3/g)	d_{BET} (nm)	Type of isotherms
ST-*co*-GMA	238.29	8.07	1.94	IV
ST-*co*-GMA-fsi nanocomposite	254.23	8.41	1.57	IV
ST-*co*-GMA-fsi-clay (1.30E) nanocomposite, 1 wt%	309.71	10.97	0.65	IV
ST-*co*-GMA-fsi-clay (1.30E) nanocomposite, 2 wt%	318.52	11.82	0.63	IV
ST-*co*-GMA-fsi-clay (1.30E) nanocomposite, 3 wt%	288.64	9.86	0.78	IV
ST-*co*-GMA-fsi-clay (1.30E) nanocomposite, 4 wt%	282.56	9.73	0.81	IV

2 wt% of clay particles in the nanocomposite, which created strong interfacial interaction between the clay and ST-*co*-GMA-fsi (Hazarika & Maji, 2014). Higher clay loadings would reduce the strong interfacial interaction, and thus the surface area and average pore volume decreased. From Table 6.1, the adsorption isotherms indicated that the pores were mesoporous. According to original IUPAC classification, the isotherm patterns of ST-*co*-GMA, ST-*co*-GMA-fsi, and ST-*co*-GMA-fsi-clay (1.30E) nanocomposites were type IV isotherms (Sing et al., 1985).

Therefore, it was proved that 2 wt% of ST-*co*-GMA-fsi-clay (1.30E) nanocomposite had the highest surface area and average pore volume with less pore size, which was reflected in the surface morphology.

6.2.4 Tensile properties

Tensile strength of ST-*co*-GMA, ST-*co*-GMA-fsi, and ST-*co*-GMA-fsi-clay (1.30E) nanocomposites are presented in Fig. 6.5. From Fig. 6.5, the tensile strength of the nanocomposites increased with the clay loading and reached a maximum at 2 wt% of clay (Surej, Praseetha, & George, 2011). The clay particles filled the spaces between the ST-*co*-GMA-fsi chains, thus giving them a rigid structure with better tensile strength value of 0.0524 and 0.0561 GPa (Ahmed, Al-Maamori, & Ali, 2015). However, the addition of clay decreased the strength as the agglomeration took place at higher clay content. The higher clay loadings such as 3 and 4 wt% occurred in

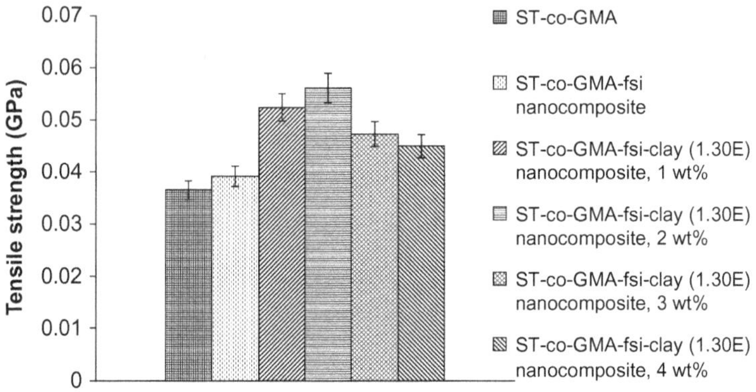

Fig. 6.5 Tensile strength of ST-*co*-GMA, ST-*co*-GMA-fsi, and ST-*co*-GMA-fsi-clay (1.30E) nanocomposites.

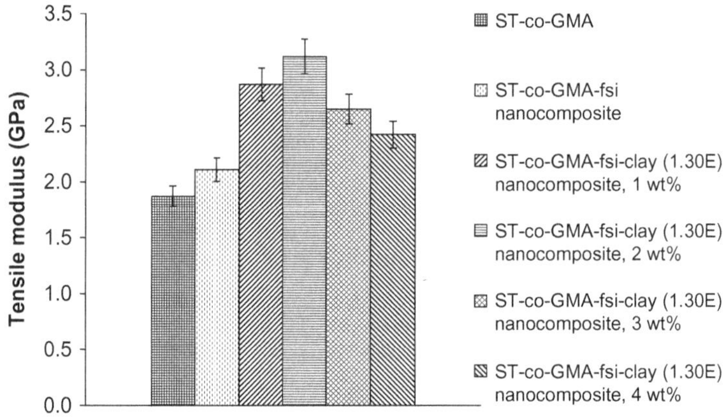

Fig. 6.6 Tensile modulus of ST-*co*-GMA, ST-*co*-GMA-fsi, and ST-*co*-GMA-fsi-clay (1.30E) nanocomposites.

aggregates in ST-*co*-GMA-fsi chains, which weakened the bonds in the nanocomposites and thus the tensile strength decreased (Abbas & Al-Husnawi, 2014).

Fig. 6.6 showed the tensile modulus of ST-*co*-GMA, ST-*co*-GMA-fsi, and ST-*co*-GMA-fsi-clay (1.30E) nanocomposites. The ST-*co*-GMA-fsi-clay (1.30E) nanocomposite with 2 wt% showed the highest tensile modulus compared to ST-*co*-GMA, ST-*co*-GMA-fsi, 1, 3, and 4 wt% of ST-*co*-GMA-fsi-clay (1.30E) nanocomposites. The clay particles at lower loading (1 and 2 wt%) were easily dispersed within ST-*co*-GMA-fsi due to the small size of the particles and therefore, the tensile modulus of the nanocomposites increased (Chang & Chang, 1996). However, the increase in quantity of clay reduced the tensile modulus. This was due to the heterogeneous distribution of the clay with ST-*co*-GMA, which led to the agglomeration of clay particles within the nanocomposites (Abbas & Al-Husnawi, 2014).

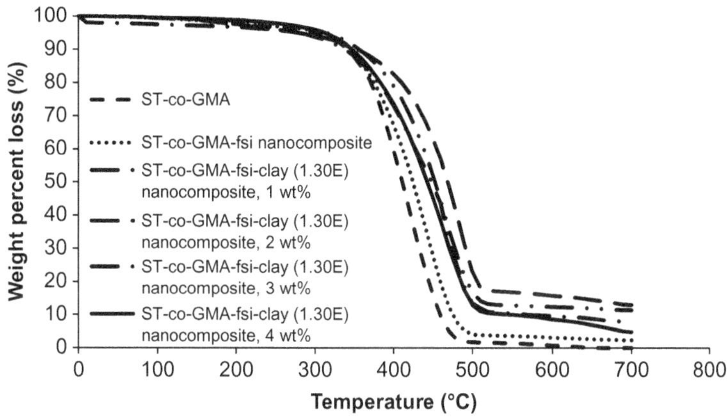

Fig. 6.7 TGA curves of ST-*co*-GMA, ST-*co*-GMA-fsi, and ST-*co*-GMA-fsi-clay (1.30E) nanocomposites.

It was concluded that 2 wt% of clay (1.30E) provided the best tensile strength and modulus improvement into ST-*co*-GMA-fsi nanocomposite. This confirmed that clay (1.30E) at lower loading provided better intercalation and compatibility toward the ST-*co*-GMA-fsi.

6.2.5 Thermogravimetric analysis

The thermal stability of ST-*co*-GMA, ST-*co*-GMA-fsi, and ST-*co*-GMA-fsi-clay (1.30E) nanocomposites are shown in Fig. 6.7. Fig. 6.7 shows that the thermal stability of ST-*co*-GMA-fsi-clay (1.30E) nanocomposites is significantly higher compared to ST-*co*-GMA and ST-*co*-GMA-fsi nanocomposites.

The nanocomposites were thermally degraded in two stages. The first step of the thermal degradation was dehydration, which occurred at about 310–380°C. This occurred at the end chain of the ST-*co*-GMA, followed by random chain scissions (Zhuang, Qian, Lv, & Wan, 2010). Based on the TGA (thermogravimetric analysis) thermograph, the weight loss was 3.8%, 3.0%, 2.5%, 2.3%, 2.7%, and 2.9% for ST-*co*-GMA, ST-*co*-GMA-fsi, and (1, 2, 3, and 4 wt%) of ST-*co*-GMA-fsi-clay (1.30E) nanocomposites respectively. The second thermal degradation of the nanocomposites took place at around 480°C where nitrogen gas evolved (Jitianu, Lammers, Arbuckle-Kiel, & Klein, 2012). The weight loss of ST-*co*-GMA, ST-*co*-GMA-fsi, and (1, 2, 3, and 4 wt%) of ST-*co*-GMA-fsi-clay (1.30E) was 96.1%, 94.7%, 86.1%, 84.8%, 87.7%, and 92.3%. From the two thermal degradation, it showed that ST-*co*-GMA-fsi-clay (1.30E) nanocomposites started the thermal degradation at a higher temperature relative to ST-*co*-GMA and ST-*co*-GMA-fsi nanocomposites. This was due to the degradation temperature of silica at 360°C, which led to fast thermal degradation on ST-*co*-GMA-fsi nanocomposite. For ST-*co*-GMA, the thermal degradation started at lower temperature among all the nanocomposites as there were weak covalent bonds between ST and GMA (Jiang,

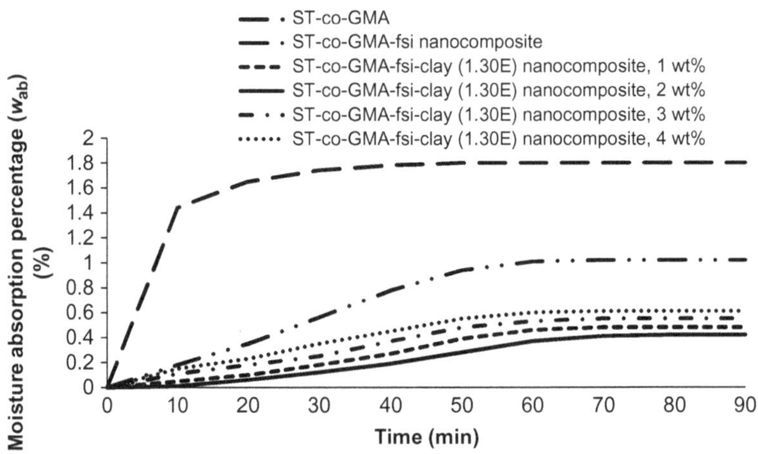

Fig. 6.8 Moisture absorption curves of ST-*co*-GMA, ST-*co*-GMA-fsi, and ST-*co*-GMA-fsi-clay (1.30E) nanocomposites.

Deng, & Yang, 2008). Therefore, the incorporation of 2 wt% of clay (1.30E) showed the best compatibility as proven by SEM result.

The activation energy of ST-*co*-GMA, ST-*co*-GMA-fsi, and ST-*co*-GMA-fsi-clay (1.30E) nanocomposites were calculated using Arrhenius equation (Chanmal & Jog, 2008). Table 6.2 shows that the activation energy of ST-*co*-GMA-fsi-clay (1.30E) nanocomposites were significantly higher than that of ST-*co*-GMA and ST-*co*-GMA-fsi nanocomposites. This showed that clay loaded nanocomposites had greater thermal stability. Overall, 2 wt% of clay (1.30E) was compatible for incorporation into ST-*co*-GMA-fsi nanocomposite to enhance the thermal properties.

6.2.6 Moisture absorption analysis

The moisture absorption of ST-*co*-GMA, ST-*co*-GMA-fsi, and ST-*co*-GMA-fsi-clay (1.30E) nanocomposites were clearly shown in Fig. 6.8. From Fig. 6.8, ST-*co*-GMA showed a sharp increment of moisture absorption in the first 10 min, and it remained constant after 50 min. ST-*co*-GMA-fsi and various clay loading of ST-*co*-GMA-fsi-clay (1.30E) nanocomposites showed the same moisture absorption pattern with a gradual increment up to 60 min.

From Fig. 6.8, ST-*co*-GMA showed the highest moisture absorption due to the lower entanglement of molecules on both sides of the interface. Thus, weak interfacial bonds were formed between ST and GMA (Liang & Yang, 2011). Silica improved the adhesion with ST-*co*-GMA, which increased the moisture absorption ability by reducing the water diffusion pathway (Zeng, Yu, Lu, & Paul, 2005). Incorporation of clay generated better tortuous pathway and increased the barrier property for water diffusion within the nanocomposites compared to incorporation of silica (Devi & Maji, 2012). Clay (1.30E) acted as barriers to the hindrance of the moisture into the ST-*co*-GMA-fsi (Leszczynska, Njuguna, Pielichowski, & Banerjee, 2007).

Table 6.2 Activation energy of ST-*co*-GMA, ST-*co*-GMA-fsi, and ST-*co*-GMA-fsi-clay (1.30E) nanocomposites determined by Arrhenius equation

Samples	T_i (°C)[a]	T_m (°C)[b]	T_f (°C)[c]	W_{Ti} (%)[d]	W_{Tm} (°C)[e]	W_{Tf} (°C)[f]	Activation energy, E_a (kJ/mol)
ST-*co*-GMA	244.5	492.5	648.7	96.2	1.8	0.1	40.9
ST-*co*-GMA-fsi nanocomposite	246.8	493.5	693.5	97.0	4.7	2.3	43.4
ST-*co*-GMA-fsi-clay (1.30E) nanocomposite, 1 wt%	249.5	500.0	693.5	97.5	17.0	11.4	67.1
ST-*co*-GMA-fsi-clay (1.30E) nanocomposite, 2 wt%	250.5	501.2	693.5	97.7	22.3	12.9	68.9
ST-*co*-GMA-fsi-clay (1.30E) nanocomposite, 3 wt%	248.5	495.6	693.5	97.3	15.8	9.6	54.9
ST-*co*-GMA-fsi-clay (1.30E) nanocomposite, 4 wt%	247.2	495.2	693.5	97.1	14.9	4.8	50.1

[a] Temperature corresponding to the beginning of decomposition.
[b] Temperature corresponding to the maximum rate of mass loss.
[c] Temperature corresponding to the end of decomposition.
[d] Mass loss at temperature corresponding to the beginning of decomposition.
[e] Mass loss at temperature corresponding to the maximum rate of mass loss.
[f] Mass loss at temperature corresponding to the end of decomposition.

The ST-*co*-GMA-fsi-clay (1.30E) nanocomposites 1 and 2 wt% showed lower moisture absorption (<0.5%) among all the nanocomposites. This was due to the nano-sized clay particles that penetrated the ST-*co*-GMA-fsi with strong interfacial bonding and reduced the ability to absorb water through the nanocomposite structure (Rozman, Kumar, Abusamah, & Saad, 1998). Higher clay loadings (3 and 4 wt%) created lots of tiny in-between spaces due to excessive clay within the nanocomposites, which created abundant surface area, allowing water molecules to enter the nanocomposites (Barshad, 1952).

Throughout this study, 2 wt% of ST-*co*-GMA-fsi-clay nanocomposites significantly reduced the diffusion of moisture into the nanocomposites and enhanced the compatibility between clay (1.30E) and ST-*co*-GMA-fsi.

6.3 Summary

ST-*co*-GMA-fsi-clay (1.30E) nanocomposites were prepared via free radical polymerization. ST-*co*-GMA-fsi bonding and fsi-clay bonding were confirmed by Si-O-C and Si-O-Si peaks in FTIR spectroscopy. From the surface morphology, it was found that the dispersion of 2 wt% of clay (1.30E) significantly enhanced the compatibility of the nanocomposites. The surface analysis proved that the 2 wt% of ST-*co*-GMA-fsi-clay (1.30E) nanocomposite had higher surface area and pore volume with smaller pore size compared to ST-*co*-GMA and other nanocomposites. The tensile strength and modulus of 2 wt% of ST-*co*-GMA-fsi-clay (1.30E) nanocomposite was the highest, which confirmed the best compatibility of clay (1.30E) into ST-*co*-GMA-fsi nanocomposite. The clay (1.30E) with 2 wt% improved the thermal stability significantly due to the strong interfacial bonding within the nanocomposite. The strong interfacial bonding formed in 2 wt% of ST-*co*-GMA-fsi-clay (1.30E) nanocomposite led to higher water resistance. It was concluded that 2 wt% of ST-*co*-GMA-fsi-clay (1.30E) nanocomposite enhanced the physical, mechanical, and morphological properties as well as thermal stability.

References

Abbas, F., & Al-Husnawi, H. (2014). *A study of the effect of zinc oxide on physical properties of NR/SBR blends*. Iraq: University of Kufa.

Ahmed, J. K., Al-Maamori, M. H., & Ali, H. M. (2015). Effect of nano silica on the mechanical properties of styrene-butadiene rubber (SBR) composite. *International Journal of Materials Science and Applications*, *4*(7), 15–20.

Barrantes, I. M., Valentin, J. L., Rodriquez, A., Garrido, I. Q., & Paris, R. (2012). Poly(styrene)/silica hybrid nanoparticles prepared via ARTP as high quality fillers in elastomeric composites. *Journal of Materials Chemistry*, *22*, 1403–1410.

Barshad, I. (1952). Adsorptive and swelling properties of clay-water system. *Clays and Clay Minerals*, *1*(1), 70–77.

Bayramoglu, G., Senkali, B. F., & Arica, M. Y. (2013). Preparation of clay-poly(glycidyl methacrylate) composite support for immobilization of cellulose. *Applied Clay Science*, *85*, 88–95.

Bellamy, L. J. (1975). *The infra-red spectra of complex molecules*. London: Chapman and Hall.

Bicak, N., Bulutcu, N., Senkal, B. F., & Gazi, M. (2001). Modification of crosslinked glycidyl methacrylate-based polymers for boron-specific column extraction. *Reactive and Functional Polymers*, *47*(3), 175–184.

Brunauer, B., Emmett, P. H., & Teller, E. (1938). Adsorption of gases in multimolecular layers. *Journal of the American Chemical Society*, *60*(2), 309–319.

Chang, C. R., & Chang, F. C. (1996). Polymer blends of polyamide-6 and poly(phenylene oxide) *co*mpatibilized by styrene-*co*-glycidyl methacrylate. *Journal of Applied Polymer Science*, *61*(13), 2411–2421.

Chanmal, C. V., & Jog, J. P. (2008). Dielectric relaxations in PVDF/$BaTiO_3$ nanocomposites. *eXPRESS Polymer Letter*, *2*(4), 294–301.

Clemons, T. (2013). *Applications of multifunctional poly(glycidyl methacrylate) (PGMA) nanoparticles in enzyme stabilization and drug delivery*. Australia: The University of Western Australia.

Conzatti, L. (2012). The clay mineral modifier as the key to sheer the properties of rubber nanocomposites. *Applied Clay Science*, *61*, 14–21.

Devi, R. R., & Maji, T. K. (2012). Chemical modification of simul wood with styrene acrylonitrile copolymer and organically modified nanoclay. *Wood Science and Technology*, *46*(1), 299–315.

Fan, J., Liu, S., Chen, G., & Qi, Z. (2002). SEM study of a polystyrene/clay nanocomposite. *Journal of Applied Polymer Science*, *83*(1), 66–69.

Flitz, C. (2008). *Preparation and characterization of recycled polypropylene based nanocomposites*. Turkey: Middle East Technical University.

Grama, S., Boiko, N., Bilyy, R., Klyuchivska, O., Antonyuk, V., Stoika, R., et al. (2014). Novel fluorescent poly(glycidyl methacrylate)-silica microspheres. *European Polymer Journal*, *56*, 92–104.

Grim, R. E. (1968). *Clay mineralogy*. New York: McGraw-Hill.

Hanifah, S. A., Hamzah, N., & Heng, L. Y. (2013). Rapid synthesis of magnetic microspheres poly(glycidyl methacrylate-*co*-styrene) by photopolymerization. *Sains Malaysiana*, *42*(4), 487–493.

Hazarika, A., & Maji, T. K. (2014). Modification of softwood by monomers and nanofillers. *Defence Science Journal*, *64*(3), 262–272.

Jiang, S., Deng, J., & Yang, W. (2008). Block copolymers prepared by free radical polymerization using α-methylstyrene-containing precopolymer as macroinitiator. *Polymer Journal*, *40*, 543–548.

Jing, S. Y., Lee, H. J., & Choi, C. K. (2002). Chemical bond structure on Si-O-C composite films with a low dielectric constant deposited by using inductively coupled plasma chemical vapour deposition. *Journal of the Korean Physical Society*, *41* (5), 769–773.

Jitianu, A., Lammers, K., Arbuckle-Kiel, G. A., & Klein, L. C. (2012). Thermal analysis of organically modified siloxane melting gels. *Journal of Thermal Analysis and Calorimetry*, *107*(3), 1039–1045.

Jlassi, K., Chandran, S., Micusik, M., Benna-Zayani, M., Yagci, Y., Thomas, S., et al. (2015). Poly(glycidyl methacrylate)-grafted clay nanofiller for highly transparent and mechanically robust epoxy composites. *European Polymer Journal*, *72*, 89–101.

Kim, K., Seo, B., Lee, J. Y., Choi, B. J., Kwag, G. H., Paik, H. J., et al. (2015). Reduced filler flocculation in the silica-filled styrene-butadiene-glycidyl methacrylate terpolymer. *Composite Interfaces*, *22*(2), 137–149.

Kunita, M. H., Girotto, E. M., Muniz, E. C., & Rubira, A. F. (2006). Polypropylene grafted with glycidyl methacrylate using supercritical CO_2 medium. *Brazilian Journal of Chemical Engineering*, *23*(2), 267–271.

Leszczynska, A., Njuguna, J., Pielichowski, K., & Banerjee, J. R. (2007). Polymer/montmorillonite nanocomposites with improved thermal properties: Part I. Factors influencing thermal stability and mechanisms of thermal stability improvement. *Thermochimica Acta*, *453*(2), 75–96.

Liang, S. J., & Yang, W. T. (2011). Functionalization of polypropylene and compatibilization of polypropylene/nylon 6 blends with copolymers containing α-methyl styrene units. *Acta Polymerica Sinica*, *11*(2), 180–185.

Lv, J. N., Fang, S. J., & Chen, L. (2009). Preparation and characterization of affinity polymer basic microspheres by soap-free emulsion polymerization. *Chinese Journal of Polymer Science*, *27*(1), 101–108.

Mirous, B. K. (2006). *Synthesis and presumptive crosslinking of stimuli–response diblock polymer brushes*. USA: The University of Akron.

Mukasa-Tebandeke, I. Z., Ssebuwufy, P. J. M., Nyanzi, S. A., Schumann, A., Nyakairu, G. W. A., Ntale, M., et al. (2015). The elemental, mineralogical, IR, DTA and XRD analyses characterized clays and clay minerals of central and Eastern Uganda. *Advances in Materials Physics and Chemistry*, *5*, 67–86.

Nemati, N., Taromi, F. A., & Mirzataheri, M. (2015). *In-situ* water based emulsion copolymerization of styrene/ethylhexyl acrylate in presence of Cloisite 15A. *Procedia Materials Science*, *11*, 542–547.

Nita, L. E., Chiriac, A. P., Cimmino, C., Duraccio, D., & Vasile, C. (2008). Polymerization in magnetic field. XIX. Thermal behavior of the copolymers of methyl methacrylate with glycidyl methacrylate synthesized in the magnetic field presence. *The Open Macromolecules Journal*, *2*(1), 26–31.

Ou, B., Yang, G., Xiao, Y., Zeng, X., Zhou, Z., Liu, Q., et al. (2013). Covalent functionalization of silica nanoparticle with poly(glycidyl methacrylate) via ATRP at ambient temperature. *Journal of Macromolecular Science, Part A: Pure and Applied Chemistry*, *50*(1), 25–28.

Rozman, H. D., Kumar, R. N., Abusamah, A., & Saad, M. J. (1998). Miscibility between natural rubber and tackifiers. II. Phase diagrams of the blends of natural rubber and petroleum resins. *Journal of Applied Polymer Science*, *67*(2), 221–229.

Sing, K. S. W., Everett, D. H., Haul, R. A. W., Moscou, L., Pierotti, R. A., Rouquerol, J., et al. (1985). Reporting physisorption data for gas/solid systems with special reference to the determination of surface area and porosity (Recommendations 1984). *International Union of Pure and Applied Chemistry*, *57*(4), 603–619.

Surej, R. C., Praseetha, P. N., & George, K. E. (2011). In situ addition polymerization of styrene using nanoclay and evaluation of properties. *International Journal of Engineering Research and Application*, *2*(4), 1045–1049.

Xu, H., Li, Y., & Yu, D. (2005). Studies on the poly(styrene-*b*-butadiene-*b*-styrene)/clay nanocomposites prepared by melt intercalation. *Journal of Applied Polymer Science*, *98*(1), 146–152.

Yang, J., Hu, J., Wang, C., Qin, Y., & Guo, Z. (2004). Fabrication and characterization of soluble multi-walled carbon nanotubes reinforced P(MMA-*co*-EMA) composites. *Macromolecular Materials and Engineering*, *289*(9), 828–832.

Yin, C., Zhang, Q., & Gong, D. (2013). Preparation and properties of silica/styrene butadiene rubber masterbatches by latex co-coagulating technology. *Polymer Composites*, *35*(6), 1212–1219.

Zanchet, A., Carli, L. N., Giocanela, M., Brandalise, R. N., & Crespo, J. S. (2012). Use of styrene butadiene rubber industrial waste devulcanized by microwave in rubber composites for automotive application. *Materials Design*, *39*, 437–443.

Zeng, Q. H., Yu, A. B., (Max) Lu, G. Q., & Paul, D. R. (2005). Clay-based polymer nanocomposites: Research and commercial development. *Journal of Nanoscience and Nanotechnology*, *5*(10), 1574–1592.

Zhuang, X., Qian, X., Lv, J., & Wan, Y. (2010). An alternative method to remove PEO–PPO PEO template in organic–inorganic mesoporous nanocomposites by sulfuric acid extraction. *Applied Surface Science*, *256*(17), 5343–5348.

Physico-mechanical and thermal properties of clay/fumed silica diffuse polylactic acid nanocomposites

7

Md. Rezaur Rahman, Josephine Lai Chang Hui
Department of Chemical Engineering and Energy Sustainability, Faculty of Engineering, University Malaysia Sarawak (UNIMAS), Kota Samarahan, Sarawak, Malaysia

7.1 Introduction

Polymer nanocomposites have gradually come into use worldwide; this is due to the remarkable thermal and mechanical properties that they exhibit with respect to conventional polymer composites (Alexandre & Dubois, 2000). The incorporation of nanoscale filler into PLA is highly recommended as the nanofiller improves the physical, mechanical, and thermal properties of the conventional PLA.

PLA is linear aliphatic polyester; it attracts the attention of both academic and industry researchers due to its renewability, biodegradability, and biocompatibility (Drumright, Gruber, & Henton, 2000). It is commonly used in biomedical and packaging applications. Due to the poor physical and mechanical properties of pure PLA, it should be further improved by introducing various types of filler.

Nowadays, fumed silica is incorporated into polymer to produce polymer-silica nanocomposites. PLA-SiO_2 hybrids via sol–gel process showed a compatible phase with addition of 3-isocyanatopropyltriethoxysilane (IPTES) as a silane coupling agent to create covalent bonds between the organic and inorganic phases (Bang & Kim, 2012). In addition, twin-screw extruded PLA-silica nanocomposites showed significant improvement in thermal stability and was qualitatively consistent in the increments of glass transition temperature (Zhang, Lou, Ilias, Krishnamachari, & Yan, 2008). The addition of silica into PLA nanocomposites successfully synthesized by sol–gel process improved the tensile strength and thermal stability of the nanocomposites. Moreover, silica acting as a compatibilizer is induced into PLA-low-density polyethylene (PLA-LDPE). It improved the compatibility between PLA and LDPE phases and thus, the toughness and crystallinity were significantly improved (Vrsaljko, Macut, & Kovacevic, 2014).

The drawbacks of pure PLA are poor thermal resistance and low mechanical properties compared to other polymer matrices (Chaloupi, 2011). The limitations of using PLA for wider industrial application can be overcome by incorporation of various forms of clay into PLA at low levels. The addition of small amounts of clay improved

Silica and Clay Dispersed Polymer Nanocomposites. https://doi.org/10.1016/B978-0-08-102129-3.00007-5

the mechanical properties of melt compounded PLA-clay nanocomposites (Jollands & Gupta, 2010). Thermal stability was significantly improved when organically modified MMT was incorporated into the PLA-cocamidopropylbetaine (PLA-CAB) blend (Balakrishnan, Hassan, Wahit, Yussuf, & Razak, 2010). Different types of clay dispersed into polymer matrices had a positive impact on the physical and morphological properties of the composites (Ray & Okamoto, 2003).

Good elastic properties of melt intercalated PLA-polycaprolactone-nanoclay (PLA-PCL-nanoclay) proved better compatibility with the nanocomposites (Eng et al., 2013). Solution cast organoclay made up of CAB-MMT was induced into PLA matrix enhanced the thermal stability of the nanocomposites up to 4% (McLauchlin & Thomas, 2009). CAB-MMT was highly dispersed into PLA matrix without a large agglomeration of particles, as MMT acted as an intercalation and nucleating agent (Ou, Ho, & Lin, 2003).

In this present work, solution-intercalation film-casting technique was used to produce PLA-fsi-clay nanocomposites using different types of clays (1.28E, 1.30E, 1.31PS, and 1.34TCN) with different weight percentage (2, 5, and 10 wt%). Physical, mechanical, morphological, and thermal properties of nanocomposites were investigated. The compatibility of different types of clay with various weight loadings PLA-fsi was also reported.

7.2 Results and discussion

7.2.1 Spectral (FTIR) analysis

FTIR analysis was used to study the chemical interaction between the PLA-fsi and clay. The IR spectra of PLA-fsi-clay (1.28E), PLA-fsi-clay (1.30E), PLA-fsi-clay (1.31PS), and PLA-fsi-clay (1.34TCN) nanocomposites are shown in Figs. 7.1 to 7.4, respectively.

Fig. 7.1(a–c) presented the FTIR spectra of PLA-fsi-clay (1.28E) nanocomposites with 2, 5, and 10 wt% clay loadings, respectively. From Fig. 7.1(a), it was observed that 2 wt% of PLA-fsi-clay (1.28E) nanocomposite had broader peaks at 1080, 1450, 1750, and 2946 cm^{-1}. On the other hand, Fig. 7.1(b–c) showed that the absorption band at 1080 cm^{-1} was associated with silanol groups at 5 and 10 wt% clay loadings. The peak at 1080 cm^{-1} overlapped the bands at 1042 cm^{-1} of C—O—C and C—CH_3 of PLA; this was attributed to the C—O stretching vibrations that were impeded in this region (Liu, Zhang, & Zhou, 2013). In addition, the IR spectrum at 1450 cm^{-1} was attributed to the characteristic of C—H stretching and bending of —CH_2 groups, which were clearly shown in PLA-fsi-clay (1.28E) nanocomposites with 5 and 10 wt% clay loadings. A strong absorption band at 1750 cm^{-1} corresponded to the C=O stretching vibration as detected in Fig. 7.1(b–c) (Bocchini et al., 2010). Moreover, the peak at 2946 cm^{-1} was also visibly detected in Fig. 7.1(b–c), assigned to the —CH groups in PLA (Kontou, Niaounakis, & Georgiopoulus, 2011). The IR spectra of PLA-fsi-clay (1.30E) with similar clay loadings are shown in Fig. 7.2(a–c). The FTIR result showed that similar patterns were observed for 2, 5, and 10 wt% of

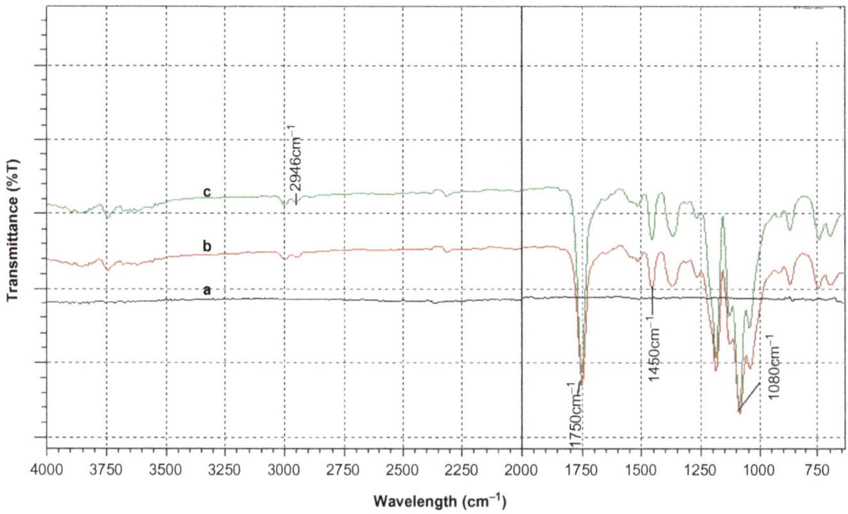

Fig. 7.1 FTIR spectra of (a) PLA-fsi-clay (1.28E) nanocomposite, 2 wt%; (b) PLA-fsi-clay (1.28E) nanocomposite, 5 wt%; and (c) PLA-fsi-clay (1.28E) nanocomposite, and 10 wt%.

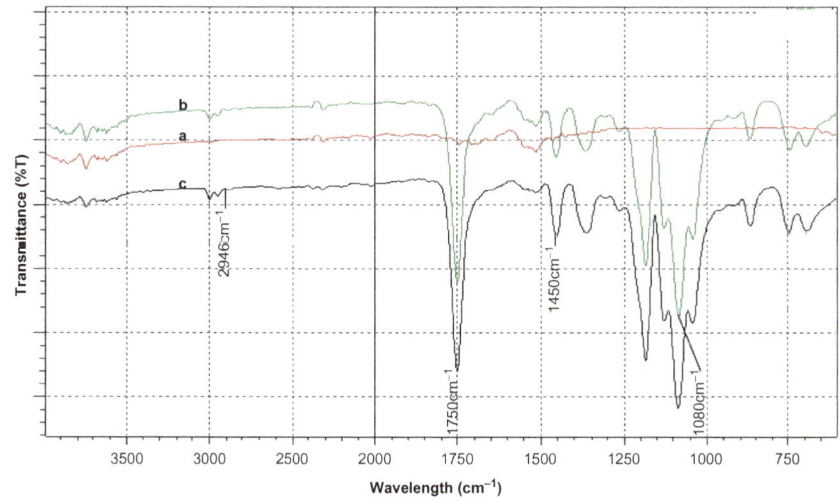

Fig. 7.2 FTIR spectra of (a) PLA-fsi-clay (1.30E) nanocomposite, 2 wt%; (b) PLA-fsi-clay (1.30E) nanocomposite, 5 wt%; and (c) PLA-fsi-clay (1.30E) nanocomposite, and 10 wt%.

PLA-fsi-clay (1.30E) nanocomposites compared to PLA-fsi-clay (1.28E) nanocomposites.

Figs. 7.3(a–c) and 7.4(a–c) showed the IR spectra of PLA-fsi-clay (1.31PS) and PLA-fsi-clay (1.34TCN) nanocomposites, respectively. The peak at 3650 cm^{-1} was detected clearly in Figs. 7.3 and 7.4. This peak was assigned to the vibration of

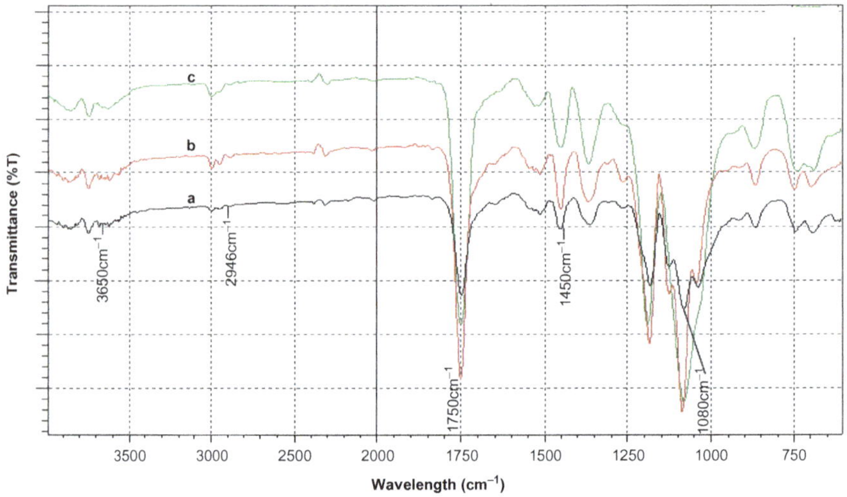

Fig. 7.3 FTIR spectra of (a) PLA-fsi-clay (1.31PS) nanocomposite, 2 wt%; (b) PLA-fsi-clay (1.31PS) nanocomposite, 5 wt%; and (c) PLA-fsi-clay (1.31PS) nanocomposite, and 10 wt%.

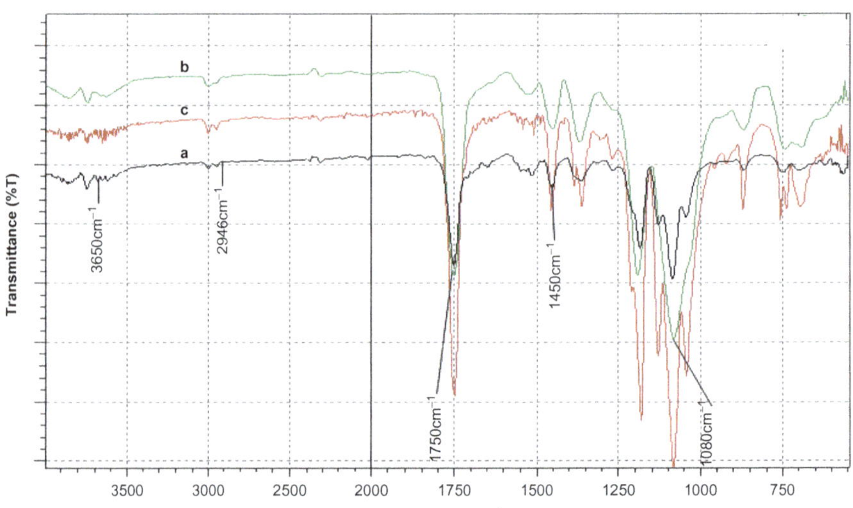

Fig. 7.4 FTIR spectra of (a) PLA-fsi-clay (1.34TCN) nanocomposite, 2 wt%; (b) PLA-fsi-clay (1.34TCN) nanocomposite, 5 wt%; and (c) PLA-fsi-clay (1.34TCN) nanocomposite, and 10 wt%.

different hydroxyl groups located at the octahedral surface of the clay and the moisture absorbed (Yuan et al., 2008). Furthermore, the FTIR spectra of 5 and 10 wt% of PLA-fsi-clay (1.31PS) and PLA-fsi-clay (1.34TCN) nanocomposites showed comparable peaks to PLA-fsi-clay (1.28E) and PLA-fsi-clay (1.30E) nanocomposites.

It was noted that IR spectra for 2 wt% of all PLA-fsi-clay nanocomposites showed much broader peaks compared to 5 and 10 wt% of the nanocomposites. From the IR spectra, it is concluded that clay (1.28E) was the most compatible to be introduced into PLA-fsi compared to clay (1.30E), clay (1.31PS), and clay (1.34TCN).

7.2.2 X-ray fluorescence analysis

X-ray fluorescence (XRF) was performed to identify the compositions of the minerals of PLA-fsi-clay nanocomposites as presented in Table 7.1. The major chemical composition analysis of the PLA-fsi-clay nanocomposites contained silicon, nitrogen, oxygen, and carbon in major quantities and other minor elements (Besling, Goossens, Meester, & Schoonman, 1998). According to the findings, the compositional analysis proved that the clay fitted into the PLA-fsi in different oxide composition (Saikia & Parthasarathy, 2010).

Table 7.1 Oxide contents of PLA-fsi-clay nanocomposites

Samples	Si	N	C	S	K	Cu	Zn	Ti
PLA-fsi-clay (1.28E) nanocomposite, 2 wt%	55.1	2.6	4.6	34.6	0.5	1.4	0.6	0.6
PLA-fsi-clay (1.28E) nanocomposite, 5 wt%	45.6	3.5	5.27	35.3	5.5	3.3	0.7	0.8
PLA-fsi-clay (1.28E) nanocomposite, 10 wt%	37.2	3.8	10.6	39.2	3.8	2.9	1.6	0.9
PLA-fsi-clay (1.30E) nanocomposite, 2 wt%	50.3	10.5	10.0	24.3	0.5	2.0	1.6	0.8
PLA-fsi-clay (1.30E) nanocomposite, 5 wt%	42.6	4.3	5.4	41.1	1.3	3.1	1.3	0.9
PLA-fsi-clay (1.30E) nanocomposite, 10 wt%	41.7	6.7	11.3	33.7	2.0	3.0	1.0	0.6
PLA-fsi-clay (1.31PS) nanocomposite, 2 wt%	42.6	4.3	5.4	41.1	1.3	3.1	1.3	0.9
PLA-fsi-clay (1.31PS) nanocomposite, 5 wt%	41.7	6.7	11.3	33.7	2.0	3.0	1.0	0.6
PLA-fsi-clay (1.31PS) nanocomposite, 10 wt%	50.3	10.5	10.1	24.3	0.5	2.0	1.6	0.8
PLA-fsi-clay (1.34TCN) nanocomposite, 2 wt%	41.7	6.7	11.3	33.7	2.0	3.0	1.0	0.6
PLA-fsi-clay (1.34TCN) nanocomposite, 5 wt%	50.3	10.5	10.1	24.3	0.5	2.0	1.6	0.8
PLA-fsi-clay (1.34TCN) nanocomposite, 10 wt%	42.6	4.3	5.4	41.1	1.3	3.1	1.3	0.9

From Table 7.1, 2 wt% of PLA-fsi-clay nanocomposites showed the highest amount of Si compound. This confirmed the presence of silanol groups in the nanocomposites. An overlapping of symmetric Si—C and Si—N groups for all nanocomposites is clearly shown in Table 7.1. This was due to C and N atoms that were bonded to Si to produce Si—C and Si—N bonds (Luo, Zhao, Cai, & Du, 2012). PLA-fsi-clay (1.28E) nanocomposite contained the highest percentage of the silicon group, followed by PLA-fsi-clay (1.30E), PLA-fsi-clay (1.31PS), and PLA-fsi-clay (1.34TCN) nanocomposites. The silanol group was assigned to Si—O—Si, as it was well-linked with PLA-fsi. From the findings, it proved that 2 wt% of clays was the optimum weight percentage, and PLA-fsi-clay (1.28E) nanocomposites were more compatible compared with other nanocomposites.

7.2.3 Scanning electron microscopy analysis

The surface morphology of PLA-fsi-clay (1.28E), PLA-fsi-clay (1.30E), PLA-fsi-clay (1.31PS), and PLA-fsi-clay (1.34TCN) nanocomposites was studied by scanning electron microscopy (SEM). Figs. 7.5A, 7.6A, 7.7A, and 7.8A was corresponded to the micrographs of PLA-fsi-clay (1.28E), PLA-fsi-clay (1.30E), PLA-fsi-clay (1.31PS), and PLA-fsi-clay (1.34TCN) at 2 wt% clay loading respectively. It was found that 2 wt% of PLA-fsi-clay nanocomposites appeared to have uniform and homogeneous dispersion without aggregation of fumed silica and clay (Mustapa, Shanks, & Kong, 2013). This indicated the strong covalent bond between PLA-fsi matrices with 2 wt% of clays. However, the smoother surface was observed for 2 wt% of PLA-fsi-clay (1.28E) nanocomposite followed by PLA-fsi-clay (1.30E), PLA-fsi-clay (1.31PS), and PLA-fsi-clay (1.34TCN) nanocomposites.

5 wt% of PLA-fsi-clay nanocomposites showed an uneven surface due to the non-homogeneous particle distribution as shown in Figs. 7.5B, 7.6B, 7.7B and 7.8B. From Figs. 7.5C, 7.6C, 7.7C, and 7.8C, agglomeration was observed in 10 wt% of PLA-fsi-clay nanocomposites. This was due to the discrete distribution of pores, and the size of pores increased substantially with higher clay loadings. Thus, PLA-fsi-clay nanocomposites had poor adhesion at higher clay loadings (5 and 10 wt%) due to their high surface energy. Therefore, 2 wt% of PLA-fsi-clay nanocomposites showed better surface morphology compared to 5 and 10 wt%.

Between the various types of clay, clay (1.28E) showed the best compatibility with PLA-fsi. There was no aggregation formed at the surface of PLA-fsi-clay (1.28E) nanocomposite. This indicated that the better compatibility between clay (1.28E) and PLA-fsi matrix had occurred. On the other hand, PLA-fsi-clay (1.30E), PLA-fsi-clay (1.31PS), and PLA-fsi-clay (1.34TCN) nanocomposites showed nonuniform dispersion of clays into PLA-fsi which produced pores and agglomerates on the surface as shown in Figs. 7.6B and C, 7.7B and C and 7.8B and C. The higher clay loadings prevented the agglomerate from breaking into a single particle due to the strong interaction that existed among the nanoparticles (Wen et al., 2011). This showed that clay (1.28E) was more compatible compared to clay (1.30E), clay (1.31PS), and clay (1.34TCN) when introduced into PLA-fsi. The higher percentage of clay loadings

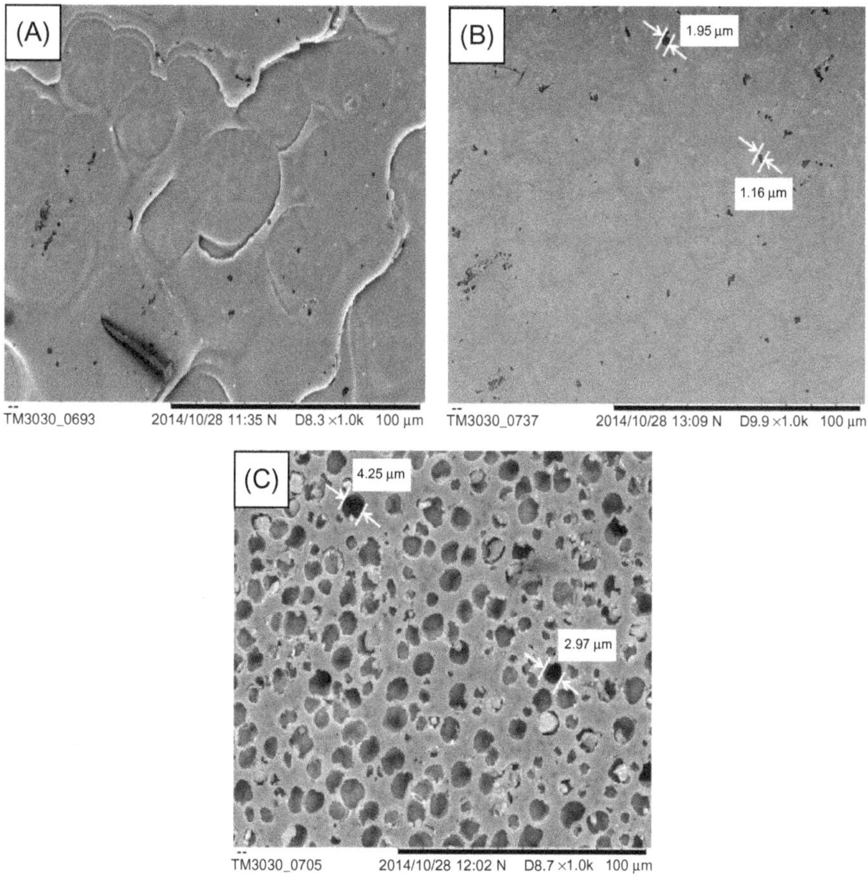

Fig. 7.5 SEM micrographs of (A) PLA-fsi-clay (1.28E) nanocomposite, 2 wt%; (B) PLA-fsi-clay (1.28E) nanocomposite, 5 wt%; and (C) PLA-fsi-clay (1.28E) nanocomposite, and 10 wt%.

produced pores that acted as a stress propagator, which impacted the physical and mechanical properties as well as thermal stability.

7.2.4 Adsorption isotherm analysis

The N_2 adsorption isotherms measures for PLA-fsi-clay nanocomposites at 77 K are shown in Figs. 7.9 to 7.12. Nitrogen uptake by inorganic oxide samples proceeded as multilayer adsorption, followed by capillary condensation (Sing, 2001). The uptake occurred at relatively higher pressure. $P/P_o > 0.2$ indicated the existence of mesoporosity (Alcaniz-Monge, Perez-Cadenas, & Lozano-Castello, 2010). Hysteresis loops observed were associated with the filling and emptying of mesopores. Specific surface area (S_{BET}) was calculated using the Brunauer-Emmett-Teller (BET) equation (Brunauer, Emmett, & Teller, 1938).

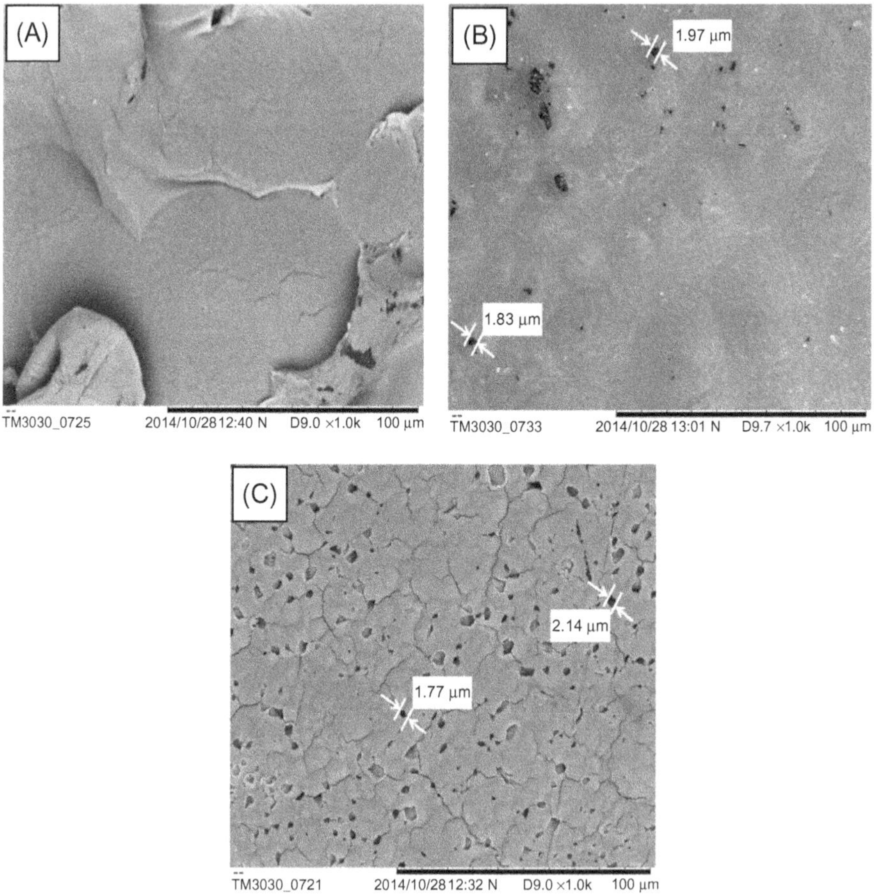

Fig. 7.6 SEM micrographs of (A) PLA-fsi-clay (1.30E) nanocomposite, 2 wt%; (B) PLA-fsi-clay (1.30E) nanocomposite, 5 wt%; and (C) PLA-fsi-clay (1.30E) nanocomposite, and 10 wt%.

Figs. 7.9 to 7.12 showed the N_2 adsorption isotherms for PLA-fsi-clay (1.28E), PLA-fsi-clay (1.30E), PLA-fsi-clay (1.31PS), and PLA-fsi-clay (1.34TCN) nanocomposites, respectively. The isotherms of all PLA-fsi-clay nanocomposites showed an initial ascending section up to $P/P_o = 0.05$. For PLA-fsi-clay (1.28E) nanocomposites, it showed a rather straight section which extended up to $P/P_o = 0.20$, 0.25, and 0.26 respectively for 10, 5, and 2 wt% clay loadings. PLA-fsi-clay (1.30E) and PLA-fsi-clay (1.31PS) nanocomposites showed a rather straight section which extended up to $P/P_o = 0.25$ for all the clay weight loadings. PLA-fsi-clay (1.34TCN) nanocomposites showed an extended straight section up to $P/P_o = 0.20$ for 10 wt% and 0.25 for 2 and 5 wt% clay loadings. Finally, the isotherm exhibited an upward sweep near saturation pressure.

It was clearly seen that PLA-fsi-clay nanocomposites showed higher BET surface area and pore volume with small pore size at 2 wt% clay loading. This proved that clay

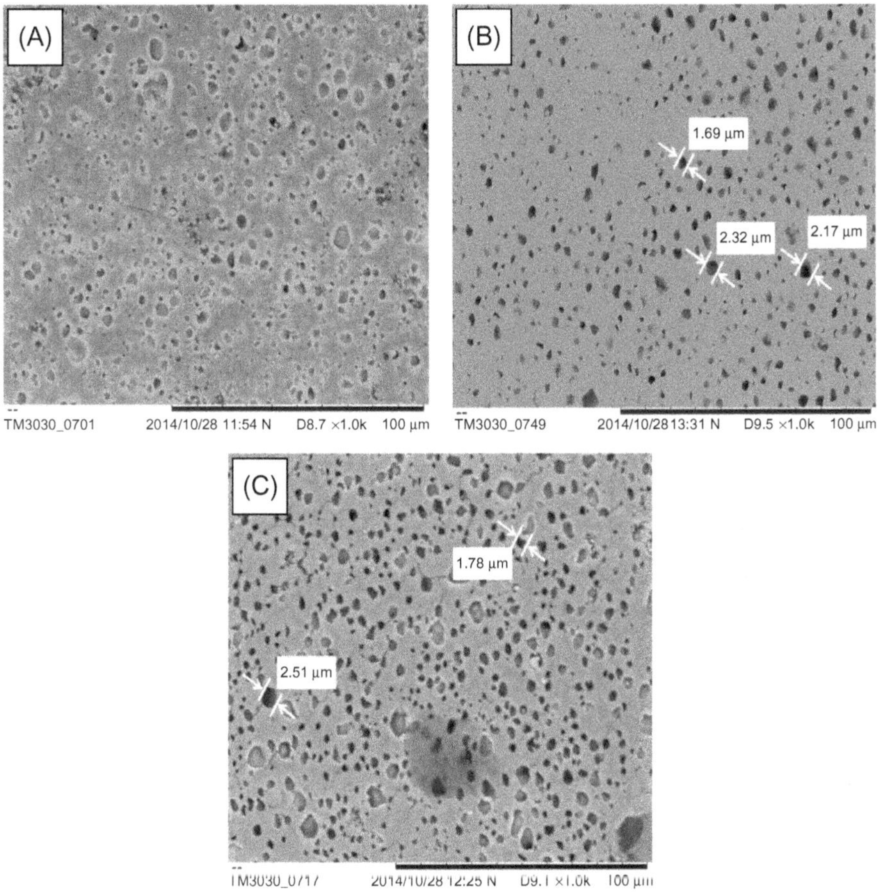

Fig. 7.7 SEM micrographs of (A) PLA-fsi-clay (1.31PS) nanocomposite, 2 wt%; (B) PLA-fsi-clay (1.31PS) nanocomposite, 5 wt%; and (C) PLA-fsi-clay (1.31PS) nanocomposite, and 10 wt%.

(1.28E), clay (1.30E), clay (1.31PS), and clay (1.34TCN) were compatible when PLA-fsi at 2 wt% clay loading as shown in Table 7.2. Clay (1.28E) showed higher BET surface area and pore volume with smaller pore size compared to clay (1.30E), clay (1.31PS), and clay (1.34TCN). This proved that clay (1.28E) was more compatible for introduction into PLA-fsi. From the isotherms, there was a hysteresis loop. Based on the IUPAC classification, the isotherm pattern was identified as type IV isotherms (Sing, 2001). Therefore, the pores of these isotherms were mesoporous. The average pore size had increased with the intercalation of clay into PLA-fsi due to the interlamellar expansion. The formation of a bidimensional porous network consisting of mesopores occurred due to the organic cation exchange process (Frost, Carmody, Xi, & Kokot, 2007).

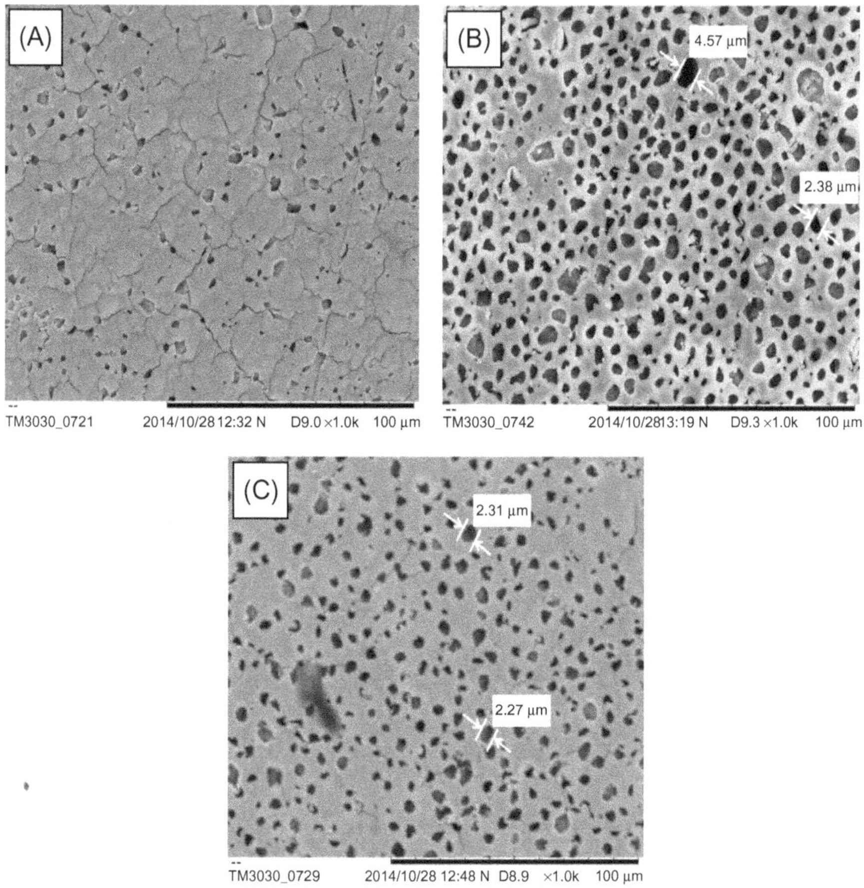

Fig. 7.8 SEM micrographs of (A) PLA-fsi-clay (1.34TCN) nanocomposite, 2 wt%; (B) PLA-fsi-clay (1.34TCN) nanocomposite, 5 wt%; and (C) PLA-fsi-clay (1.34TCN) nanocomposite, and 10 wt%.

Overall, 2 wt% clay loading showed the optimum result compared to 5 and 10 wt% of PLA-fsi-clay nanocomposites. PLA-fsi-clay (1.28E) nanocomposite showed higher surface area and average pore volume with smaller pore size compared to other nanocomposites, which was reflected in the surface morphology.

7.2.5 Tensile properties

The tensile strength and tensile modulus of PLA-fsi-clay nanocomposites are shown in Figs. 7.13 and 7.14, respectively. It was observed that various clays and clay loadings had an effect on tensile strength and tensile modulus of PLA-fsi-clay nanocomposites.

Fig. 7.13 shows a comparison of the tensile strength between four types of clays (1.28E, 1.30E, 1.31PS, and 1.34TCN) with three different weight percentages (2,

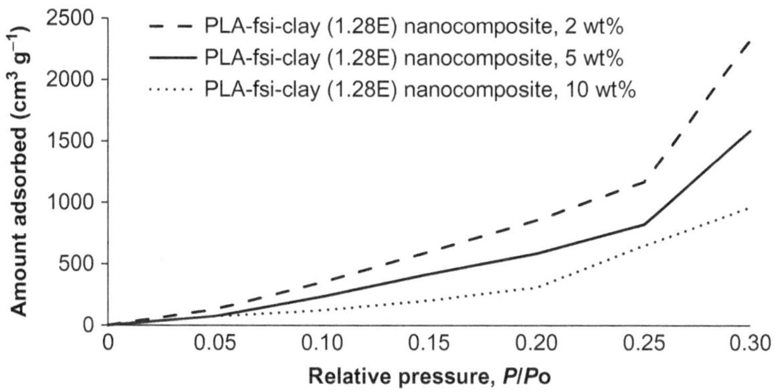

Fig. 7.9 N_2 adsorption isotherms of PLA-fsi-clay (1.28E) nanocomposites with different clay loadings.

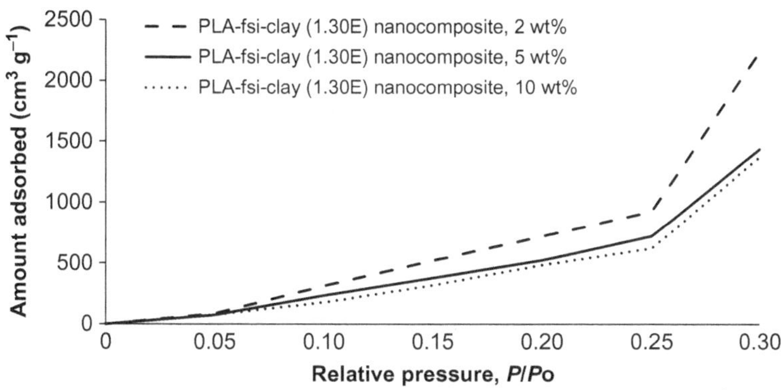

Fig. 7.10 N_2 adsorption isotherms of PLA-fsi-clay (1.30E) nanocomposites with different clay loadings.

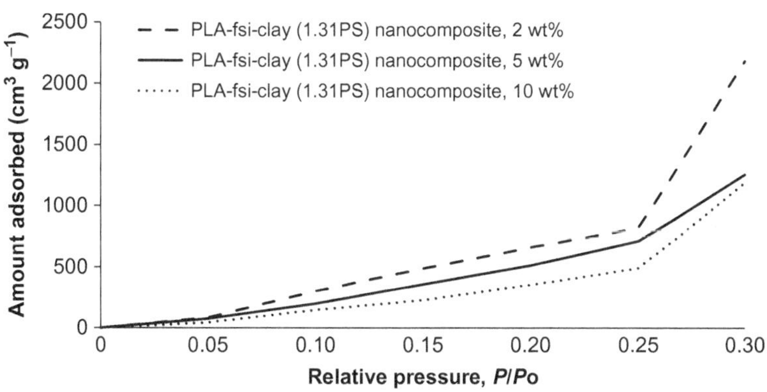

Fig. 7.11 N_2 adsorption isotherms of PLA-fsi-clay (1.31PS) nanocomposites with different clay loadings.

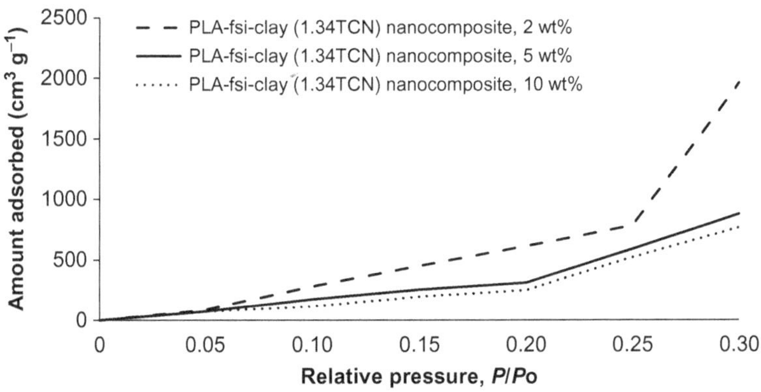

Fig. 7.12 N_2 adsorption isotherms of PLA-fsi-clay (1.34TCN) nanocomposites with different clay loadings.

5, and 10 wt%). Overall, it was clearly shown that 2 wt% of all types of clays had higher tensile strength compared to 5 and 10 wt% of PLA-fsi-clay nanocomposites. 2 wt% of PLA-fsi-clay (1.28E) nanocomposite exhibited the highest tensile strength (0.057GPa) compared to 10 wt% of PLA-fsi-clay (1.34TCN) with the lowest tensile strength value (0.023GPa). When the clay loading increased, agglomeration occurred and created more pores which increased the pore size that affect the tensile strength of the nanocomposites.

Fig. 7.14 represents the tensile modulus of the PLA-fsi-clay nanocomposites. 2 wt% of PLA-fsi-clay (1.28E) nanocomposite showed the highest tensile modulus (2.69GPa) while 10 wt% of PLA-fsi-clay (1.34TCN) nanocomposite showed the lowest tensile modulus with a value of 0.13GPa. This clearly shows that lower clay loading (2 wt%) significantly improved the tensile modulus compared to the higher clay loadings (5 and 10 wt%). It was evident that the presence of a small amount (2 wt%) of clay largely improved the tensile modulus of the nanocomposites and reached the maximum tensile value.

The improvements in tensile strength and modulus were ascribed to the presence of clay nanoparticles. Clay nanoparticles provide good compatibility and interfacial bonding strength toward PLA-fsi. In addition, a stable hindrance layer was formed between clay particles, which inhibited the agglomeration and thus improved the dispersion of the clay within PLA-fsi significantly (Zheng, Zheng, & Ning, 2003). Favorable technique used to enhance the nanosized clay and formed a strong barrier between the voids. This produced an ultra-high interfacial area and ionic bonds between the clay and PLA-fsi, which resulted in large increases of tensile strength and tensile modulus (Alexandre & Dubois, 2000).

Overall, PLA-fsi-clay (1.28E) nanocomposites showed the best tensile properties as the clay (1.28E) contained trimethyl stearyl ammonium, which imparted anti-wetting properties and caused the surface of PLA-fsi-clay (1.28E) nanocomposites to become hydrophobic (Seeni Meera, Sankar, Murali, Janiskar, & Mandal, 2012). This proved that clay (1.28E) was a compatible addition to PLA-fsi. Furthermore,

Table 7.2 Physical properties detected from N_2 adsorption at 77 K on PLA-fsi-clay nanocomposites

Samples	Specific surface area, S_{BET} (m^2/g)	Average pore volume, V_m ($10^{-4} cm^3/g$)	d_{BET} (nm)	Type of isotherm
PLA-fsi-clay (1.28E) nanocomposite, 2 wt%	587.9	30.9	1.8	IV
PLA-fsi-clay (1.28E) nanocomposite, 5 wt%	147.8	19.5	2.0	IV
PLA-fsi-clay (1.28E) nanocomposite, 10 wt%	53.2	12.7	2.3	IV
PLA-fsi-clay (1.30E) nanocomposite, 2 wt%	298.6	27.3	1.8	IV
PLA-fsi-clay (1.30E) nanocomposite, 5 wt%	109.4	18.8	2.0	IV
PLA-fsi-clay (1.30E) nanocomposite, 10 wt%	108.6	17.9	2.0	IV
PLA-fsi-clay (1.31PS) nanocomposite, 2 wt%	183.8	25.9	1.8	IV
PLA-fsi-clay (1.31PS) nanocomposite, 5 wt%	74.6	16.3	2.2	IV
PLA-fsi-clay (1.31PS) nanocomposite, 10 wt%	61.4	13.5	2.2	IV
PLA-fsi-clay (1.34TCN) nanocomposite, 2 wt%	181.5	24.5	1.8	IV
PLA-fsi-clay (1.34TCN) nanocomposite, 5 wt%	49.6	11.6	2.9	IV
PLA-fsi-clay (1.34TCN) nanocomposite, 10 wt%	27.2	9.8	3.8	IV

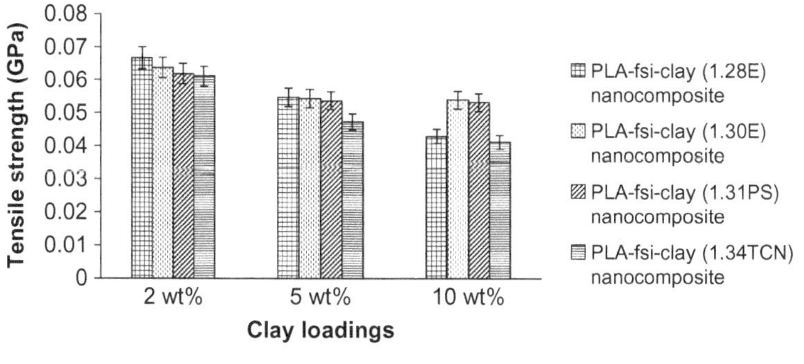

Figure 7.13 Tensile strength of PLA-fsi-clay nanocomposites.

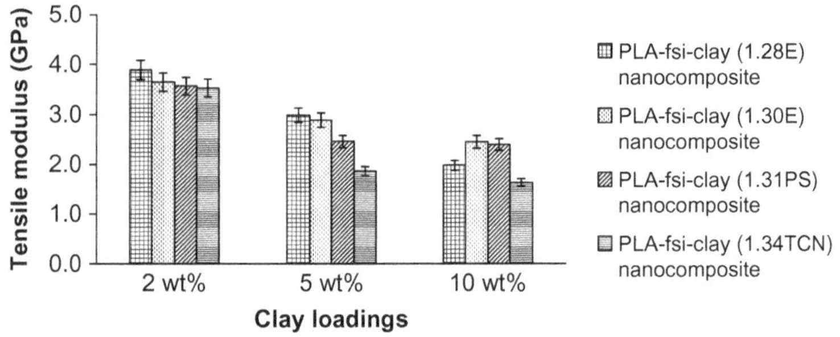

Figure 7.14 Tensile modulus of PLA-fsi-clay nanocomposites.

2 wt% of PLA-fsi-clay nanocomposites enhanced the tensile strength and tensile modulus of nanocomposites compared to 5 and 10 wt%.

7.2.6 Thermogravimetric analysis

The impact of fumed silica and clay on the thermal degradation of PLA-fsi-clay nanocomposites was analyzed by thermogravimetric analysis (TGA). There were two steps in the thermal decomposition of nanocomposites, namely the moisture loss of trapped solvent and weight loss due to dehydration process (below 300°C) (Jiang, Zhang, & Wolcott, 2007).

The first step degradation occurred at 200°C, while the second step, degradation, occurred at 350°C for all the nanocomposites. According to Fig. 7.15, PLA-fsi-clay (1.28E) nanocomposite exhibited a small weight percent loss of 3.8% below 300°C. This clearly showed the removal of absorbed water on the surface of the nanocomposites. At second stage, 2 wt% of PLA-fsi-clay (1.28E) nanocomposite showed the highest weight percentage loss (12.05%) among all the nanocomposites, while 10 wt% of PLA-fsi-clay (1.34TCN) nanocomposite showed the lowest weight percent loss with value of 1.53%, as shown in Figs. 7.15 and 7.18. Figs. 7.16 and 7.17 showed

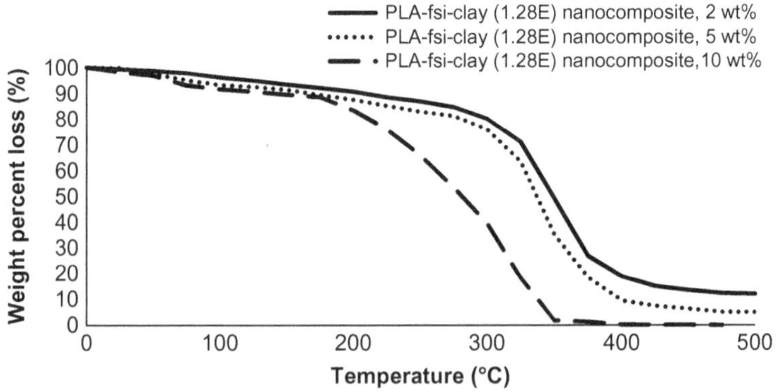

Fig. 7.15 TGA curves of PLA-fsi-clay (1.28E) nanocomposites.

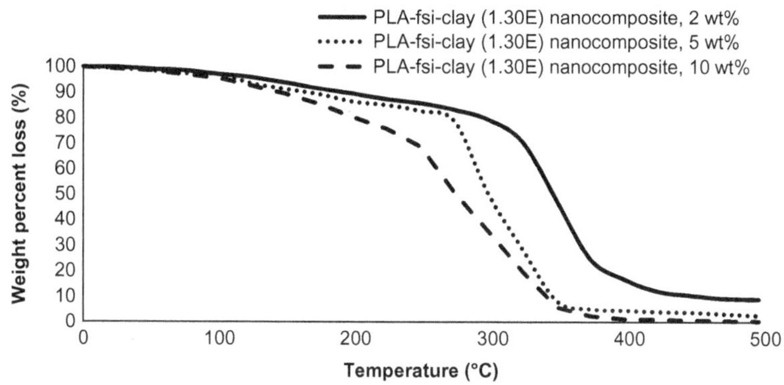

Fig. 7.16 TGA curves of PLA-fsi-clay (1.30E) nanocomposites.

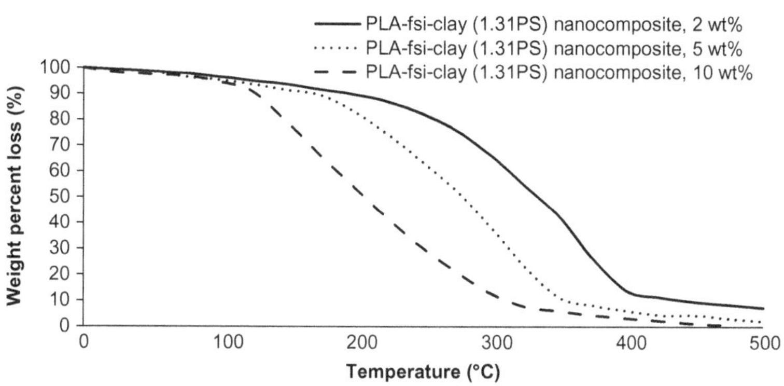

Fig. 7.17 TGA curves of PLA-fsi-clay (1.31PS) nanocomposites.

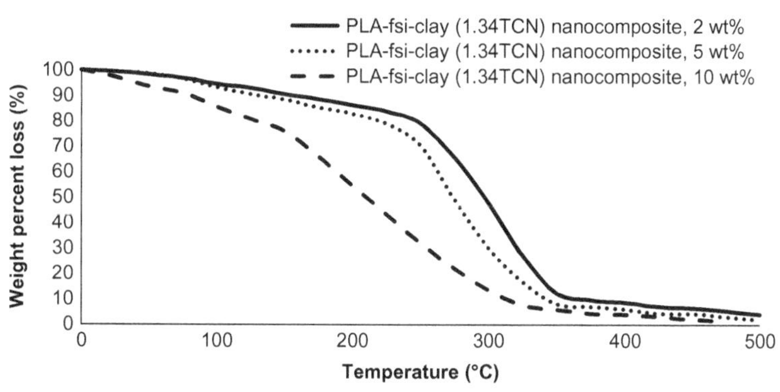

Fig. 7.18 TGA curves of PLA-fsi-clay (1.34TCN) nanocomposites.

that the PLA-fsi-clay (1.30E) and PLA-fsi-clay (1.31PS) nanocomposites had average weight percentage loss. The corresponding temperature of 2 wt% of PLA-fsi-clay nanocomposites shifted to higher temperatures, which led to better thermal stability. The improved thermal stability of the 2 wt% of PLA-fsi-clay nanocomposites was attributed to the clay acting as a crosslinking agent that retarded the motion of polymer chain (Benlikaya, Alkan, & Kaya, 2009). It was concluded that the 2 wt% of clays was compatible with PLA-fsi and thus, increased the thermal stability compared to 5 and 10 wt% of PLA-fsi-clay nanocomposites.

Table 7.3 shows the thermal characteristics such as initial temperature (T_i), maximum rate loss temperature (T_m), and final decomposition temperature (T_f), as well as the activation energy. It was found that the activation energy of PLA-fsi-clay (1.28E) nanocomposites was significantly higher than PLA-fsi-clay (1.30E), PLA-fsi-clay (1.31PS), and PLA-fsi-clay (1.34TCN) nanocomposites. Clay (1.28E) at lower loading increased the thermal stability, as they created a physical protective barrier on the surface of the nanocomposites (Wu, Wu, & Zhang, 2006). The complex structure of surface modified clay (1.28E) at lower loading delayed the volatilization in the nanocomposites (Gilman et al., 1998).

Overall, the lower clay loading (2 wt%) enhanced the thermal stability of PLA-fsi-clay nanocomposites. Moreover, the incorporation of clays, especially clay (1.28E), into PLA-fsi successfully improved thermal stability of PLA-fsi-clay nanocomposites, which was reflected in other results.

7.2.7 Moisture absorption analysis

The moisture absorption of the PLA-fsi-clay (1.28E), PLA-fsi-clay (1.30E), PLA-fsi-clay (1.31PS), and PLA-fsi-clay (1.34TCN) nanocomposites with 2 wt%, 5 wt%, and 10 wt% clay loadings are clearly shown in Figs. 7.19 to 7.22. All PLA-fsi-clay nanocomposites showed a sharp increase in moisture absorption at the first 15 min, and remained constant for 75 min. Fick's law was used to determine the moisture absorption of PLA-fsi-clay nanocomposites where the mass of water absorbed increased linearly with the square root of time until it reached an equilibrium state (Vilay, Mariatti, Mat Taib, & Todo, 2008).

From Figs. 7.19 to 7.22, 2 wt% of PLA-fsi-clay nanocomposites showed lower moisture absorption ($<2.0\%$) compared to 5 and 10 wt%. The strong PLA-fsi and clay adhesion at 2 wt% led to fewer pores in the nanocomposites, and thus the water diffusion pathways were inhibited. The hydroxyl groups of clay created strong hydrogen bonds with the hydroxyl groups in the PLA matrix, and thus the interactions between inter molecules were improved and decreased the water sensitivity.

In addition, the clays, especially clay (1.28E), formed strong covalent bonds with PLA-fsi and acted as a water barrier against the diffusion of moisture into the PLA-fsi (Taghizadeh & Sabouri, 2013). Moreover, 2 wt% of PLA-fsi-clay nanocomposites reduced the diffusion of moisture into the nanocomposites and the compatibility between the clay and PLA-fsi was improved.

Table 7.3 Activation energy of PLA-fsi-clay nanocomposites determined by Arrhenius equation

Samples	T_i (°C)[a]	T_m(°C)[b]	T_f (°C)[c]	W_{T_i} (%)[d]	W_{T_m}(°C)[e]	W_{T_f}(°C)[f]	Activation Energy, E_a (kJ/mol)
PLA-fsi-clay (1.28E) nanocomposite, 2 wt%	108.9	316.6	421.1	96.2	80.4	15.2	6670.5
PLA-fsi-clay (1.28E) nanocomposite, 5 wt%	95.3	313.7	419.3	93.3	73.0	9.5	3163.5
PLA-fsi-clay (1.28E) nanocomposite, 10 wt%	91.4	250.3	364.9	91.6	65.9	7.8	1946.8
PLA-fsi-clay (1.30E) nanocomposite, 2 wt%	103.2	304.0	418.1	97.9	79.4	12.2	5523.5
PLA-fsi-clay (1.30E) nanocomposite, 5 wt%	98.5	286.0	364.0	97.2	78.6	8.3	2815.3
PLA-fsi-clay (1.30E) nanocomposite, 10 wt%	79.5	240.5	355.0	97.1	68.9	5.9	2601.2
PLA-fsi-clay (1.31PS) nanocomposite, 2 wt%	97.3	211.6	389.8	96.5	74.7	13.6	4878.6
PLA-fsi-clay (1.31PS) nanocomposite, 5 wt%	95.0	173.5	347.6	95.2	72.9	11.1	2263.9
PLA-fsi-clay (1.31PS) nanocomposite, 10 wt%	75.5	151.1	310.9	96.9	68.5	7.5	2030.4
PLA-fsi-clay (1.34TCN) nanocomposite, 2 wt%	91.0	245.3	361.4	94.5	79.1	12.3	2902.5
PLA-fsi-clay (1.34TCN) nanocomposite, 5 wt%	89.3	233.9	343.6	93.3	72.5	8.1	1749.4
PLA-fsi-clay (1.34TCN) nanocomposite, 10 wt%	66.4	150.4	317.7	90.9	76.0	7.6	1561.5

[a]Temperature corresponding to the beginning of decomposition.
[b]Temperature corresponding to the maximum rate of mass loss.
[c]Temperature corresponding to the end of decomposition.
[d]Mass loss at temperature corresponding to the beginning of decomposition.
[e]Mass loss at temperature corresponding to the maximum rate of mass loss.
[f]Mass loss at temperature corresponding to the end of decomposition.

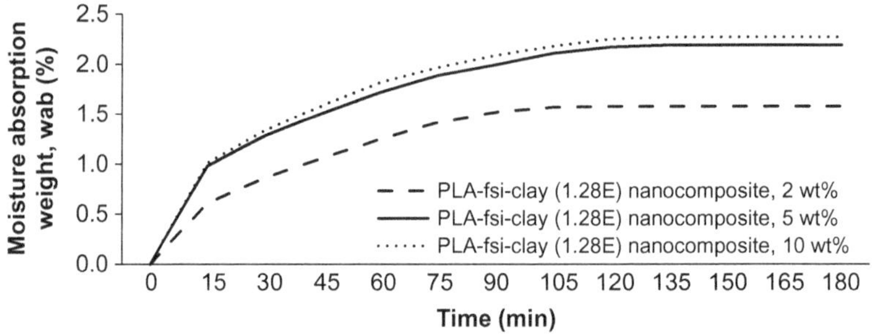

Fig. 7.19 Moisture absorption curves of PLA-fsi-clay (1.28E) nanocomposites.

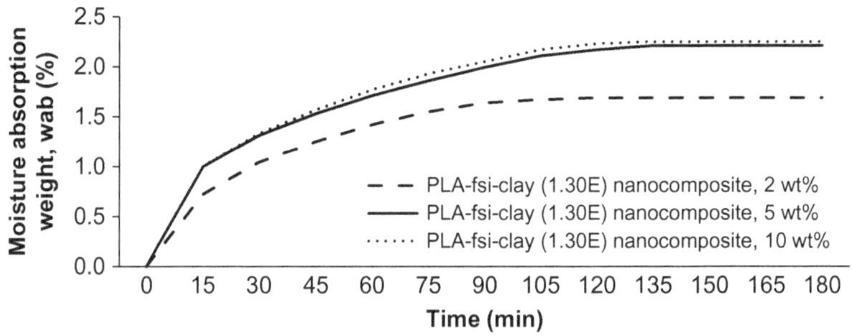

Fig. 7.20 Moisture absorption curves of PLA-fsi-clay (1.30E) nanocomposites.

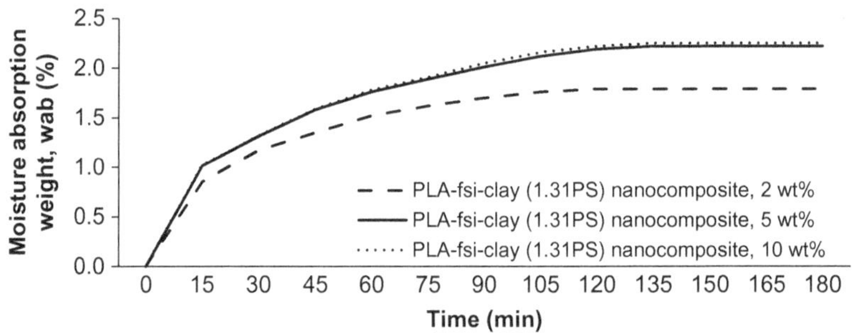

Fig. 7.21 Moisture absorption curves of PLA-fsi-clay (1.31PS) nanocomposites.

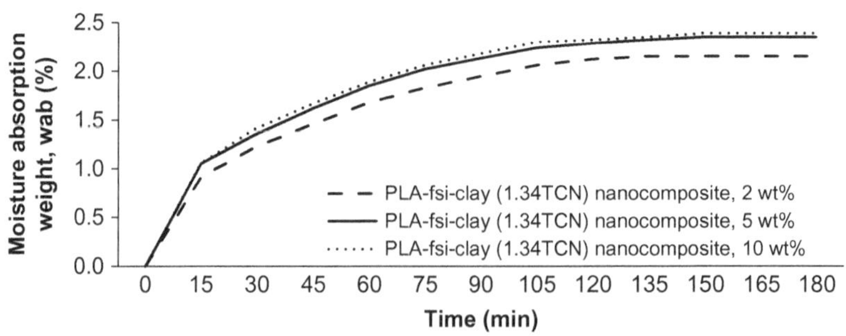

Fig. 7.22 Moisture absorption curves of PLA-fsi-clay (1.34TCN) nanocomposites.

7.3 Summary

The resultant nanocomposites were characterized by FTIR, XRF, BET, SEM, tensile tests, TGA, and moisture absorption tests. The FTIR spectrum indicated that PLA-fsi-clay (1.28E) nanocomposite with 2 wt% of clay showed broader peaks (especially the vibration of C—O—C group confined by silica network) than PLA-fsi-clay (1.30E), PLA-fsi-clay (1.31PS) and PLA-fsi-clay (1.34TCN) nanocomposites with 2 wt% of clay as proven by the moisture absorption tests. XRF results confirmed that PLA-fsi-clay (1.28E) with 2 wt% of clay contained the highest percentage of silicon groups compared to other nanocomposites. The incorporation of 2 wt% of clay (1.28E) showed the best compatibility with PLA-fsi followed by 2 wt% of clay (1.30E), clay (1.31PS), and clay (1.34TCN). The BET result showed PLA-fsi-clay (1.28E) nanocomposite with 2 wt% of clay had a higher surface area and average pore volume with smaller pore size. 2 wt% of clay (1.28E) loaded nanocomposites largely improved the tensile strength and modulus. TGA results revealed that the thermal stability with higher activation energy of PLA-fsi-clay (1.28E) nanocomposite at 2 wt% was significantly improved. 2 wt% clay loading was shown to be the optimum formulation that produced PLA-fsi-clay nanocomposites with the best physico-mechanical, morphological, and thermal properties.

References

Alcaniz-Monge, J., Perez-Cadenas, M., & Lozano-Castello, D. (2010). Influence of pore size distribution on water adsorption on silica gels. *Journal of Porous Materials*, *17*(4), 409–416.

Alexandre, M., & Dubois, P. (2000). Polymer-layered silicate nanocomposites: preparation, properties and uses of a new class of materials. *Materials Science and Engineering: R: Reports*, *28*(1–2), 1–63.

Balakrishnan, H., Hassan, A., Wahit, M. U., Yussuf, A. A., & Razak, S. B. A. (2010). Novel toughened polylactic acid nanocomposite: mechanical, thermal and morphological properties. *Materials and Design*, *31*(7), 3289–3298.

Bang, G., & Kim, S. W. (2012). Biodegradable poly(lactic acid)-based hybrid coating materials for food packaging films with gas barrier properties. *Journal of Industrial and Engineering Chemistry*, *18*(3), 1063–1068.

Benlikaya, R., Alkan, M., & Kaya, I. (2009). Preparation and characterization of sepiolite poly(ethyl methacrylate) and poly(2-hydroxyethyl methacrylate) nanocomposites. *Polymer Composites*, *30*(11), 1585–1594.

Besling, W. F. A., Goossens, A., Meester, B., & Schoonman, J. (1998). Laser-induced chemical vapor deposition of nanostructured silicon carbonitride thin films. *Journal of Applied Physics*, *83*(1), 544–553.

Bocchini, S., Fukushima, K., Di Blasio, A., Fina, A., Frache, A., & Geobaldo, F. (2010). Polylactic acid and polylactic acid-based nanocomposite photooxidation. *Biomacromolecules*, *11*(11), 2919–2926.

Brunauer, B., Emmett, P. H., & Teller, E. (1938). Adsorption of gases in multimolecular layers. *Journal of the American Chemical Society*, *60*(2), 309–319.

Chaloupi, N. N. (2011). *Development of polylactide-clay nanocomposites for food packaging applications*. Canada: University of Montreal.

Drumright, R. E., Gruber, P. R., & Henton, D. E. (2000). Polylactic acid technology. *Advanced Materials*, *12*(23), 1841–1846.

Eng, C. C., Ibrahim, N. A., Zainuddin, N., Ariffin, H., Yunus, W. M. Z. W., Then, Y. L., Teh, C. C. (2013). Enhancement of mechanical and thermal properties of polylactic acid/polycaprolactone blends by hydrophilic nanoclay. Indian Journal of Materials Science, 2013(2013): 1–11.

Frost, R., Carmody, O., Xi, Y., & Kokot, S. (2007). Surface characterization of selected sorbent materials for common hydrocarbon fuels. *Surface Science*, *601*(9), 2066–2076.

Gilman, J., Kashiwagi, T., Giannelis, E., Manias, E., Lomakin, S., Lichtenhan, J., et al. (1998). Nanocomposites: radiative gasification and vinyl polymer flammability. *Special Publication Royal Society Chemistry*, *224*, 203–221.

Jiang, L., Zhang, J., & Wolcott, M. P. (2007). Comparison of polylactide/nano-sized calcium carbonate and polylactide/montmorillonite composites: reinforcing effects and toughening mechanisms. *Polymer*, *48*(26), 7632–7644.

Jollands, M., & Gupta, R. K. (2010). Effect of mixing conditions on mechanical properties of polylactide/montmorillonite clay nanocomposites. *Journal of Applied Polymer Science*, *118*(3), 1489–1493.

Kontou, E., Niaounakis, M., & Georgiopoulus, P. (2011). Comparative study of PLA nanocomposites reinforced with clay and silica nanofillers and their mixtures. *Journal of Applied Polymer Science*, *122*(3), 1519–1529.

Liu, M., Zhang, Y., & Zhou, C. (2013). Nanocomposites of halloysite and polylactide. *Applied Clay Science*, *75-76*, 52–59.

Luo, Y., Zhao, J., Cai, Y., & Du, S. (2012). Effect of amino-functionalization on the interfacial adhesion of multi-walled carbon nanotubes/epoxy nanocomposites. *Materials Design*, *33*, 405–412.

McLauchlin, A. R., & Thomas, N. L. (2009). Preparation and thermal characterisation of poly(lactic acid) nanocomposites prepared from organoclays based on an amphoteric surfactant. *Polymer Degradation and Stability*, *94*(5), 868–872.

Mustapa, I. R., Shanks, R. A., & Kong, I. (2013). Poly(lactic acid)-hemp-nanosilica hybrid composites: thermomechanical, thermal behavior and morphological properties. *International Journal of Advanced Science and Engineering Technology*, *3*(1), 192–199.

Ou, C. F., Ho, M. T., & Lin, J. R. (2003). The nucleating effect of montmorillonite on crystallization of PET/montmorillonite nanocomposite. *Journal of Polymer Research*, *10*(2), 127–132.

Ray, S. S., & Okamoto, S. M. (2003). Polymer/layered silicate nanocomposites: a review from preparation to processing. *Progress in Polymer Science*, *28*(11), 1539–1641.

Saikia, B. J., & Parthasarathy, G. (2010). Fourier transform infrared spectroscopic characterization of kaolinite from Assam and Meghalaya, northeastern India. *Journal of Modern Physics*, *1*(4), 206–210.

Seeni Meera, K. M., Sankar, R. M., Murali, A., Janiskar, S. N., & Mandal, A. B. (2012). Sol–gel network silica/modified montmorillonite clay hybrid nanocomposites for hydrophobic surface coatings. *Colloids and Surfaces B: Biointerfaces*, *90*(1), 204–210.

Sing, K. (2001). The use of nitrogen adsorption for the characterization of porous materials. *Colloids and Surfaces A: Physicochemical and Engineering Aspects*, *187-188*, 3–9.

Taghizadeh, M., & Sabouri, N. (2013). Biodegradation behaviors and water adsorption of poly(vinyl alcohol)/starch/carboxymethyl cellulose/clay nanocomposites. *International Nano Letters*, *3*(1), 1–8.

Vilay, V., Mariatti, M., Mat Taib, R., & Todo, M. (2008). Effect of fiber surface treatment and fiber loading on the properties of bagasse fiber-reinforced unsaturated polyester composites. *Composites Science and Technology*, *68*(3–4), 631–638.

Vrsaljko, D., Macut, D., & Kovacevic, V. (2014). Potential role of nanofillers as compatibilizers in immiscible PLA/LDPE blends. *Journal of Applied Polymer Science.*, *132*(6). https://doi.org/10.1002/app.41414.

Wen, X., Zhang, K., Wang, Y., Han, L., Han, C., Zhang, H., et al. (2011). Study of the thermal stabilization mechanism of biodegradable poly(L-lactide)/silica nanocomposites. *Polymer International*, *60*(2), 202–210.

Wu, D., Wu, L., & Zhang, M. (2006). Rheology and thermal stability of polylactide/clay nanocomposites. *Polymer Degradation and Stability*, *91*(12), 3149–3155.

Yuan, P., Southon, P. D., Liu, Z., Green, M. E. R., Hook, J. M., Antill, S. J., et al. (2008). Functionalization of halloysite clay nanotubes by grafting with gamma aminopropyl triethoxysilane. *Journal of Physical Chemistry C*, *112*(40), 15742–15751.

Zhang, J., Lou, J., Ilias, S., Krishnamachari, P., & Yan, J. (2008). Thermal properties of poly(lactic acid) fumed silica nanocomposites: Experiments and molecular dynamics simulations. *Polymer*, *49*(9), 2381–2386.

Zheng, Y., Zheng, Y., & Ning, R. (2003). Effects of nanoparticles SiO_2 on the performance of nanocomposites. *Materials Letters*, *57*(19), 2940–2944.

Biomedical and packaging application of silica and various clay dispersed nanocomposites

8

Akshay Kakar, Elammaran Jayamani*, Muhammad Khusairy Bin Bakri*, Md. Rezaur Rahman†*
*Faculty of Engineering, Computing and Science, Swinburne University of Technology Sarawak Campus, Kuching, Sarawak, Malaysia, †Department of Chemical Engineering and Energy Sustainability, Faculty of Engineering, University Malaysia Sarawak (UNIMAS), Kota Samarahan, Sarawak, Malaysia

8.1 Introduction

Conventionally, polymer materials are reinforced with synthetic or natural inorganic compounds for either enhancing their properties, or to make the material more economical. Most common fillers include:

- Particles, such as calcium carbonate
- Fibers, such as glass fibers
- Plate-shaped particles, such as mica

Although these commonly reinforced polymeric materials are extensively used in a number of fields, it has been reported that incorporation of these fillers in polymers tend to have certain shortcomings that include increases in (Alexandre & Dubois, 2000; Fischer, 2003; Giannelis, 1996; Lagaly, 1999; Varlot, Reynaud, Kloppfer, Vigier, & Varlet, 2001):

- weight
- brittleness
- opacity

However, nanocomposites belong to a whole new class of composites in which the dispersed particles are in the nano scale, in at least one of its dimensions. Nanocomposites can therefore be classified into three categories based on the number of dimensions that are in the nanometer scale (Alexandre & Dubois, 2000; Fischer, 2003; Giannelis, 1996; Lagaly, 1999):

- Iso-dimensional: All three dimensions are in the nano scale.
- Nanotubes or whiskers: Two dimensions are in the nanoscale, while the third dimension is larger.
- Layered crystals or clays: These typically exist as sheets that have their thickness in the nanometer scale, whereas its length and width are at a larger scale.

Silica and Clay Dispersed Polymer Nanocomposites. https://doi.org/10.1016/B978-0-08-102129-3.00008-7

Among these types of nanocomposites, the clay and layered silicates have been studied most extensively. This is probably because they are easily available and their intercalation chemistry has been thoroughly explored over a long period of time (Gorrasi, Tortora, Vittoria, Galli, & Chiellini, 2002).

Polymer-layered silicate nanocomposites are produced by integrating finely distributed layered silicate fillers in a polymer matrix (Fischer, 2003). However, because of their face-to-face stacking in agglomerated tactoids, the nano layers are difficult to disperse in most polymers. It is well known that hydrophilic fillers are incompatible with hydrophobic matrix. Due to this unharmonious combination, dispersion of the tactoids becomes even more challenging. The layered silicates therefore need to be modified organically to produce clay (organoclay) that is compatible with polymer. In fact, there is ample evidence which shows that alkyl ammonium surfactants can enhance compatibility of the clay, and a hydrophobic polymer matrix of the surface chemistry by replacing the native clay silicate structure's inorganic exchange cations in the cavities or "galleries" (Lebaron, Wang, & Pinnavaia, 1999). Diverse methods may then be used to combine the ion-exchanged layered silicates in polymer matrix by (Pavlidou & Papaspyrides, 2008):

- polymerization
- solution intercalation
- simple melt mixing

Regardless of the method used, nanoparticles are added to the matrix or matrix precursors as 1–100 μm powders containing associated nanoparticles. Engineering the apt interfacial chemistry between nanoparticles filler and the polymer matrix is crucial, but not adequate to alter the micron-scale heterogeneity of the composition with the initial powder and into nanoscale homogenization of nanoparticles in the polymer's nanocomposite (Vaia & Wagner, 2004). Therefore, suitable conditions have to be developed during the preparation stage of the nanocomposite.

The polymer-layered silicates with nanometer-sized showed extremely highly dispersed surface area, and hold exclusive properties. These properties are not usually found in the general microscopic filler composites. (Alexandre & Dubois, 2000; Fischer, 2003; Giannelis, 1996; Lagaly, 1999). It is well known that by incorporating just a minor fraction of clay in the polymer matrix and without affecting the optical homogeneity of the composite, drastic improvements can be achieved in the physical properties, such as (Pavlidou & Papaspyrides, 2008):

- tensile strength and modulus
- heat distortion temperature (HDT)
- gas permeability

The most noteworthy are the surprising properties obtained from the inclusion of stiff filler in a polymer matrix, for example the often reported maintenance (or even enhancement) of the impact strength. The weight fraction of the inorganic additive is usually <10 wt%. This makes the materials lighter than most common composites (Balazs, Singh, Zhulina, & Lyatskaya, 1999; Fischer, 2003; Ginzburg, Singh, & Balazs, 2000; Lincoln, Vaia, Wang, & Hsiao, 2001; Osman, Mittal, & Lusti,

2004). These exclusive characteristics make the nanocomposites perfect materials for products ranging from high-barrier packaging for food, resilient biomedical tools, device components, and electronics, to strong and heat-resistant automotive parts (Balazs et al., 1999). Moreover, polymer-layered silicate nanocomposites have been suggested as exemplary systems to inspect polymer structure and dynamics in restrained environments (Lincoln et al., 2001; Vaia & Giannelis, 2001). Nevertheless, in spite of the latest advancement in polymer nanocomposite technology, there are many essential questions that have not been answered. For example (Fornes, Yoon, Keskkula, & Paul, 2001):

- How are general composite properties affected by the changes in polymer crystalline structure induced by the clay?
- How is it possible to modify organoclay chemistry to increase exfoliation reproducibility for a polymer system?
- How are composites affected by the properties' process parameters?

Additional studies are required to answer those questions.

8.2 Structure and characteristics of layered silicates and clays

Layered silicates used in the fabrication of nanocomposites are either synthetic or natural minerals. The layers are generally bound together with counterions. Tetrahedral sheets are their rudimentary building blocks. In this, silicon is enveloped by octahedral sheets and four oxygen atoms, where metals such as aluminum are enclosed by eight atoms of oxygen, which is shown in Fig. 8.1.

Therefore, in 1:1 layered structures (e.g., in kaolinite) a tetrahedral sheet is bonded with an octahedral sheet, where sharing of the oxygen atoms occurs (Chin, Thurn-Albrecht, Kim, Russell, & Wang, 2001). Whereas, the crystal lattice of 2:1 layered silicates (or 2:1 phyllosilicates) consists of two-dimensional layers. Here a central octahedral sheet of alumina is attached to two external silica tetrahedra by the tip. Due to this, the oxygen ions of the octahedral sheet also belong to the tetrahedral sheets. The thickness of the layer is approximately 1 nm. The adjacent dimensions may vary from 300 Å to several microns, and even larger, which is dependent on the particulate silicate, the clay source, and the preparation method (e.g., clays prepared by milling typically have lateral platelet dimensions of approximately 0.1–1.0 μm). Therefore, the aspect ratio of these layers (ratio length/thickness) is particularly high, with values >1000 (Alexandre & Dubois, 2000; Beyer, 2002; McNally, Raymond Murphy, Lew, Turner, & Brennan, 2003; Solomon, Almusallam, Seefeldt, Somwangthanaroj, & Varadan, 2001). The basic 2:1 structure with silicon in the tetrahedral sheets and aluminum in the octahedral sheet, without any substitution of atoms, is called pyrophyllite. Since the layers do not expand in water, pyrophyllite has only an external surface area and no internal one.

When silicon in the tetrahedral sheet is substituted by aluminum, the resulting structure is called "mica." Due to this substitution, the mineral, which is balanced

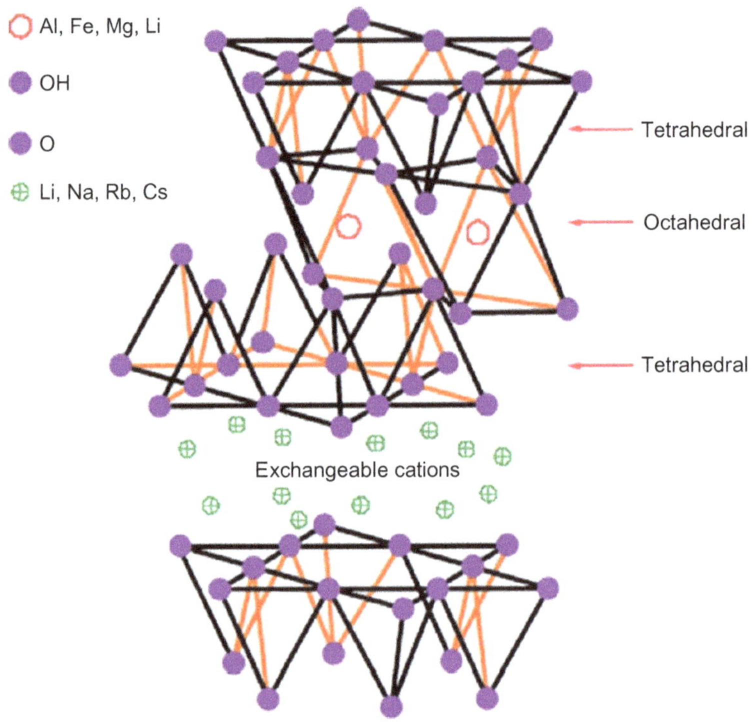

Fig. 8.1 The structure of a 2:1 layered silicate (Pavlidou & Papaspyrides, 2008). Originally from Beyer, G. (2002). Nanocomposites: a new class of flame retardants for polymers. *Plastics, Additives and Compounding*, *4*(10), 22–28.

by interlayer potassium cations, is characterized by a negative surface charge. However, because the size of the potassium ions matches the hexagonal hole created by the Si/Al tetrahedral layer, it is able to fit very tightly between the layers. Consequently, the interlayers collapse and the layers are held together by the electrostatic attraction between the negatively charged tetrahedral layer and the potassium cations. Therefore, micas do not swell in water and, like pyrophyllite, have no internal surface area (Dixon, 1991). On the other hand, if in the original pyrophyllite structure the trivalent Al-cation in the octahedral layer is partially substituted by the divalent Mg-cation, the structure of montmorillonite is formed, which is the best-known member of a group of clay minerals, called "smectites" or "smectite clays." In this case, the overall negative charge is balanced by sodium and calcium ions, which exist hydrated in the interlayer (Manias et al., 2001). The resulting structure do not fit in the tetrahedral layer. Additionally, the layers are bonded by water, relatively weak forces and other polar molecules can go in between the unit layers, enlarging the lattice (Chin et al., 2001).

Along with montmorillonite, hectorite, and saponite are the layered silicates that are most frequently used in nanocomposite materials. Their potentially high aspect ratio and the unique intercalation/exfoliation characteristics are the driving reasons making these materials very popular (Cho & Paul, 2001). Overall, it is well recognized that structural flawlessness is close to perfection, as the reinforcing components become smaller and that the ultimate properties of reinforcing composite components may be probable if their dimensions reach atomic or molecular scales. For instance, carbon nanotubes demonstrate the current highest known values of elastic modulus (1.7 TPa). Similarly, individual clay sheets, being only 1 nm thick, demonstrate an impeccable crystalline structure. The smaller the reinforcing components are the larger is their internal surface. Therefore, they have a tendency to combine rather than to scatter homogeneously in a matrix (Fischer, 2003). In reality, the silicate layers have the inclination to arrange themselves to form stacks with a regular van der Waals gap between them, called an "interlayer" or "gallery" (Alexandre & Dubois, 2000; Beyer, 2002; McNally et al., 2003). The interlayer dimension is determined by the crystal structure of the silicate (for dehydrated Na-montmorillonite, this dimension is approximately 1 nm) (Solomon et al., 2001). Analysis of layered silicates has shown that there are numerous levels of organization within the clay minerals. The tiniest particles, primary particles, are in the scale of 10 nm and consist of stacks of parallel lamellae. Micro-aggregates are formed by lateral joining of several primary particles, and aggregates are composed of several primary particles and micro-aggregates (Ishida, Campbell, & Blackwell, 2000).

8.3 Organic modification of layered silicates

Layered silicates only form a homogenous mixture with hydrophilic polymers, such as poly(vinyl alcohol) and poly(ethylene oxide). To create a homogenous mixture with other polymers, the alkali counterions must be exchanged for a cationic-organic surfactant, which shown in Fig. 8.2. Alkyl ammonium ions are mostly used, although other "onium" salts can be used, such as phosphonium and sulfonium (Alexandre &

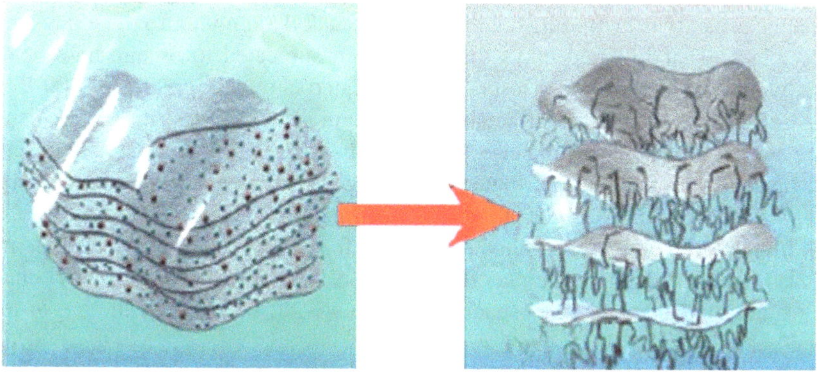

Fig. 8.2 Schematic picture of an ion-exchange reaction (Pavlidou & Papaspyrides, 2008). Originally from Fischer, H. (2003). Polymer nanocomposites: from fundamental research to specific applications. *Materials Science and Engineering C*, *23*(6–8), 763–772.

Dubois, 2000; Manias et al., 2001; Pavlidou & Papaspyrides, 2008). This can be readily achieved through ion-exchange reactions that render the clay organophilic (Kornmann, Lindberg, & Berglund, 2001). In order to obtain the exchange of the onium ions with the cations in the galleries, water swelling of the silicate is needed. For this reason, alkali cations are preferred in the galleries because 2-valent and higher valent cations prevent water swelling. The hydrate formation of monovalent intergallery cations is the main cause for water swelling. Natural clays may comprise divalent cations (i.e., calcium) and need exchange processes with sodium, before the further treatment with onium salts (Kornmann et al., 2001). The alkali cations, as they are not structural and may be easily substituted by other positively charged atoms or molecules. Thus they are called exchangeable cations (Xie et al., 2001). The organic cations reduce the surface energy of the silicate surface and enhance wetting with the polymer matrix (Giannelis, 1996; Kornmann et al., 2001). Furthermore, the lengthy organic chains of such surfactants, with positively charged ends, are tied to the surface of the negatively charged silicate layers, causing an increase in the height of the gallery (Kim, Lee, Hoffmann, Kressler, & Stöppelmann, 2001). It then becomes plausible for organic species (i.e., polymers or prepolymers) to diffuse between the layers and ultimately detach them (Kornmann et al., 2001; Zerda & Lesser, 2001). Sometimes, the alkyl ammonium cations may even offer functional groups that can react with the polymer or start polymerization of monomers (Chin et al., 2001). The microchemical environment in the galleries is hence apt for the intercalation of polymer molecules (Huang, Zhu, Yin, Qian, & Sun, 2001). Finally, the surface modification both increases the basal spacing of clays and functions as a compatibilizer between the hydrophilic clay and the hydrophobic polymer (Zerda & Lesser, 2001).

Depending on the functionality, packing density, and length of the organic modifiers, the organo-modified layered silicates (OMLSs, organosilicates or organoclays) may be engineered to optimize their compatibility with a given polymer (Vaia & Giannelis, 1997; Xie et al., 2001). It is worth noticing that, on the basis of the cation exchange capacity of the clay, the content of the surfactant is usually about 35–45 wt% (Yu, Yang, Zhang, Zhao, & Mai, 2003). Actually, one way to measure the clay cation exchange capacity is by finding out the amount of alkyl ammonium salt held up by the organoclays. That is, dried clays that have been subjected to organo-modification along with a sample of the corresponding untreated clay, are ignited at 1000°C.

In general, the longer the surfactant chain length, and the higher the charge density of the clay, the further apart the clay layers will be forced. This is anticipated because both of these parameters contribute to growing the volume occupied by the intragallery surfactant (Lebaron et al., 1999). For example, Wang, Choi, Koo, Choi, and Chung (2001) prepared organoclays with different alkyl ammonium chain lengths and also used an organophilic clay, Cloisite 20A, which has two long alkyl chains. They found that the interlayer spacing increases with the increase in size of alkyl amine chain length.

At first, the orientation of surfactant chains was comprehended from infrared and XRD measurements. Based on which the organic chains have been long thought to lie either parallel to the silicate layer, forming mono or bilayers or, depending on the packing density and the chain length, to radiate away from the surface, forming mono or even a bimolecular tilted "paraffinic" arrangement (Alexandre & Dubois, 2000; Lebaron et al., 1999). A more accurate explanation has been offered by (Vaia & Wagner, 2004) based on FTIR experiments. By observing frequency changes of the asymmetric CH_2 stretching and bending vibrations, they showed that alkyl chains can differ from liquid-like to solid-like. The liquid-like structure dominates because the interlayer density or chain length decreases, or if the temperature increases. When the available surface area per molecule is within a certain range, the chains are not entirely chaotic, but maintain some orientation order similar to that of the liquid crystalline state. As the chain lengthens, the interlayer structure appears to evolve in a stepwise fashion, from a chaotic to a more ordered mono layer, then "jumping" to a more disordered pseudo-bilayer. Moreover, an NMR study by Wang et al. (2001) indicated the concurrence of ordered trans and disordered gauche conformations (Sinha Ray & Bousmina, 2005).

Summarizing this section, there are two particular characteristics of layered silicates that are exploited in polymer-layered silicate nanocomposites. The first is the ability of the silicate particles to disperse into individual layers. Since dispersing a layered silicate can be pictured like opening a book, an aspect ratio as high as 1000 for fully dispersed individual layers can be obtained (contrast that to an aspect ratio of about 10 for undispersed or poorly dispersed particles). The second characteristic is the ability to fine-tune their surface chemistry through ion exchange reactions with organic and inorganic cations. These two characteristics are, of course, interrelated because the degree of dispersion in a given matrix that, in turn, determines aspect ratio, depends on the interlayer cation (Giannelis, 1996; Ishida et al., 2000).

8.4 Nanocomposite structure

Any physical mixture of a polymer and silicate (or inorganic material in general) does not necessarily form a nanocomposite. The situation is analogous to polymer blends. In most cases, separation into discrete phases normally takes place. In immiscible systems, the poor physical attraction between the organic and the inorganic components leads to relatively poor mechanical properties. Furthermore, particle agglomeration tends to reduce strength and produce weaker materials (Giannelis, 1996). Thus, when the polymer is unable to intercalate between the silicate sheets, a phase-separated composite is obtained, whose properties are in the same range as traditional microcomposites (Alexandre & Dubois, 2000; Beyer, 2002). Beyond this traditional class of polymer-filler composites, two types of nanocomposites can be obtained, depending on the preparation method and the nature of the components used, including polymer matrix, layered silicate, and organic cation (Alexandre & Dubois, 2000; Beyer, 2002).

It is generally accepted that exfoliated systems give better mechanical properties than intercalated ones, which are shown in Fig. 8.3 (Chin et al., 2001; Varlot et al., 2001). The complete dispersion of clay nanolayers in a polymer optimizes the number of available reinforcing elements for carrying an applied load and deflecting cracks. The coupling between the tremendous surface area of the clay and the polymer matrix

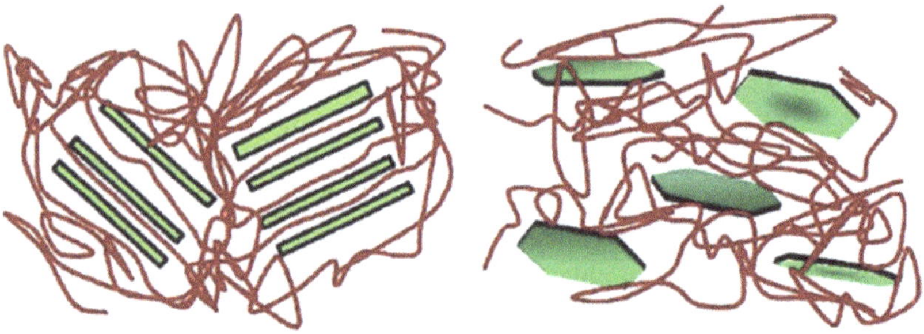

Fig. 8.3 Schematic illustration of two different types of thermodynamically achievable polymer/layered silicate nanocomposites (Pavlidou & Papaspyrides, 2008).
Originally from Sinha Ray, S., & Bousmina, M. (2005). Biodegradable polymers and their layered silicate nanocomposites: in greening the 21st century materials world. *Progress in Materials Science*, *50*(8) 962–1079.

facilitates stress transfer to the reinforcement phase, allowing for mechanical property improvements (Beyer, 2002; Wu et al., 2001). However, it is not easy to achieve complete exfoliation of clays and, indeed with few exceptions, the majority of the polymer nanocomposites reported in the literature were found to have intercalated or mixed intercalated-exfoliated nanostructures (Chin et al., 2001). This is because the silicate layers are highly anisotropic, with lateral dimensions ranging from100 to 1000nm, and even when separated by large distances (i.e., when delaminated) cannot be placed completely randomly in the sea of polymer. Furthermore, the majority of the polymer chains in the hybrids are tethered to the surface of the silicate layers. Thus, it can be expected that there are domains in these materials, even above the melting temperature of the constituent polymers, wherein some long-range order is preserved and the silicate layers are oriented in some preferred direction. This long-range order and domain structure is likely to become better defined at the higher silicate contents, where the geometrically imposed mean distance between the layers becomes less than the lateral dimensions of the silicate layers, thus forcing some preferential orientation between the layers. However, there might be considerable poly-dispersity effects in terms of the orientation and the distance between the silicate layers. Many such randomly oriented grains make up the entire sample, leading to the presence of disordered material. Thus, in general the material possesses a layered structure with grains wherein the silicate layers are oriented in a preferred direction, leading to the presence of grain boundaries and concomitant defects (Krishnamoorti & Giannelis, 1997).

8.5 Nanocomposite structural characterization

Two complementary techniques are generally used to characterize the structures of nanocomposites: XRD and transmission electron microscopy (TEM) (Alexandre & Dubois, 2000; Beyer, 2002; Porter, Metcalfe, & Thomas, 2000; Sinha Ray &

Okamoto, 2003). Due to its ease of use and availability, XRD is most commonly used to probe the nanocomposite structure and occasionally to study the kinetics of polymer melt intercalation (Sinha Ray & Okamoto, 2003). This technique allows the determination of the spaces between structural layers of the silicate utilizing Bragg's law. By monitoring the position, shape, and intensity of the basal reflections from the distributed silicate layers, the nanocomposite structure may be identified (Sinha Ray & Okamoto, 2003). For immiscible polymer/OMLS mixtures, the structure of the silicate is not affected, and thus, the characteristics of the OMLS basal reflections do not change. On the other hand, in comparison with the spacing of the organoclay used, the intercalation of the polymer chains increases the interlayer spacing, leading to a shift of the diffraction peak toward lower angle, according to Bragg's law. In such intercalated nanocomposites, the repetitive multilayer structure is well preserved, allowing the interlayer spacing to be determined. In contrast, the extensive layer separation associated with exfoliated structures disrupts the coherent layer stacking and results in a featureless diffraction pattern. Thus, for exfoliated structures, no more diffraction peaks are visible in the XRD diffractograms either because of too much spacing between the layers (i.e., exceeding 8 nm in the case of ordered exfoliated structure) or because the nanocomposite does not present ordering (Alexandre & Dubois, 2000; Beyer, 2002; Yu et al., 2003).

Some layered silicates initially do not exhibit well-defined basal reflections. Thus, peak broadening and intensity decreases are very difficult to study systematically. Therefore, conclusions concerning the mechanism of nanocomposite formation and structure based solely on XRD patterns are only tentative. On the other hand, TEM allows a qualitative understanding of the internal structure and can directly provide information in real space, in a localized area, on morphology and defect structures (Ma, Xu, Ren, Yu, & Mai, 2006; Morgan & Gilman, 2003).

8.6 Nanocomposite mechanical properties

The first mechanism that has been put forward to explain the reinforcing action of layered silicates is one also valid for conventional reinforcements, such as fibers. That is, rigid fillers are naturally resistant to straining due to their high moduli. Therefore, when a relatively softer matrix is reinforced with such fillers, the polymer, particularly that adjacent to the filler particles, becomes highly restrained mechanically. This enables a significant portion of an applied load to be carried by the filler, assuming, of course, that the bonding between the two phases is adequate (Fornes & Paul, 2003). From this mechanism, it becomes obvious that the larger the surface of the filler in contact with the polymer, the greater the reinforcing effect will be. This could partly explain why layered silicates, having an extremely high specific surface area (on the order of $800\,m^2/g$) impart dramatic improvements of modulus even when present in very small amounts in a polymer. In fact, the low silicate loading required in nanocomposites to affect significant property improvements, is probably their most distinguishing characteristic. In most conventionally filled polymer systems, the modulus increases linearly with the filler volume fraction, whereas for nanocomposites, much lower filler concentrations increase the modulus sharply and to a much larger extent (Sinha Ray & Okamoto, 2003). However, some authors have argued that the dramatic improvement of modulus for such extremely low clay concentrations (i.e., 2–5 wt %) cannot be attributed simply to the introduction of the higher modulus inorganic filler layers. A proposed theoretical approach assumes a layer of affected polymer on the filler surface, with a much higher modulus than the bulk equivalent polymer. This affected polymer can be thought of as a region of the polymer matrix that is physisorbed on the silicate surface, and is thus stiffened through its affinity for and adhesion to the filler surface. Obviously, for such high aspect ratio fillers as the layered silicate layers, the surface area exposed to the polymer is huge and, therefore, the significant increases in the modulus with very low filler content are not surprising. Furthermore, beyond the percolation limit, the additional silicate layers are incorporated in polymer regions that are already affected by other silicate layers, and thus it is expected that the enhancement of modulus will become much less dramatic (Shia & Hui, 1998). In order to prove the effect of degree of exfoliation on nanocomposite mechanical properties, Fornes and Paul (2003) used an analytical approach to elucidate how incomplete exfoliation influences nanocomposite stiffness.

In cross-linked polyester/OMLS nanocomposites, the modulus decreases with increasing clay content; in fact, the drop for the 2.5 wt% nanocomposite was greater than expected. To explain this phenomenon, it was proposed that the intercalation and exfoliation of the clay in the polyester resin serve to effectively decrease the number of crosslinks from a topological perspective. The origin of the greater drop in properties of the 2.5 wt% nanocomposites may be traced to the morphology; for example, it was observed that the sample showed exfoliation on a global scale compared to the nanocomposite containing 10 wt% clay, indicating that the cross-linking density is inversely proportional to the degree of exfoliation (Bharadwaj et al., 2002). Apart from the modulus, the addition of OMLS in a polymer matrix usually also increases the tensile strength compared to that of the neat polymer material. For example, Shelley et al. (Shelleya, Matherb, & DeVries, 2001) reported a 175% improvement in yield stress accompanied by a 200% increase in tensile modulus for a nylon 6 nanocomposite containing 5 wt% clay. Most polymer-clay nanocomposite studies report tensile properties, such as modulus, as a function of clay content. This plot of Young's modulus of nylon 6 nanocomposite vs. filler weight content, shows a constant large rate of increase of modulus up to ca. 10 wt% of nanoclay, whereas above this threshold the aforementioned leveling-off of Young's modulus is observed. This change corresponds to the passage from totally exfoliated structure (below 10 wt%) to partially exfoliated-partially intercalated structure (for 10 wt% clay and above), as determined by XRD and TEM (Alexandre & Dubois, 2000; Sinha Ray & Okamoto, 2003).

In another study, Liu and Wu (2002) studied the mechanical performance of PA66 nanocomposites prepared via melt intercalation, using epoxy co-intercalated clay. The tensile strength increases rapidly from 78 MPa for PA66 up to 98 MPa for PA66CN5, but the increasing amplitude decreases when the clay content is above 5 wt%. A similar phenomenon is observed in the dependence of tensile modulus of PA66CN on clay content. The smaller increase in amplitude observed with a clay loading above 5 wt% was again attributed to the inevitable aggregation of the layers at high clay content. Another example shows both the tensile modulus and the yield strength of neat PA12 and the nanocomposites, which increased steadily with increasing organoclay loading. Compared to the virgin polymer, the tensile modulus of PA12/clay systems was improved by about 40% upon adding only 5 wt% of clay, while limited improvement of the tensile strength was observed by incorporating clay in the matrix. Again, it was suggested that there is an optimum clay concentration for nanocomposite tensile strength improvement. With a further increase in clay loading, a moderate decrease of tensile strength was observed, suggesting that the relative amount of intercalation/exfoliation of the clay morphology gradually increases with increasing clay content, because the tensile strength is usually sensitive to the degree of dispersion (Phang et al., 2005).

Similarly, other factors that influence the degree of exfoliation, apart from the clay content, also have an impact on nanocomposite modulus and strength.

The addition of organoclay leads to a substantial improvement in stiffness for the composites based on each of the three PA6 samples examined, for example, LMW, MMW, and HMW (low, medium, and high molecular weight, respectively). Interestingly, the stiffness increases with increasing matrix molecular weight at any given concentration, even though the moduli of the neat PA6s are all quite similar. Yield strength increases with MMT content; however, while the HMW and MMW-based nanocomposites show a steady increase in strength with clay content, the LMW-based nanocomposites show a less pronounced effect. These differences with respect to molecular weight are attributed to the better exfoliation achieved for the higher molecular weight matrices (Fornes et al., 2001). Other factors that may play a crucial role in improvement of nanocomposite mechanical properties include the organic modification of the clay and the addition of compatibilizers to the polymer matrix. As a representative example, Young's modulus values of PP/PP-MA nanocomposites are listed in and compared with the corresponding microcomposite, as well as simple PP-MA/PP polymer blends. It is readily observed that increasing the amount of PP-MA increases the modulus, while comparison of PP with the simple PP-MA/PP blends rules out any possible effect of matrix modification due to the presence of increasing amounts of PP-MA (Alexandre & Dubois, 2000). In another study, Hotta and Paul (2004) performed tensile tests on various PE and PE-MA nanocomposites based on organoclays with one or two alkyl tails. The increase in modulus with addition of MMT is much stronger for the organoclay with two alkyl tails than for the one with a single tail, as would be expected on the basis of the much better dispersion of clay platelets for the surfactant with two alkyl tails. Similar trends were observed also for nanocomposite yield strength. Interestingly, the authors noted that there is no advantage in adding PE-MA for building modulus or strength at low MMT content ($\leq$2.5 wt%), in spite of the morphological differences seen. On the contrary, there is a clear advantage in adding PE-MA at higher MMT contents. Even though the benefit for modulus is not as great as might be expected, in the absence of PE-MA, the yield strength actually decreases on addition of MMT beyond 2.5 wt%.

Both strength and modulus are higher in the case of the reactive intercalating agent, owing to the better dispersion of the organoclay (Zhao et al., 2005). The ultimate strength increased dramatically with clay content and reached a maximum at 5 wt% MMT, where the ultimate strength of the nanocomposites increased by about 450% for C16-MMT and 600% for MO-MMT, compared with that of pure PU, indicating that the improved mechanical strength depends on the characteristics of the modifier (Xiong, Liu, Yang, & Wang, 2004). The extent of improvement of nanocomposite mechanical properties will also depend directly upon the average length of the dispersed clay particles, because this determines their aspect ratio and, hence, their surface area (Kojima, Usuki, Kawasumi, Okada, & Fukushima, 1993; Sinha Ray & Okamoto, 2003). At this point, we note that several authors have also pointed out factors that have an adverse effect on nanocomposite modulus and/or strength and need to be taken into consideration when preparing nanocomposite materials. Quite interestingly, Gopakumar, Lee, Kontopoulou, and Parent (2002) found that the exfoliation of 5 and 10 wt% clay in PE-MA increased Young's modulus by 30% and 53%, respectively, whereas the tensile stress at yield showed only a marginal increase, up to a maximum of 15% for the 10 wt% clay composition. The authors noted that the greatly enhanced interfacial area derived from exfoliation of the clay improves the mechanical reinforcement potential of the filler. However, given that the mechanical properties of a filled system depend on two principal factors, for example, crystallinity of the polymer matrix and the extent of filler reinforcement, the degree of crystallinity must also be considered. In another study dealing with the effect of matrix variations on mechanical properties of nanocomposites, Chaudhary, Prasad, Gupta, and Bhattacharya (2005) studied the tensile properties of nanocomposites based on ethyl-vinyl-acetate copolymers with various VA contents and two alternative organoclays. Because in ethyl-vinyl-acetate copolymer with increasing VA content the crystallinity of the polymer decreases (and will lower the stiffness), while the polarity increases (and will increase the intercalation), the authors suggested that in their system, the stiffness and toughness responses would reflect an interplay of two factors: (a) an increase in the "rigid" amorphous phase due to polymer-clay intercalation and (b) an increase in the "mobile" amorphous phase due to the increasing VA content. Experimental results showed that the influence of increasing clay concentration on the tensile behavior of ethyl-vinyl-acetate copolymer matrices was significant only with a lower moderately polar ethyl-vinyl-acetate copolymer matrix (9% and 18% VA). Thus, a linear proportionality was found between clay concentration and tensile modulus for ethyl-vinyl-acetate copolymer-9 and ethyl-vinyl-acetate copolymer-18, a relation not observed with ethyl-vinyl-acetate copolymer-28. In fact, it is very difficult to compare the extent of the improvement of the mechanical properties of different ethyl-vinyl-acetate copolymer/clay nanocomposites reported so far, because ethyl-vinyl-acetate copolymers of different vinyl acetate contents have been processed into the nanocomposites with different clays and different modifying agents by different methods (Srivastava, Pramanik, & Acharya, 2006). In the case of high Tg thermosets, it has been suggested that neither intercalated nor exfoliated nanosilicates lead to an improvement of the tensile stress at break, but rather make the materials more brittle. This effect appears to be generally more pronounced for intercalated structures than for exfoliated ones (Alexandre & Dubois, 2000). The results of tensile tests conducted by Hackman and Hollaway (2006) on epoxy nanocomposites conventionally prepared under low-shear (stirring for 5 h at 90°C). The tensile modulus increased by 9.0% and 19.9% with 5 and 10 wt% clay loading, respectively. However, the ultimate tensile stress decreased with increasing clay content, although the variation was large. The authors attributed this phenomenon to the fact that large clay particles act as impurities and increase stress concentrations. The flexural modulus and ultimate flexural stress increased by 19.6% and 7.7%, respectively, for specimens containing 10 wt% clay. For nanocomposites processed under high shear, the tensile modulus and ultimate tensile stress increased by 18.7% and 9.3%, respectively, when 5 wt% clay loading was applied. In this case, the improvement in ultimate tensile strength was attributed to the smaller particles not generating stress concentrations leading to premature failure. A summary of the tensile properties of soft (SPU) and hard (HPU) polyurethane elastomers and of the corresponding nanocomposites, prepared by either solvent casting or melt compounding. As can be seen upon silicate addition, large improvements in stiffness were observed, which, however, were accompanied by a decrease in tensile strength and elongation (Finnigan, Martin, Halley, Truss, & Campbell, 2004). Similar trends have been reported by (Tortora et al., 2002). Both exfoliated and intercalated PU/o-MMT nanocomposites showed an improvement in the elastic modulus upon increasing the clay content, but a decrease in the stress and strain at break. In general, it has been argued that in the presence of polar or ionic interactions between the polymer and the silicate layers, the stress at break is usually increased, whereas when there is lack of interfacial adhesion, no or very slight tensile strength enhancement is recorded (Alexandre & Dubois, 2000). Pegoretti, Kolarik, Peroni, and Migliaresi (2004) found that the yield strength was not reduced by the addition of clay to recycled PET and considered this to be a sign of good interfacial adhesion; however, in the same study, a slight decrease of stress at break and a dramatic reduction of strain were reported. On the other hand, in PS intercalated nanocomposites the ultimate tensile stress was found to decrease compared to that of the PS matrix and dropped further at higher filler content. This lack of strength was attributed to the fact that only weak interactions exist at the PS/clay interface, contrary to other compositions in which polar interactions may prevail, strengthening the matrix interface (Noh & Lee, 1999). It should be noted that several authors have reported their inability to measure nanocomposite yield stress because the materials often become brittle and fail before reaching the yield point. Such remarks were made by Gorrasi et al. (2002), who conducted tensile tests on PCL nanocomposite, containing 30 wt% clay, and also on blends of this nanocomposite with HMW PCL. For the blend containing 15 wt% clay, only the elastic modulus could be evaluated because the sample did not reach the yield point, while lower clay concentrations in the blend led to better mechanical properties in terms of flexibility and drawability. For the initial nanocomposite, however, it was not even possible to draw the sample because of its brittleness. An interesting study was performed by Chang et al.

Chang, Jong, Lak, and Im (2004) prepared PET-based nanocomposites through in situ intercalative polymerization, and subsequently produced nano-hybrid fibers by extrusion through the die of a capillary rheometer. The hot extrudates were stretched through the die of a capillary rheometer at 270°C and immediately drawn to various draw ratios (DR). As is evident, the tensile properties of the fibers formed increased with the growing amount of organoclay at DR = 1. When the organoclay was increased from 0 to 3 wt% in hybrids at DR = 1, the strength linearly improved from 46 to 71 MPa, and the modulus from 2.21 to 4.10 GPa. On the other hand, it is quite interesting to note the effect of DR on the tensile strength and modulus of PET and PET nanocomposite fibers. For pure PET, the strength and modulus increased from 46 to 51 MPa, and 2.21 to 2.39 GPa, respectively, as the DR was increased from 1 to 16. However, the ultimate strength and modulus of the hybrid fibers decreased markedly with increasing DR. An increase in the mechanical tensile strength with increasing DR is very common for engineering plastics and is usually observed in flexible coil-like polymers. However, nanocomposite fibers did not follow this trend. Chang et al. suggested that higher stretching of the fiber leads to debonding and creation of voids in the hybrid, which reduce the tensile mechanical properties. This study clearly illustrates that nanocomposite materials may have a different response to mechanical loads than the corresponding neat polymer matrices. Finally, even though nanocomposite researchers are generally interested in the tensile properties of the final materials, there are a few reports concerning the flexural properties of polymer-layered silicate nanocomposite nanocomposites (Ray & Okamoto, 2003; Ray, Yamada, Okamoto, Ogami, & Ueda, 2003).

8.7 Toughness and strain

The brittle behavior often exhibited by nanocomposites probably originates from the formation of micro voids due to debonding of clay platelets from the polymer matrix upon failure. This has been testified through careful inspection of fracture surfaces and is also correlated to observations by in situ deformation experiments using transmission electron microscopy (Morawiec, Pawlak, Slouf, Galeski, & Krasnikowa, 2005; Phang et al., 2005). In fact, the observation of nanocomposite fracture surfaces is quite interesting. No distinct clay agglomerates are observed by scanning electron microscopy (SEM) even at high magnification, which is shown in Fig. 8.4. For 1 wt% clay addition, the fracture surface became smoother compared to neat PA12 and 5 wt% clay loaded composites. Careful inspection of the fracture surface at higher magnification of nanocomposite with 5 wt% clay verifies the formation of microvoids due to the debonding of clay platelets from the matrix. Usually, microvoids are formed around the large homogeneities, which become evident, especially at high clay loadings. These micro voids will coalesce with the formation of larger cracks causing embrittlement, ultimately resulting in reduced toughness (Phang et al., 2005).

In the case of nylon 12 nanocomposites, the Izod impact strength monotonically decreases as the clay concentration increases. The toughness (representing the energy absorption during the fracture process) decreases by about 25% with 5 wt% of clay. Similar observations of reduction in impact strength are also reported in nylon 6/clay nanocomposites and PE-based nanocomposites, indicating that the incorporation of clay into semi crystalline thermoplastics usually results in toughness reduction, for

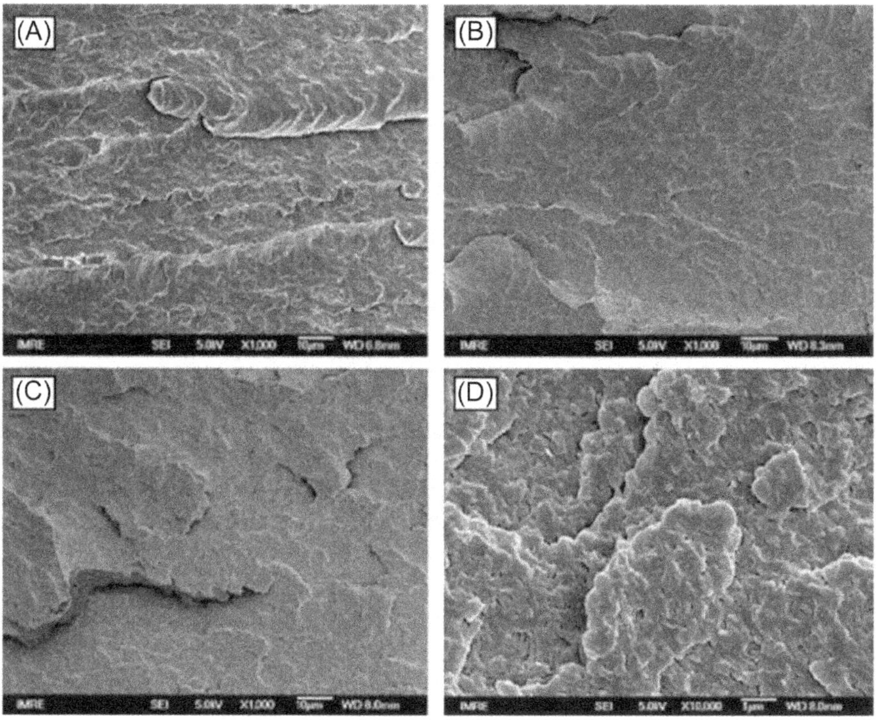

Fig. 8.4 SEM images showing fracture surfaces after impact tests: (A) neat PA12; (B) and (C) PA12 nanocomposites containing 1 and 5 wt% clay, respectively; (D) high magnification of (C) (Pavlidou & Papaspyrides, 2008).
Originally from Phang, I. Y., Liu, T., Mohamed, A., Pramoda, K. P., Chen, L., Shen, L., et al. (2005). Morphology, thermal and mechanical properties of nylon 12/organoclay nanocomposites prepared by melt compounding. *Polymer International*, *464*, 456–464.

example, the aforementioned embrittlement effect from clay addition (Phang et al., 2005). On the other hand, some studies report little or no change of toughness upon clay intercalation/exfoliation. For example, while the tensile strength and modulus of PP nanocomposites increased rapidly with increasing clay content from 0 to 5 wt%, the notched Izod impact strength was constant, within experimental error, in the clay content range between 0 and 7 wt% (Liu & Wu, 2001). Another study reports the impact properties for exfoliated nylon 6-based nanocomposites prepared either by in situ intercalative polymerization or by melt intercalation. In that study, marginal reductions in impact properties are reported, whatever the exfoliation process used. In the case of in situ intercalative polymerization, the Izod impact strength is reduced from 20.6 to 18.1 J/m when 4.7 wt% clay is incorporated. Charpy impact tests show similar reduction in the impact strength, with a drop from 6.21 kJ/m^2 for the filler free matrix, down to 6.06 kJ/m^2 for the 4.7 wt% nanocomposite. The decrease in the Izod impact strength of melt-intercalated nylon 6 nanocomposites is not very pronounced over a relatively large range of filler content (Liu, Qi, & Zhu, 1999).

Furthermore, toughness improvements upon clay dispersion have also been reported—a remarkable result, considering that conventional polymer-clay composites, containing aggregated nanolayer tactoids ordinarily improve rigidity, but sacrifice toughness and elongation (Lebaron et al., 1999). As an example, Liu and Wu (2002) observed a toughening effect in PA66CN. The notched Izod impact strength increased from 96 to 146 J/m upon 5 wt% clay addition, and remained higher than that of PA66, even with higher clay content. Such results are particularly surprising, considering that from length-scale arguments it is known that toughening occurs over a specific size range; effective toughening necessitates a filler size >0.1 μm and may not be energetically favorable at nano-length scales. Also, the sizes of the nanoparticles are generally too small to provide toughening through a crack-bridging mechanism and cannot effectively enhance crack-trajectory tortuosity. Therefore, the extremely reduced scale of a fully exfoliated nanocomposite does not lend itself to a toughening application. However, in an intercalated system, there is considerable interaction between silicate layers that might alleviate this concern (Zerda & Lesser, 2001). For example, Zerda and Lesser (2001) showed that the gross yielding behavior of a glassy thermoset was substantially modified upon the formation of intercalated nanocomposites, with void formation within clay aggregates leading to the evolution of a visible shear-banded zone in compression samples.

The fracture behavior appears to be most dramatically improved in the intercalated system. The fracture energy of the composites was increased by 100% at clay concentration of 5 wt%. By investigating the surface roughness and crack propagation under subcritical loading, it has been hypothesized that the creation of additional surface area on crack propagation is the primary means for toughening intercalated systems. The morphology of the system plays an important role in the toughening mechanism, because the spacing of regions of intercalated clay is important to toughening. It is believed, therefore, that the intercalated morphology can afford some property improvements that are unavailable to the fully exfoliated systems. Concerning the fracture behavior of ethyl-vinyl-acetate copolymer-based nanocomposites, Peeterbroeck, Alexandre, and Je (2005) concluded that it is independent of the origin of the clay, while it appears to be related to the nature of the clay organo-modifier and the state of nanocomposite dispersion. On the other hand, Kim et al. (2001) attributed the enhanced toughness they observed for intercalated PA12 nanocomposites to the fact that some amount of applied energy is dissipated by splitting, sliding, or opening of the separated bundles in the stacked layers. Also, Le Pluart, Duchet, and Sautereau (2005) reported that the incorporation of a benzyl dimethyl tallow alkyl ammonium montmorillonite in rubbery and glassy epoxy matrices leads to promising improvement of mechanical properties. They obtained an interesting stiffness/toughness balance for very low filler contents and without reducing the Tg of the matrix, which is particularly interesting considering how the brittleness of epoxies limits their use in technological areas where their high Tg is often highly appreciated.

Thellen et al. (2005) conducted tensile tests on PLA-based nanocomposite blown films and recorded improvements up to 40% for both the modulus and the elongation. Yao, Song, Hourston, and Luo (2002) also reported improvements in strain at break. The authors suggested that the improved elasticity is due in part to the plasticizing effect of gallery onium ions, which contribute to dangling chain formation in the matrix. Accordingly, (Chen & Evans, 2006) observed a dramatic improvement in tensile elongation at break in the presence of clay. At low clay loadings, test pieces underwent yielding during tension, similar to pristine PCL, but with dramatic increases in ductility, quite the opposite of the usual effect of adding a particulate filler to a polymer. When the clay loading was high, typically higher than 20 wt%, the composites became brittle and did not reach the yield point. Finally, it is worth summarizing the work of Hong et al. (2005) on PP-based RTPO/clay nanocomposites, prepared by using PP-MA as a compatibilizer. PP-based RTPO (or in reactor made TPO) is a blend of PP and poly(ethylene-*co*-propylene) (EPR), produced by the bulk polymerization of propylene, followed by gas-phase copolymerization of ethylene and propylene driven by the $TiCl_4/MgCl_2$-based catalyst system. Such materials, like the conventional blends of PP/EPR prepared by mechanical blending, exhibit improved flexibility and toughness compared to neat PP. Moreover, because the rubber phase can be dispersed uniformly and reach a high degree of dispersion in these in situ blends, it is possible to achieve more intimate interaction between the matrix and the rubber phase.

Above the glass transition temperature, when materials become soft, the reinforcement effect of the clay particles becomes more prominent, due to the restricted movement of the polymer chains. This results in the observed enhancement of G (Sinha Ray & Okamoto, 2003). For example, an epoxy-based nanocomposite, containing 4 vol% silicates, showed a 60% increase in G' in the glassy region, compared to the unfilled epoxy, while the equivalent increase in the rubbery region was 450% (Messersmith & Giannelis, 1994). Similar results have also been reported in the case of PP- (Hoai, Maiti, Okamoto, Kotaka, & Hasegawa, 2001), PCL- (Jimenez, Ogata, Kawai, & Ogihara, 1997), SBS- (Laus & Sandrolini, 1997), PA, PLA, and epoxy-based nanocomposites (Messersmith & Giannelis, 1994). Enhancement of the loss modulus, G', has also been reported for nanocomposite materials, however this aspect of dynamic mechanical performance is far less discussed in the literature. Finally, the tan δ values are affected in different ways by nanocomposite formation, depending on the polymer matrix. For example, in PS based nanocomposites, a shift of tan δ to higher temperatures has been observed, accompanied by a broadening of this transition (Noh & Lee, 1999), while the opposite effect was reported in the case of PP-based nanocomposites (Hoai et al., 2001).

Some authors observed a decrease of tan δ peaks, and considered this indicative of a glass transition suppression by the presence of the clay. However, (Fornes & Paul, 2003) pointed out that this conclusion is a misinterpretation, because the low values for the nanocomposites are simply the result of dividing the relatively constant loss modulus values in the Tg region, by larger and larger storage modulus values. Quite surprisingly, dioctadecyldimethyl ammonium chloride showed that above Tg, the moduli for the pure PU and the PU/o-MMT nanocomposites show no obvious difference, while below Tg, addition of o-MMT strongly influences the modulus values. Interestingly, the authors found that E' and E'' of the PU/o-MMT decrease in comparison with values for the PU, for unclear reasons. On the other hand, significant enhancements of E' and E'' were seen for the nanocomposite prepared using a particular modified clay (Xiong et al., 2004). In the case of PLA-based nanocomposites, it was observed that PLACNs with a very small amount of o-PCL as a compatibilizer exhibited a very large enhancement of mechanical properties compared to that of PLACN with comparable clay loading (Sinha Ray, Maiti, Okamoto, Yamada, & Ueda, 2002). (Krikorian & Pochan, 2003) also studied the dynamic mechanical properties of neat PLA and nanocomposites prepared with OMLS. These authors found that at high temperatures, the reinforcement effect of OMLS weakens, and suggested that this indicates a weakening of the thermomechanical stability of these materials at high temperature.

8.8 Flame retardance

Polymers are being used in more applications for flame retardant behavior. Traditionally, flame retardancy has been achieved either by using intrinsically flame retardant polymers, such as fluoropolymers or PVC, or by incorporating flame retardants (FRs), such as aluminum trihydrate, magnesium hydroxide, organic brominated compounds or intumescent systems. However, such FRs exhibit significant disadvantages. For example, aluminum trihydrate and magnesium hydroxide need to be applied at very high loadings to be effective, resulting in high density and lack of flexibility of the end-products, as well as low mechanical properties and problems in compounding. On the other hand, concerns over the environmental impact have made halogen containing materials a less popular option in many countries. Moreover, the addition of many FRs increases the production of soot and carbon monoxide during combustion. Finally, intumescent systems are relatively expensive and electrical requirements can restrict their application (Beyer, 2002; Porter et al., 2000).

Among test methods developed for the evaluation of flame retardant properties, the one most commonly used by researchers is cone calorimetry, as it provides valuable information and may even indicate the flame retardancy mechanism. The measuring principle in this test, which is standardized as ASTME 1354 and ISO 5660, is that of oxygen consumption. This states that there is a constant relationship between the mass of oxygen consumed from the air and the amount of heat released during polymer combustion. In a typical cone calorimeter experiment, the sample is exposed to a defined heat flux, usually 35 or 50 kW/m^2; properties such as heat release rate (HRR), peak of heat release (PHRR), time to ignition (TTI), total heat released (THR), mass loss rate (MLR), mean CO yield, and mean specific extinction area can be simultaneously measured (Beyer, 2002; Zanetti, Lomakin, & Camino, 2000). The HRR is considered to be the most important variable characterizing a fire. A high HRR causes fast ignition and flame spread, while the PHRR represents the point in a fire where heat is likely to propagate further, or ignite adjacent objects (Beyer, 2002; Morgan, 2006).

This difference comes into effect shortly after the initial combustion, when the nanocomposite has had time to form a char on its surface (Porter et al., 2000). In fact, it is generally considered that the formation of a thermal insulating and low permeability char on the outer surface of a nanocomposite during combustion is the key to understanding improvement in flame retardant properties (Beyer, 2002; Porter et al., 2000). More specifically, heat transfer from an external source or from a flame promotes thermal decomposition of the organoclay and the polymer. This results in accumulation and reassembly of clay platelets on the surface of the burning material (Zanetti, Bracco, & Costa, 2004). Therefore, the carbonaceous char formed superficially during combustion is rich in silicates and can be viewed as a sort of ceramic char-layered silicate nanocomposite (Zanetti et al., 2004). In fact, XRD and TEM examination of such residues has revealed intercalated structures. Actually, it is exactly its own nanocomposite structure that allows the residue formed to act as a protective barrier by reducing the heat and mass transfer between the flame and the polymer. That is, the char insulates the underlying polymer from heat and also slows oxygen uptake and the escape of volatile gases produced by polymer degradation. Both these actions interfere with the combustion cycle by reducing the amount of fuel available for burning (Beyer, 2002; Porter et al., 2000). This mechanism has been put forward in most studies reporting on the flame retardant properties of nanocomposites. For example, combustion experiments on ethyl-vinyl-acetate copolymer-based nanocomposites showed that the HRR and MLR were reduced by 70%–80% in a nanocomposite with low silicate loadings (2%–5%) presumably because of a refractory char-clay surface layer formed by the reassembling of the clay layers and catalyzed charring of the polymer. (Duquesne et al., 2003) also reported on the fire retardancy of ethyl-vinyl-acetate copolymer-based nanocomposites, using two montmorillonites: Cloisite Na+and Cloisite 30B. While the PHRR was clearly reduced when either clay was added to the polymer (relative decreases of 25% for Cloisite Na+and 50% for Cloisite 30B), the authors noted that the TTI was also reduced and that the THR was similar for pure polymer and the clay-containing polymer. Comparison of the residues after the cone calorimeter experiment demonstrates the different behavior of the three materials. The pure polymer does not give a residue, whereas the ethyl-vinyl-acetate copolymer/Na + − 5 gives a powdery gray "ash," and the ethyl vinyl-acetate copolymer/30B-5 system gives a fragile, carbonaceous residue around 3 mm thick. In the case of PE/clay nanocomposites, significant reductions of PHRR were reported, whereas the TTI results were somewhat more complicated. For low clay contents, TTI was increased, due to the barrier effect of the clay. However, with increasing clay content, a reduction of TTI was attributed to catalytic effects. Again, a large reduction of flammability in nanocomposites compared to pure PE was attributed to the formation of char (Zhao et al., 2005).

However, the TTI was similar or slightly reduced for the nanocomposites and the initial rate of weight loss/heat release was higher, probably due to early decomposition of the organic modifier of the clay and subsequent clay-catalyzed polymer degradation. As shown by the "fire performance index" (PHRR/TTI), the nanocomposites perform much better than pure PUs. However, they showed higher rates of smoke emission than their pure PU counterparts, presumably because of a greater degree of secondary condensed phase chemistry in an oxygen-depleted environment, leading to the formation and evolution of aromatic and carbonaceous species. Finally, Berta, Lindsay, Pans, and Camino (2006) conducted vertical burn tests, where a flame was applied twice for 10s to the base of the test specimens and the burning behavior was observed. Pure PU materials dripped heavily upon ignition, whereas for the PU nanocomposites, dripping was strongly suppressed or eliminated. HRR plots of PP based RTPO neat resin, microcomposites, and nanocomposites. The PHRR of the nanocomposite containing 10 wt% clay is 37% lower than that of neat resin, while the microcomposite and the neat resin show very similar behavior. At the end of combustion, the neat resin leaves no residue and the microcomposite leaves only a little powder, while the nanocomposite leaves a consistent char-like residue. Generally, PP thermally degrades to volatile products above 250°C through a radical chain process, propagated by carbon centered radicals created by carbon-carbon bond scission. In neat resin, thermal degradation continues during the combustion, and all intermediate degradation products volatilize completely. On the other hand, there may be physico-chemical adsorption of the intermediate degradation products on clay surfaces in clay nanocomposites. This adsorption may delay volatilization of the degradation products and promote accumulation of the incomplete degradation products on the clay (Hong et al., 2005). Zanetti and Costa also found that the nanodispersed morphology in PE/ethyl-vinyl-acetate copolymer nanocomposites leads to a substantial decrease of combustion rate of the polymer matrix, whereas the microcomposite shows a smaller decrease of the rate of combustion. However, a quite interesting observation is that although the clay must be nanodispersed to affect the flammability of nanocomposites, it does not need to be completely delaminated. In other words, it has been argued that the flame retardancy obtained in nanocomposites when merely intercalation has occurred is at least as good as when complete exfoliation is achieved. In fact, excellent performance has been observed when the clay layers have remained separated by only approximately 3 nm, which is considered to be in the "intercalation" zone (Messersmith & Giannelis, 1994; Sinha Ray & Okamoto, 2003).

At this point, it is worth summarizing a very interesting study by Kashiwagi et al., aiming to further elucidate the nanocomposite flame retardancy mechanism. The authors performed cone calorimeter tests on PA6 and nanocomposites containing 2% and 5% clay. They noticed that the MLR (burning) curve of each sample is proportional to the HRR curve. Thus, the specific heat of combustion (Hc), obtained from the HRR divided by the MLR remains unchanged for the three samples, implying that the observed reduction in HRR (and MLR) tends to be due to chemical and physical processes in the condensed phase, rather than in the gas phase. To prove this conjecture, the authors exposed samples to the same external flux as in the cone calorimeter, but in nitrogen, to avoid any gas phase effects. Despite the quantitative difference in MLR between the two cases, the overall differences among the three samples were very similar between the burning case and the gasification case. Given that the burning behavior depends on processes in both the gas and the condensed phase, while the gasification behavior depends only on those in the condensed phase, the authors suggested that the observed improvement in the flammability resistance of nanocomposites is due to chemical and physical processes in the condensed phase. During gasification, the nanocomposites appeared to be more viscous than the neat PA6 sample and dark floccules appeared on their surfaces, grew with time, and were left at the bottom of the container at the end of the test. Similar carbonaceous floccules were also observed in the residues of the burnt samples tested in the cone calorimeter.

Further analysis showed that up to 80% of the protective floccules consist of clay particles, while the remaining 20% consist of thermally stable organic components with possible graphitic structure. The authors attributed the accumulation of clay particles on the burning/gasifying sample surface to two possible mechanisms. One is the recession of the polymer resin from the surface by pyrolysis with dewetted clay particles left behind, and the other is the transportation of clay particles pushed by numerous rising bubbles of degradation products. The development of the aforementioned radiative gasification technique is expected to increase the understanding of the condensed phase decomposition processes of pyrolysis, because the mechanism of nanocomposite flame retardancy can be looked at in more detail through such techniques. With apparatus similar to the two cone calorimeter, this method allows pyrolysis in a nitrogen atmosphere at heat fluxes similar to those found in fires (Beyer, 2002; Porter et al., 2000).

Nanocomposites could offer significant advantages in the area of polymers flame retardancy and these advantages become even more evident when nanocomposites are compared with conventional FRs. First of all, only very low concentrations of silicate are necessary in nanocomposites, resulting in commercial advantages such as low density, lower cost and ease of preparation. Moreover, these materials are an environmentally friendly alternative to some types of fire retardants, as they contain no halogens, phosphates, or aromatics other than those that may be present in the polymer matrix; and they do not produce increases in the carbon monoxide and soot levels during combustion like those associated with conventional FRs. Also, while traditional fillers very often severely degrade the physical properties of the polymer or discolor it, an important feature of nanocomposites is the simultaneous improvement in many physical properties, without any change in the polymer color (Beyer, 2002; Lagaly, 1999; Porter et al., 2000). Finally, unlike some FRs, silicates provide physical integrity to the material burning in configurations (e.g., vertical upward combustion) in which dripping of flaming material could occur, which constitutes an additional hazard of fire propagation to surrounding materials. However, even though flame retardancy is one of the most promising properties of nanocomposites, and despite the fact that these materials have shown to perform very well in laboratory tests concerning their resistance to fire, much more research needs to be carried out to establish their applications and limitations. The potential synergy between traditional flame retardants and nanocomposites, which is discussed in the following paragraph, could further extend the uses of nanocomposites in this field (Porter et al., 2000).

8.9 Nanocomposites: Advantages and applications

As described above, polymer nanocomposite materials often exhibit properties superior to conventional composites, such as strength, stiffness, thermal and oxidative stability, barrier properties, as well as flame retardant behavior. These improved properties are generally attained at lower filler content in comparison with conventionally filled systems. Therefore, polymer-layered silicate nanocomposites are far lighter in weight than a conventional composite, which makes them quite competitive for specific applications (Sinha Ray & Okamoto, 2003). Moreover, for systems with favorable thermodynamics of mixing, the organoclay can be incorporated in the final stages of polymer processing (e.g., extrusion, injection or compression molding) to obtain nanocomposite materials. Thus, polymer nanocomposites are amenable to most of the common processing techniques in today's industrial practice—which will lower the barriers to their commercialization. Another unique aspect of nanocomposites is the lack of property tradeoffs. Traditionally, blend or composite formulations require tradeoffs between desired performance, mechanical properties, cost, and processability. However, polymer nanocomposite technology provides a route around these traditional limitations, and offers, for the first time, the opportunity to design materials without the compromises typically found in conventionally filled polymers. The aforementioned attractive characteristics of polymer-layered silicate nanocomposites already suggest a variety of possible biomedical and packaging applications:

1. Fuel tanks: it will reduce the number to fuel trucks exploding on the streets while transporting fuel.
2. Fuel pumps: reduce the risk to explosions and injury in gas stations
3. Packaging and transportation of highly inflammable materials: this would include nonfuel materials that range from fire crackers to military grade explosives.
4. Vehicle body casing for firetrucks, military vehicles and space shuttles
5. Casing for consumer vehicle's gas tank
6. Food packaging
7. Various types of splints (ranging from finders to hands and legs)
8. Or material for medical devices that tend to heat higher than 41°C (limit set by FDA).

It is for this reason that many companies have taken a strong interest and have invested in developing nanoclays and polymer nanocomposites. A TPO nanocomposite with as little as 2.5% layered silicate is as stiff as and much lighter than parts with 10 times the amount of conventional talc filler. Thus, the weight savings can reach 20%, depending on the part and the material that is being replaced by the TPO nanocomposite. William Windscheif, Basell's Global Business Vice President for Advanced Polyolefins, called this application "a small step, but a giant one for nanocomposites" and added that it heralds a broader shift to nano-PP in automotive technology. The automotive industry has well adopted the nanocomposite technology in their products mainly because of their reliability on consistency of strength and low density, which makes their products lighter and stronger. Because these materials have a proven track record under constant load, these materials will create a buzz in the biomedical device manufacturing industry.

It is worth noticing that the weight advantage of polymer nanocomposites could have a significant impact on environmental protection and material recycling. It is predicted that widespread use of polymer nanocomposites would save 1.5 billion liters of gasoline over the life of 1 year's production of vehicles and reduce related CO_2 emissions by >5 billion kilograms.

In fact, the excellent barrier properties of clay-based polymer nanocomposites could result in considerable enhancement of shelf life for many types of packaged food. Meanwhile, the optical transparency of polymer nanocomposite film is generally similar to their pristine counterparts, which is impossible with conventional polymer composites. Therefore, the above property advantages would make them widely acceptable in packaging industries as wrapping films and beverage containers. For example, Bayer has developed a new grade of plastic films for food packaging, which are made from PA6 exfoliated nanocomposites. Honeywell is also developing nano-PA materials that can beat the cost of high-barrier plastics, or even glass. Nanocor's Imperm compound supplements the inherent gas barrier of amorphous MDX6 nylon from Mitsubishi Gas Chemical with the addition of nanoclay. Used as the core of a three-layer PET bottle, Imperm reportedly ensured a 28.5-week beer shelf life. Imperm is said to adhere to PET without tie layers, while sufficient clarity is retained to meet requirements for the amber bottle. Meanwhile, Ube America is developing nanocomposite barriers for automotive fuel systems, using up to 5% nanoclay in PA6 and PA6/66 blends. Also quite interesting are potential applications of nanocomposites based on biodegradable polymers. Such polymers have become indispensable in a wide range of applications. However, despite their attractive degradation characteristics and significant demand for such materials, the lack of structural and functional stability prevents currently available biodegradable polymers from having widespread commercial impact. Therefore, biodegradable polymer-based nanocomposites appear to have a very bright future for a wide range of applications as high-performance biodegradable materials (Sinha Ray & Okamoto, 2003). Layered silicate nanoparticles, when distributed within the matrix of a fiber-reinforced polymer (FRP), can retard the diffusion of environmental moisture and other chemicals to the fiber-matrix interface where their presence can result in delamination and fiber weakening. Thus, the use of nanoparticles helps to preserve the integrity of FRPs and to prolong the service life of composites when these are used in outdoor applications such as bridges and utility poles.

Other potential nanocomposite applications include nano-pigments that are believed to be an environment-friendly alternative to toxic cadmium and palladium pigments, as well as controlled drug delivery systems, etc. However, despite the current optimism surrounding polymer-nanocomposite materials, the mechanical performance benchmarks set by highly loaded short-fiber composites are still beyond the capabilities of low concentration nanocomposites. Manufacturing costs also remain a significant factor restricting the growth of polymer-nanocomposite applications. In fact, some early application development programs have lapsed for cost reasons. Such casualties include the timing-belt cover based on nylon 6 nanocomposite. According to Silberglitt, there are two possible paths or trends—a high-growth path, under which nanocomposite materials are pervasively applied throughout society, and a low-growth path, under which the use of nanocomposites leads to incremental improvements in specific technology areas.

There are observations by Giannelis and co-workers that their preparation is possible by simple melt-mixing of the polymer with the layered silicate. Other preparation routes include intercalation of polymer or prepolymer from solution, in situ intercalative polymerization, and template synthesis. In most cases, layered silicates first need to be modified with cationic-organic surfactants, in order to become miscible with polymeric matrices. Then, whether a nanocomposite will form or not, and whether this will be intercalated or exfoliated, depends on a variety of factors. These include the type of polymer, layered silicate and organic modifier, the preparation technique, and processing conditions. In general, nanocomposite materials, particularly those with exfoliated structures, present significant improvements of modulus and strength, whereas contradictory results are reported concerning their elongation and toughness. Improvements of storage and loss moduli are also reported by many authors. Other interesting characteristics of this class of materials include improved barrier properties, thermal stability, and flame retardance. Despite some contradictory results reported in the literature and presented here, concerning certain aspects of polymer-layered silicate nanocomposite technology, we hope this review will be a useful tool for those conducting research in this field.

References

Alexandre, M., & Dubois, P. (2000). Polymer-layered silicate nanocomposites: preparation, properties and uses of a new class of materials. *Materials Science & Engineering R: Reports*, *28*(1), 1–63.

Balazs, A. C., Singh, C., Zhulina, E., & Lyatskaya, Y. (1999). Modeling the phase behavior of polymer/clay nanocomposites. *Accounts of Chemical Research*, *32*(8), 651–657.

Berta, M., Lindsay, C., Pans, G., & Camino, G. (2006). Effect of chemical structure on combustion and thermal behaviour of polyurethane elastomer layered silicate nanocomposites. *Polymer Degradation and Stability*, *91*(5), 1179–1191.

Beyer, G. (2002). Nanocomposites: a new class of flame retardants for polymers. *Plastics, Additives and Compounding*, *4*(10), 22–28.

Bharadwaj, R. K., Mehrabi, A. R., Hamilton, C., Trujillo, C., Murga, M., Fan, R., et al. (2002). Structure—property relationships in cross-linked polyester—clay nanocomposites. *Polymer*, *43*, 3699–3705.

Chang, J., Jong, S., Lak, Y., & Im, S. (2004). Poly (ethylene terephthalate) nanocomposites by in situ interlayer polymerization: the thermo-mechanical properties and morphology of the hybrid fibers. *Polymer*, *45*, 919–926.

Chaudhary, D. S., Prasad, R., Gupta, R. K., & Bhattacharya, S. N. (2005). Clay intercalation and influence on crystallinity of EVA-based clay nanocomposites. *Thermochimica Acta*, *433*, 187–195.

Chen, B., & Evans, J. R. G. (2006). Poly (−caprolactone)—clay nanocomposites: structure and mechanical properties. *Macromolecules*, *39*, 747–754.

Chin, I. J., Thurn-Albrecht, T., Kim, H. C., Russell, T. P., & Wang, J. (2001). On exfoliation of montmorillonite in epoxy. *Polymer*, *42*(13), 5947–5952.

Cho, J. W., & Paul, D. R. (2001). Nylon 6 nanocomposites by melt compounding. *Polymer*, *42*, 1083–1094.

Dixon, J. B. (1991). Roles of clays in soils. *Applied Clay Science*, *5*(5–6), 489–503.

Duquesne, S., Jama, C., Le Bras, M., Delobel, R., Recourt, P., & Gloaguen, J. M. (2003). Elaboration of EVA–nanoclay systems—characterization, thermal behaviour and fire performance. *Composites Science and Technology*, *63*(8), 1141–1148.

Finnigan, B., Martin, D., Halley, P., Truss, R., & Campbell, K. (2004). Morphology and properties of thermoplastic polyurethane nanocomposites incorporating hydrophilic layered silicates. *Polymer*, *45*, 2249–2260.

Fischer, H. (2003). Polymer nanocomposites: From fundamental research to specific applications. *Materials Science and Engineering C*, *23*(6–8), 763–772.

Fornes, T. D., & Paul, D. R. (2003). Modeling properties of nylon 6/clay nanocomposites using composite theories. *Polymer, 44*. Available at https://doi.org/10.1016/S0032-3861(03)00471-3.

Fornes, T., Yoon, P., Keskkula, H., & Paul, D. (2001). Nylon 6 nanocomposites: the effect of matrix molecular weight. *Polymer, 42*(25), 09929–09940.

Giannelis, E. P. (1996). Polymer layered silicate nanocomposites. *Advanced Materials, 8*(1), 29–35. WILEY-VCH Verlag GmbH.

Ginzburg, V. V., Singh, C., & Balazs, A. C. (2000). Theoretical phase diagrams of polymer/clay composites: the role of grafted organic modifiers. *Macromolecules, 33*(3), 1089–1099.

Gopakumar, T. G., Lee, J. A., Kontopoulou, M., & Parent, J. S. (2002). Influence of clay exfoliation on the physical properties of montmorillonite/polyethylene composites. *Polymer, 43*, 5483–5491.

Gorrasi, G., Tortora, M., Vittoria, V., Galli, G., & Chiellini, E. (2002). Transport and mechanical properties of blends of poly(ε-caprolactone) and a modified montmorillonite-poly (ε-caprolactone) nanocomposite. *Journal of Polymer Science, Part B: Polymer Physics, 40*(11), 1118–1124.

Hackman, I., & Hollaway, L. (2006). Epoxy-layered silicate nanocomposites in civil engineering. *Composites: Part A, 37*, 1161–1170.

Hoai, P., Maiti, P., Okamoto, M., Kotaka, T., & Hasegawa, N. (2001). A hierarchical structure and properties of intercalated polypropylene/clay nanocomposites. *Polymer, 42*, 9633–9640.

Hong, C. H., Lee, Y. B., Bae, J. W., Jho, J. Y., Nam, B. U., Nam, G. J., et al. (2005). Tensile and flammability properties of polypropylene-based RTPO/clay nanocomposites for cable insulating material. *Journal of Applied Polymer Science, 97*, 2375–2381.

Hotta, S., & Paul, D. R. (2004). Nanocomposites formed from linear low density polyethylene and organoclays. *Polymer, 45*, 7639–7654.

Huang, J.-C., Zhu, Z., Yin, J., Qian, X., & Sun, Y.-Y. (2001). Poly (etherimide)/montmorillonite nanocomposites prepared by melt intercalation: morphology, solvent resistance properties and thermal properties. *Polymer, 42*(3), 873–877.

Ishida, H., Campbell, S., & Blackwell, J. (2000). General approach to nanocomposite preparation. *Chemistry of Materials, 12*(5), 1260–1267.

Jimenez, G., Ogata, N., Kawai, H., & Ogihara, T. (1997). Structure and thermal/mechanical properties of poly (e-caprolactone)—clay blend. *Journal of Applied Polymer Science, 64*, 2211–2220.

Kim, G.-M., Lee, D.-H., Hoffmann, B., Kressler, J., & Stöppelmann, G. (2001). Influence of nanofillers on the deformation process in layered silicate/polyamide-12 nanocomposites. *Polymer, 42*(3), 1095–1100.

Kojima, Y., Usuki, A., Kawasumi, M., Okada, A., & Fukushima, Y. (1993). Mechanical properties of nylon 6-clay hybrid. *Journal of Materials Research, 8*(5), 1185–1189.

Kornmann, X., Lindberg, H., & Berglund, L. a. (2001). Synthesis of epoxy—clay nanocomposites. Influence of the nature of the curing agent on structure. *Polymer, 42*, 1–7.

Krikorian, V., & Pochan, D. J. (2003). Poly (l-lactic acid)/layered silicate nanocomposite: fabrication, characterization, and properties. *Chemistry of Materials, 15*(22), 4317–4324. American Chemical Society.

Krishnamoorti, R., & Giannelis, E. P. (1997). Rheology of end-tethered polymer layered silicate nanocomposites. *Macromolecules, 30*(14), 4097–4102. American Chemical Society.

Lagaly, G. (1999). Introduction: from clay mineral-polymer interactions to clay mineral-polymer nanocomposites. *Applied Clay Science, 15*, 1–9.

Laus, M., & Sandrolini, F. (1997). New hybrid nanocomposites based on an organophilic clay and poly (styrene-b-butadiene) copolymers. *Journal of Materials Research*, *12*(11), 3134–3139.

Le Pluart, L., Duchet, J., & Sautereau, H. (2005). Epoxy/montmorillonite nanocomposites: Influence of organophilic treatment on reactivity, morphology and fracture properties. *Polymer*, *46*, 12267–12278.

Lebaron, P. C., Wang, Z., & Pinnavaia, T. J. (1999). Polymer-layered silicate nanocomposites: an overview. *Applied Clay Science*, *15*(1–2), 11–29.

Lincoln, D. M., Vaia, R. A., Wang, Z., & Hsiao, B. S. (2001). Secondary structure and elevated temperature crystallite morphology of nylon-6/layered silicate nanocomposites. *Polymer*, *42*(4), 1621–1631.

Liu, L., Qi, Z., & Zhu, X. (1999). Studies on nylon 6/clay nanocomposites by melt-intercalation process. *Journal of Applied Polymer Science*, *71*, 1133–1138.

Liu, X., & Wu, Q. (2001). PP/clay nanocomposites prepared by grafting-melt intercalation. *Polymer*, *42*, 10013–10019.

Liu, X., & Wu, Q. (2002). Polyamide 66/clay nanocomposites via melt intercalation. *Macromolecular Materials and Engineering*, *287*, 180–186.

Ma, J., Xu, J., Ren, J., Yu, Z., & Mai, Y. (2006). A new approach to polymer/montmorillonite nanocomposites. *Polymer*, *44*(2003), 4619–4624.

Manias, E., Touny, A., Wu, L., Strawhecker, K., Lu, B., & Chung, T. C. (2001). Polypropylene/montmorillonite nanocomposites. review of the synthetic routes and materials properties. *Chemistry of Materials*, *13*(10), 3516–3523.

McNally, T., Raymond Murphy, W. Y., Lew, C. J., Turner, R., & Brennan, G. (2003). Polyamide-12 layered silicate nanocomposites by melt compounding. *Polymer*, *44*, 2761–2772.

Messersmith, P. B., & Giannelis, E. P. (1994). Synthesis and characterization of layered silicate-epoxy nanocomposites. *Chemistry of Materials*, *6*, 1719–1725.

Morawiec, J., Pawlak, A., Slouf, M., Galeski, A., & Krasnikowa, N. (2005). Preparation and properties of compatibilized LDPE/organo-modified montmorillonite nanocomposites. *European Polymer Journal*, *41*, 1115–1122.

Morgan, A. B. (2006). Flame retarded polymer layered silicate nanocomposites: a review of commercial and open literature systems. *Polymers for Advanced Technologies*, *17*(4), 206–217. John Wiley & Sons, Ltd.

Morgan, A. B., & Gilman, J. W. (2003). Characterization of polymer-layered silicate (clay) nanocomposites by transmission electron microscopy and X-ray diffraction: a comparative study. *Journal of Applied Polymer Science*, *87*, 1329–1338.

Noh, M. W., & Lee, D. C. (1999). Synthesis and characterization of PS-clay nanocomposite by emulsion polymerization. *Polymer Bulletin*, *42*, 619–626.

Osman, M. A., Mittal, V., & Lusti, H. R. (2004). The aspect ratio and gas permeation in polymer-layered silicate nanocomposites. *Macromolecular Rapid Communications*, *25* (12), 1145–1149.

Pavlidou, S., & Papaspyrides, C. D. (2008). A review on polymer-layered silicate nanocomposites. *Progress in Polymer Science (Oxford)*, *33*(12), 1119–1198.

Peeterbroeck, S., Alexandre, M., & Je, R. (2005). Poly (ethylene-co-vinyl acetate)/clay nanocomposites: effect of clay nature and organic modifiers on morphology, mechanical and thermal properties. *Polymer Degradation and Stability*, *90*, 288–294.

Pegoretti, A., Kolarik, J., Peroni, C., & Migliaresi, C. (2004). Recycled poly (ethylene terephthalate)/layered silicate nanocomposites: morphology and tensile mechanical properties. *Polymer*, *45*, 2751–2759.

Phang, I. Y., Liu, T., Mohamed, A., Pramoda, K. P., Chen, L., Shen, L., et al. (2005). Morphology, thermal and mechanical properties of nylon 12/organoclay nanocomposites prepared by melt compounding. *Polymer International*, *464*, 456–464.

Porter, D., Metcalfe, E., & Thomas, M. J. K. (2000). Nanocomposite fire retardants—a review. *Fire and Materials*, *24*(1), 45–52. John Wiley & Sons, Ltd.

Ray, S. S., & Okamoto, M. (2003). Biodegradable polylactide and its nanocomposites: opening a new dimension for plastics and composites. *Macromolecular Rapid Communications*, *24*, 815–840.

Ray, S. S., Yamada, K., Okamoto, M., Ogami, A., & Ueda, K. (2003). New polylactide/layered silicate nanocomposites. 3. High-performance biodegradable materials. *Chemistry of Materials*, *15*, 1456–1465.

Shelleya, J., Matherb, P., & DeVries, K. (2001). Reinforcement and environmental degradation of nylon-6/clay nanocomposites. *Polymer*, *42*(13), 5849–5858. Elsevier.

Shia, D., & Hui, C. Y. (1998). An Interface model for the prediction of young' s modulus of layered silicate-elastomer nanocomposites. *Polymer Composites*, *19*(5), 608–617.

Sinha Ray, S., & Bousmina, M. (2005). Biodegradable polymers and their layered silicate nanocomposites: In greening the 21st century materials world. *Progress in Materials Science*, *50*(8), 962–1079.

Sinha Ray, S., Maiti, P., Okamoto, M., Yamada, K., & Ueda, K. (2002). New polylactide/layered silicate nanocomposites. 1 Preparation, characterization, and properties. *Macromolecules*, *35*(8), 3104–3110. American Chemical Society.

Sinha Ray, S., & Okamoto, M. (2003). Polymer/layered silicate nanocomposites: a review from preparation to processing. *Progress in Polymer Science (Oxford)*, *28*(11), 1539–1641.

Solomon, M. J., Almusallam, A. S., Seefeldt, K. F., Somwangthanaroj, A., & Varadan, P. (2001). Rheology of polypropylene/clay hybrid materials. *Macromolecules*, *34*(6), 1864–1872.

Srivastava, S. K., Pramanik, M., & Acharya, H. (2006). Ethylene/vinyl acetate copolymer/clay nanocomposites. *Journal of Polymer Science: Part B: Polymer Physics*, *44*, 471–480.

Thellen, C., Orroth, C., Froio, D., Ziegler, D., Lucciarini, J., Farrell, R., et al. (2005). Influence of montmorillonite layered silicate on plasticized poly (L-lactide) blown films. *Polymer*, *46*, 11716–11727.

Tortora, M., Gorrasi, G., Vittoria, V., Galli, G., Ritrovati, S., & Chiellini, E. (2002). Structural characterization and transport properties of organically modified montmorillonite/polyurethane nanocomposites. *Polymer*, *43*, 6147–6157.

Vaia, R. A., & Giannelis, E. P. (1997). Polymer melt intercalation in organically-modified layered silicates: model predictions and experiment. *Macromolecules*, *30*(25), 8000–8009. American Chemical Society.

Vaia, R. a., & Giannelis, E. P. (2001). Liquid crystal polymer nanocomposites: direct intercalation of thermotropic liquid crystalline polymers into layered silicates. *Polymer*, *42*, 1281–1285.

Vaia, R., & Wagner, H. (2004). Framework for nanocomposites. *Materials Today*, *7*, 32–37.

Varlot, K., Reynaud, E., Kloppfer, M. H., Vigier, G., & Varlet, J. (2001). Clay-reinforced polyamide: preferential orientation of the montmorillonite sheets and the polyamide crystalline lamellae. *Journal of Polymer Science, Part B: Polymer Physics*, *39*(12), 1360–1370.

Wang, K. H., Choi, M. H., Koo, C. M., Choi, Y. S., & Chung, I. J. (2001). Synthesis and characterization of maleated polyethylene/clay nanocomposites. *Polymer*, *42*, 9819–9826.

Wu, S.-H., Wang, F.-Y., Ma, C.-C. M., Chang, W.-C., Kuo, C.-T., Kuan, H.-C., et al. (2001). Mechanical, thermal and morphological properties of glass fiber and carbon fiber reinforced polyamide-6 and polyamide-6rclay nanocomposites. *Materials Letters*, *49* (July), 327–333.

Xie, W., Gao, Z., Liu, K., Pan, W. P., Vaia, R., Hunter, D., et al. (2001). Thermal characterization of organically modified montmorillonite. *Thermochimica Acta*, *367*(368), 339–350.

Xiong, J., Liu, Y., Yang, X., & Wang, X. (2004). Thermal and mechanical properties of polyurethane/montmorillonite nanocomposites based on a novel reactive modifier. *Polymer Degradation and Stability*, *86*, 549–555.

Yao, K. J., Song, M., Hourston, D. J., & Luo, D. Z. (2002). Polymer/layered clay nanocomposites: 2 polyurethane nanocomposites. *Polymer Communication*, *43*, 1017–1020.

Yu, Z.-Z., Yang, M., Zhang, Q., Zhao, C., & Mai, Y.-W. (2003). Dispersion and distribution of organically modified montmorillonite in nylon-66 matrix. *Journal of Polymer Science Part B: Polymer Physics*, *41*(11), 1234–1243. Wiley Subscription Services, Inc., A Wiley Company.

Zanetti, M., Bracco, P., & Costa, L. (2004). Thermal degradation behaviour of PE/clay nanocomposites. *Polymer Degradation and Stability*, *85*(1), 657–665.

Zanetti, M., Lomakin, S., & Camino, G. (2000). Polymer layered silicate nanocomposites. *Macromolecular Materials and Engineering*, *279*(1), 1–9. WILEY-VCH Verlag GmbH.

Zerda, A. S., & Lesser, A. J. (2001). Intercalated clay nanocomposites: Morphology, mechanics, and fracture behavior. *Journal of Polymer Science Part B: Polymer Physics*, *39*(11), 1137–1146. John Wiley & Sons, Inc.

Zhao, C., Qin, H., Gong, F., Feng, M., Zhang, S., & Yang, M. (2005). Mechanical, thermal and flammability properties of polyethylene/clay nanocomposites. *Polymer Degradation and Stability*, *87*, 183–189.

Durability and sustainability of the silica and clay and its nanocomposites

9

Akshay Kakar, Elammaran Jayamani*, Muhammad Khusairy Bin Bakri*, Md. Rezaur Rahman*[†]
*Faculty of Engineering, Computing and Science, Swinburne University of Technology Sarawak Campus, Kuching, Sarawak, Malaysia, [†]Department of Chemical Engineering and Energy Sustainability, Faculty of Engineering, University Malaysia Sarawak (UNIMAS), Kota Samarahan, Sarawak, Malaysia

9.1 Introduction

One of the most popular current developments and research is on the field of technical discipline of nanotechnology. Obviously, this also includes the broad range of polymer science and technology topics. One of the examples of progress in the field is the development of microelectronics and nanoelectronics, which currently now have a critical dimension scale below 100 nm. Other areas include nanocomposites, polymer-based biomaterials, fuel cell electrode polymer bound catalysts, nanoparticle drug delivery, layer-by-layer self-assembled polymer films, mini-emulsion particles, imprint lithography, polymer blends, and electro spun nanofibers (Paul & Robeson, 2008). In the specific field of nanocomposites, the topics were focused on the bactericidal properties, flame resistance, cosmetic applications, barrier properties, electro-optical properties composite reinforcement, and flame resistance (Paul & Robeson, 2008).

In polymer science, nanotechnology is not considered as something new. However, prior to the advance in age and involvement of nanoscale dimensions, it is currently referred to as nanotechnology. According to Paul and Robeson (2008), in blends and composites, interfacial phenomena involved nanoscale dimension. While mini-emulsion particles usually fall below 100 nm, nanoscale voids are often for asymmetric membranes, block copolymer at nanoscale level, and phase separated polymer blends at nanoscale phase dimensions. Even with nanocomposites, for decades, carbon black reinforcement of elastomers, naturally occurring fiber (e.g., asbestos-nanoscale fiber diameter), and colloidal silica modification are subjects to be investigated. Not mentioned for decades, organic–inorganic nanocomposites based on sol–gel chemistry were also investigated (Mark, Jiang, & Tang, 1984; Wen & Wilkes, 1996; Wilkes, Orler, & Huang, 1985). In essence, there is a transition zone between molecular level and macromolecular level in the nanoscale dimensions. More studies and interest were observed, especially on those involving exfoliated clay, carbon

Silica and Clay Dispersed Polymer Nanocomposites. https://doi.org/10.1016/B978-0-08-102129-3.00009-9

nanotubes, exfoliated graphite (graphene), silica, nano-silica, carbon nanofibers, nano-crystalline metals, and nanoscale inorganic filler addition or fiber modification.

As communities of researchers around the world are currently focusing on the unique use of inorganic and organic additions in polymer composites to improve its properties. One improvement is by using clay and silica (which include nanoclay, nanosilica, and mesoporous silica). Polymer added with clay and silica cause improvement in thermal, corrosion resistance, mechanical, and tribological properties, which also saw improved sustainability and durability of the materials. Thus, research was done on both experimentally and theoretically to find the improved properties in the nanocomposites, in terms of mechanical, physical, chemical, thermal, and morphological. Most of the researchers focused extensively to determine the fracture toughness, deformation behavior, Young's modulus, and its strength of the composites (Akbari & Bagheri, 2007, 2013; Boo et al., 2007; Chan, Wu, Li, & Cheung, 2002; Chavarria & Paul, 2004; Chen, Justice, Schaefer, & Baur, 2008; Choi, Yee, & Laine, 2004; Dasari et al., 2006; Gloaguen & Lefebvre, 2001; He, Liu, Tjiu, Sue, & Yee, 2008; Jiang, Zhang, & Wolcott, 2007; Khosh, Bagheri, & Zokaei, 2008; Kim, Qin, Fang, Sun, & Mather, 2003; Koval'chuk et al., 2008; Liu & Wu, 2001; Ma, Mai, Rong, Ruan, & Zhang, 2007; Marouf, 2009; Pearson, Dittanet, Marouf, & Siotong, 2010; Reichert et al., 2000; Reynaud, Jouen, Gauthier, Vigier, & Varlet, 2001; Szazdi, Pozsgay, & Pukanszky, 2007; Uddin & Sun, 2010).

Nano-indentation were also being used, which focused on the improvement of the advanced work in such materials at present (Li et al., 2004; Seltzer, Kim, & Mai, 2011; Shen, Wang, Liu, & He, 2006; Wang, Liu, Yu, & Mai, 2009) On the other hand, some researchers try to adopt the theoretical models in their studies, especially on the stiffness and strength properties of the of the composites, which are influenced by its mechanical interface region and nano-fillers in polymers (Anthoulis & Kontou, 2008; Brune & Bicerano, 2002; Cannillo et al., 2006; Ciprari, Jacob, & Tannenbaum, 2006; Desai & Haque, 2005; Fornes & Paul, 2003; Frankland, Harik, Odegard, Brenner, & Gates, 2003; Luo & Daniel, 2003; Miltner, Assche, Pozsgay, Pukanszky, & Mele, 2006; Odegard, Clancy, & Gates, 2005; Pukanszky, 2005; Sheng et al., 2004; Szazdi et al., 2007).

The addition of fillers like silica and clay are commonly used in polymer formulations to improve the mechanical properties, which is the strength and modulus of the materials. The interaction between filler particles surfaces and plastics or polymers ensured the formation of an anticipated interfacial layer which yields the stronger and improved properties (Redhwi, Siddiqui, Andrady, & Hussain, 2013). Thus, for designing better composites, one way is by decreasing the average particle size of the fillers, which consequently causes increase in a specific surface area and also increases the interfacial volume of the nano-filler and polymer. Most of the nanocomposites, which are made from thermoplastics showed superior performance at lower loading of the filler in terms of mechanical properties, compared with conventional composites which are usually made from thermoset plastics (Redhwi et al., 2013). During outdoor exposure, the usage of nanoscale fillers efficiently created a light shield on the nanocomposites, which improved the weatherability (Redhwi et al., 2013). The higher the nanofiller specific areas, the more efficient the shielding created on the nanocomposites compared with conventional fillers at the same volume fraction (Redhwi et al., 2013).

Superior weatherability of PP/double hydroxide nanocomposites (Peng & Qu, 2006) and PP/ZnO nanocomposites (Zao & Li, 2006) were shown in the results obtained. Through coatings with ZnO, rutile-based poly (methyl methacrylate) (Zan, Liu, Fa, & Peng, 2006) and polyurethane (Chen, Wang, Liao, Mai, & Zhang, 2007) remarkably improved weather-ability was shown as ZnO served as a protective lacquer on pine wood surface, which gave a level of stabilization superior to that afforded by even hindered-amine light stabilizers (HALs) or the conventional UV absorbers (Weichelt et al., 2009). However, PP/montmorillonite (MMT) clay based system had poorer weatherability compared with unfilled PP matrix nanocomposites (Kumar, Depan, Tomer, & Singh, 2009; Qin et al., 2005). Polyethelene was also reported to have similar observations (Alexandre, Dubois, Sun, Garces, & Jerome, 2002).

In low-cost design of nanocomposites, the most widely used thermoplastics are polyethylene, which have good interaction with polymer matrix. In the laboratory, to produce such composites successfully, the presence of nanofiller is needed. It is reported that the polyethylene-layered silicate nanocomposites were the synthesis to get such good properties (Alexandre et al., 2002). While, others reported a similar technique were achieved using PE/silica (Cheng, Miao, Peng, Zou, & Zhang, 2009) and PE/clay hybrid (Shin, Simon, Soares, & Scholz, 2003) nanocomposites. On the other hand, both of the low-density polyethylene (Gopakumar, Lee, Kontopoulou, & Parent, 2002; Zhang & Wilkie, 2003) and clay nanocomposites (Kawasumi, Hasegawa, Kato, Usuki, & Okada, 1997), as well as polypropylene were also reported using a similar technique and achieved required properties.

Increases in matrix-resin compatibility were usually due to chemical treatment often helped due to dispersion of the nanofillers. For instance, to prepare LLDPE-MAH/Clay nanocomposites, grafter polar monomer maleic anhydride was used to the backbone of polyethylene (Wang et al., 2002; Wang, Chung, Jang, Keum, & Song, 2002). Unfortunately, in situ polymerization is not cost-effective and practical to be used in the commercial production of composites, which caused melt mixing conventional process, such as Brabender and compounding extruders were given a priority. Melt blending was successfully achieved using silica and clay and was shown by Wei, Tang, and Huang (2004) and Zhang, Rong, Zhang, and Friedrich (2003). But large improvement in the mechanical can only be achieved if full dispersion of filler is obtained critically for the nanocomposites. For example, polyethylene/MMT blends were studied by Sanchez-Valdes, Lopez-Quintanilla, Ramirez-Vargas, Medellin-Rodriquez, and Gutierrez-Rodriguez (2006). The LDPE nanocomposites were prepared by melting the blends in films. While a low-density of polyethylene LDPE grafted with maleic anhydride (LDPE-g-MA) was used as compatibilizer to ensure fill dispersion.

Nanofiller shape is important in ensuring better adhesion interface between filler and polymer. It also provides better aspects in its durability properties, such as permeability, pores structure, resistance, particle size, etc. Respectively, nanofillers were categorized according to their shapes: one-dimensional nanoscale, two-dimensional platelets, and three-dimensional, opposed to tubes and fibers (Schadler, Brinson, & Sawyer, 2007). Graphene nano-platelets are examples of 1D, halloysite and carbon nanotubes are examples of 2D, and silica and rubber are example of 3D nanofiller. The materials' two-phase interface region between two interface size decreases with the volume fraction, even though the bulk polymer matrix properties is different than the polymer matrix properties.

Broadly, in the past few years, thermoplastics and thermosets polymer nanocomposites were again gaining attention, which evolved around the preparation, processing, creating new properties (mechanical, thermal, thermomechanical, functional, etc.), and diverse applications (Alexandre & Dubois, 2000; Azeez, Rhee, Park, & Hui, 2013; Paul & Robeson, 2008; Tjong, 2006; Vaia & Wagner, 2004; Young, Kinloch, Gong, & Novoselov, 2012). In engineering applications, taking one of the examples of influenced properties, such as fracture toughness, it is known to have the capability to store energy and have high energy absorption for polymer nanocomposites. The current progress focus on 1D, 2D, and 3D on the fracture behaviors of the addition of nanofiller, showed high demand of nanocomposites, such as epoxy, in the varieties of industries (i.e., energy, adhesives, construction, transportation, etc.).

Piggott and Chua (1987) and Cech, Palesch, and Lukes (2013) showed that there is a thin interphase layer covered around the particles of polymer matrix, which differs from those counterparts in the bulk polymer matrix composites. However, the small interphase region, especially in the microcomposites does not much influence the mechanical properties. But, the volume fraction region significantly affected the mechanical properties. This concluded the size, shape, and loading of the nanofiller directly influenced by the volume fraction interfacial. Taking into consideration the similar loading, 3D nanofiller with smaller size have bigger volume fraction than those 2D nanofiller. Thus, this showed that the 3D nanofiller are far more effective than 2D nanofiller, if the size is changed drastically (Schadler et al., 2007). Furthermore, positive interfacial areas were shown on the mechanical, triobological, and thermomechanical of the polymer nanocomposites properties (Schadler et al., 2007).

Nano-filler such as clay, silica, and others are used to toughen nanocomposites, whether it is thermoplastics, or thermoset (Rebizant, Abetz, Tournilhac, Court, & Leibler, 2003; Rebizant et al., 2004; Wu, Thio, & Bates, 2005; Hydro & Pearson, 2007; Deng, Ye, & Friedrich, 2007; Liu et al., 2008; Liu et al., 2009; Hameed, Guo, Xu, Hanley, & Mai, 2010; Kishi et al., 2011; Chen & Taylor, 2012; Wu et al., 2012; Wu, Guo, Kraska, Stuhn, & Mai, 2013; Wu, Guo, Zhang, & Mai, 2013; Luo et al., 2013). The addition of nanofiller like silica and clay in nanocomposites made the materials more durable and sustainable, as with changes in structures and ductility of the materials make it more adoptable in industrial practices.

This book chapter will discuss the polymer matrix based nanocomposites with addition of clay and silica, and the effect of the modification on the durability and sustainability of the materials. Focus was given to the reinforcement aspects of the nanocomposites, especially on those interested in this primary area. Thus, properties and potential applications are important, which include the flammability resistance, membrane properties, electrical and electronic properties, barrier properties, and polymer blend compatibilization.

9.2 The role of nanofillers on nanocomposites

Generally, composites are made from different combinations of materials and sometimes more than two different materials. Usually it is divided into matrix (i.e., polymer, plastics, etc.) and reinforced (i.e., particles, fibers, film sheets, etc.) materials. Most of the reinforced materials are stronger than the existing materials and have lower densities, while the matrix polymer is ductile or tough, or vice versa according to applications. Combining both of these two different materials, with consideration of design and fabrication technique, desirable improved version of properties can be attained.

Nanocomposites are solid materials that have at least one of the domains in nanostructure scale and multiple phases. Depending on the interfacial characteristics, morphology, and type of the materials component, novel chemical and physical properties can also be achieved. In addition, nanocomposites were well-thought-out as extended class of broad materials, in which the microstructures were modulated in zero to three dimensions length scales less than 100nm. To create better grain and block in the materials, the arrangement of the atoms in nanosize clusters is crucial, as this will indirectly create an increase in area of dispersion, particularly within the host matrix polymer composites.

Significantly, as the filler size decreases, the desirability of the nano-sized particles has been found to improv drastically. Two ideal preluded results can be achieved, one with the same properties at much lower filler and one with better absolute properties with lighter product (generally, compared with polymer, fillers have higher specific gravity). In term of basic success, in which it undergoes fair judgment, most of the nanocomposites applications are in there additive range of 5%–10%, whereas the market size fillers are much smaller in terms of weight when compared with conventional ones (Rothon, 2016).

Most of the introduced nanofillers in recent year had a limited number of issues related to commercial success. Some of the problem involves nanosize, as smaller reduction in size will always costs more money. Furthermore, when the size is smaller and reduced, the technique used is going to be different, as the mixture process will become more difficult, especially in dispersing the particles, which indirectly will significantly increase the cost. Lastly, surface structure and properties of the filler when reduced in size will cause several effects and issues on the deactivation stabilizer and affect the structure algorithm. Thus, significantly as increased in the coupling agent amount used due to high amount of surfaces, and the use of an expensive coupling agent, will also directly increas the cost (Rothon, 2016). In addition to maximum particles size attainable, it decreases and prevents the conventional composites from achieving stronger properties.

9.3 Clay

Clay is a soil material or natural finely-grained rock that form through combination of one or more types of minerals with traces of organic matter, and sometimes metal oxides. Most of the clay mineral particles used in the nanocomposites are smectitic clays. This includes hectorire, saponite, and montmorillonite (Alexandre & Dubois, 2000; Esteves, Barros-Timmons, & Trindade, 2004). These types of clay belong to the family of philossilicate 2:1. It is usually formed by combined layers in such a way that octahedral layers, which contain aluminum, are sandwiched between the two tetrahedral layers of silicon (as shown in Fig. 9.1). The layers of a and b are continued in direction and stacked in the c direction. Deposits of the geologic clay are usually composed of phyllosilicate minerals with trapped water within the mineral structure. Clays were also considered as plastics form due to its water content and hardened structure, which is brittle and non-plastic upon drying or burn heated. Depending on the soil's content where the clay is being extracted; the various colors that can be obtained ranged from brown or dull gray to deep orange to red.

Depending on the clay, the thickness can vary around 1 nm and the dimensions side can vary from several micrometers to 30 nm. Weak electrostatics and layer stacking by Van der Waals forces originated interlayer spaces or the galleries (Anadao, 2012).

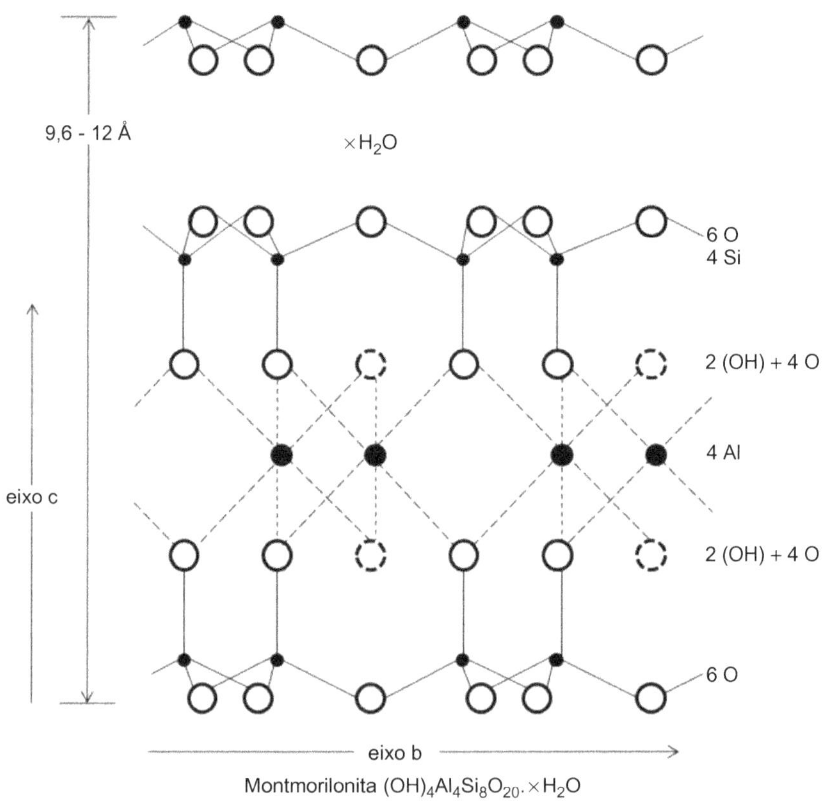

Fig. 9.1 Schematic representation of montmorillonite.
Source taken from Anadao, P. (2012). Polymer/clay nanocomposites: concepts, researches, applications and trends for the future. In F. Ebrahimi (Ed.), Nanocomposites—New trends and development. Intechopen.

Within the intercalated layer, iron and magnesium can be used to replace aluminum ions, while lithium's ions also can be used to replace magnesium ions, and thus, this will neutralize the negative charge through alkaline and terrous–alkalineceations between the intercalated layers. Moreover, polar molecules and water molecules were allowed to enter the space which caused the expansion in the c direction between these layers. Cation exchange capacity (CEC) is known as the resulting surface charge and it is usually expressed as mequiv/100g. The charge varies and highlighted according to layer and is mostly considered to be an average value in the whole crystal (Ke & Stroeve, 2005; Ray & Bousmina, 2005).

9.4 Silica

Silica is called silicone dioxide based on its chemical compound. Silica is made from the richest elements in earth's crust, which are oxygen and silicon. 56% of the covered earth contains silica, and 95% of it is rocks. Main crystalline silica are divided into three; cristobalite, tridymite, and quartz. Other varieties of silica are coesite, keatite, and lechatelierite. In applications, most of silica sand was used and mixed with Portland cement, sandstone, concrete to build houses, roads, and mortar. Silica was also used for grinding glass and stone (i.e., glass, silicon carbide, and ceramics in manufacture, as refractory materials, and also as gemstones in foundry molds.) Other than that, it also was used to remove moisture as a desiccant in the form of silica gel.

Currently, most research had benn done on the wide application of silica particle as catalysis, imaging, and drug delivery (Trewyn, Nieweg, Zhao, & Lin, 2007). With current facilities and development, the development of silica was expanded in the form of nanotechnologies, which enabled creation of nanosilica and mesoporous silica. The difference between nanosilica and mesoporous silica is the size diameter and pores size, whereas the mesoporous silica containing pores size between 2 and 50nm, while nanosilica have size less than 100nm. The most common mesoporous silica used are MCM-41 and SBA-15 (Katiyar, Yadav, Smirniotis, & Pinto, 2006). The compounded mesoporous silica was first patented around 1970 (Chiola, Ritsko, & Vanderpool, 1971). But the impact of the research was not noticed until it was reintroduced and reproduced back in 1997 (Di Renzo, Cambon, & Dutartre, 1997; Xu, Pang, Yu, Huo, & Chen, 2007). Japanese researcher, synthesized mesoporous silica nanoparticles independently. Six years after that, University of California, Santa Barbara produced much larger scale silica particles around 4.6 to 30nm (Zhao et al., 1998). The names of the silica itself were named for the university location, Santa Barbara Amorphous (SBA-15). The hexagonal arrays of pores were shown on the particles itself. These particle types allowed researcher to use it as molecular sieves.

In today's wide applications, nanoparticles of mesoporous silica can be used as imaging, thermal energy storage, medicine, and biosensors (Mitran, Berger, Munteanu, & Matei, 2015). Most of nanoparticles mesoporous silica is synthesized reacting with tetraethyl orthosilicate with a template made by micellar rods. The collection cause regular pores arrangement to be filled with nano-sized spheres and rods. According to Trewyn, Slowing, Giri, Chen, and Lin (2007), the template can be removed by washing it with the proper amount of solvent. Stoner process, simple sol–gel method, or a drying spray method can also be used to synthesize the mesoporous silica particles (Nandiyanto, Iskandar, & Okuyama, 2008; Nandiyanto, Kim, Iskandar, & Okuyama, 2009). Tetraethyl orthosilicate also can be used with additional polymer monomer as a template. When comparing TEOS and MPTMS, advantages were shown to MPTMS due to effectiveness for synthesizing such particles. Thus, uniform spheres were produced and reduction was seen on the aggregation.

9.5 Durability and sustainability of clay nanocomposites and silica nanocomposites

9.5.1 *Spectral and thermal properties*

Redhwi et al. (2013) compares the FTIR spectrum of control LDPE sample with LDPE-ZnO and LDPE-SiO_2 nanocomposites. In the former spectrum, when comparing control LDPE sample with LDPE-ZnO nanocomposite, the absorption band at 2364 cm^{-1} appears to be more intense for LDPE-ZnO. This peak may be attributed to CO_2 absorption onto the nanoparticles. The peak at 500 cm^{-1} likely to be apparent, which is likely due to the prismatic microstructure of the filler obtained at peak 512 cm^{-1} (Abdullah, Ariyanto, Shaari, Yuliarto, & Junaidi, 2009). More intense peak at 3730 cm^{-1} was seen with LDPE-nano-SiO_2 nanocomposites due to the contribution of O—H incorporated with silica filler. While the vibration band of silica powder was seen at 1117 cm^{-1}. Two intense sharp peaks at 3724 and 3703 cm^{-1} were shown by LDPE-MMT nanocomposite due to the N—H functional groups in the amine functionalized clay.

However, with nanosilica, interaction of strong hydrogen-bond (Carteret, 2009) via silanol group (Si-OH) on the nanoparticle surface is expected (peak band around 3700 cm^{-1}, which is due to silanol group). Thus, this also showed that the silica agglomeration network structure yields a 3D structure particularly efficient in reinforcing.

No significant differences on the DSC studies at the overlaid exemplary thermograms of crystalline melting cycles between the nanocomposites thermal behaviors. For each of the cases, the first heating usually started early to yield a peak at about 112–113°C. Significant lack of difference in LDPE nanocomposites were seen, which is in contrast to that HDPE (Gopakumar et al., 2002; Wang, Choi, et al., 2002; Wang, Chung, et al., 2002) and PP (23) nanocomposites. In examples, increased nucleation facilitated by the nanoclay particles was attributed due to the change in thermal behavior. Compared with LDPE, both PP and HDPE are readily crystallizable and not much significant increase in nucleation is present of MMT in the composites. Redhwi et al. (2013) finding showed consistency with the reported data on LDPE/clay nanocomposites, but with a much higher percentage (~30 wt%) of the clay (24). There was minimal volume fraction of nanofiller used in these compositions primarily due to the minimal change in the thermal behavior of present nanocomposites and the small fraction consequently from the interfacial volume created in the composite.

9.5.2 Mechanical properties

The amount of filler contained in the composites affected the mechanical properties mostly in the polymer matrix. Deformations were observed on the test pieces, in the uniform direction, and strains up to the yield point of the gauge length. In the test piece, a neck was formed and further deformation was localized in the region where the length expanded further with strain. Redhwi et al. (2013) mentioned that the tensile curve shapes remained similar, even though there are variations in the volume fraction and filler type. Under the 10%, the extensibility of the tensile was not affected, even with different filler used and it is usually less than the unfilled LDPE. However, in particular, the tensile modulus is sensitive with the degree of reinforcement achieved in the LDPE nanocomposites. Results obtained by Redhwi et al. (2013) showed that the average modulus of LDPE increased the base value from about 35.4 MPa with 25% for nanosilica, 150% for nano-ZnO, and 150% for MMT clay. However, the reinforcement efficiency was observed in decreased order of Silica > MMT clay > ZnO through certain calculation (Redhwi et al., 2013).

Hybridization between carbon (Rattanasom, Saowapark, & Deeprasertkul, 2007) and mica (Maldas & Kokta, 1993) showed promising results. It decreased the absorption of water and swell on thickness observed when nanoclay with reed flours hybridization, even though it upgraded the whole system tensile properties. The effect of coupling agent on the physical and mechanical properties of reed flour/PP/nanoclay hybrid composite were also shown by certain researchers (Najafi, Kord, Abdi, & Ranaee, 2012). Other than that, integration of nanoclay in HDPE/rice husk system greatly enhanced the mechanical properties (Kord, 2011). Hemp fiber/PP based hybrid composites also showed increase in stiffness and flexibility due to the involving effects of clay nanoparticles (Kord, 2012).

Thickness enlargement due to swelling increased by nanoparticles loading and substantial reductions in tendency of water absorption were also seen in some cases. Emerging as an attention-grabbing work, the effect of the incorporation and compatibilization of organoclay in wood flour/HDPE composites systems affected the thermal and mechanical properties of the composites (Zhong, Poloso, Hetzer, & De Kee, 2007). 5 wt% of nanoclay hybridization with micro-crystalline cellulose in the reinforced micro-crystalline cellulose ethylene-propylene (EP) copolymer showed remarkable changes in the Young's modulus from 1.04 to 1.24 GPa (Kumar & Singh, 2007). Researchers also found out that the dispersion of nanoclay on wood-plastic-nanoclay hybrid composites gave better enhancement for its physical and mechanical properties (Kord & Kiakojouri, 2011). The thermal and mechanical performances were also seen improved in terms of compatibilization, especially when organoclay were incorporated within pine cone fiber hybrid polymer composites (Arrakhiz, Benmoussa, Bouhfid, & Qaiss, 2013). Notably observed and contrary to fiber, there are reductions in the stiffness/modulus of the system when additions of clay were added in. In a different study, interfacial interaction and good dispersion between pine cone fiber and clay within it polymer composites were revealed by SEM for clay/pine cone fiber hybrid polymer nanocomposites.

The nanoclay incorporated into the formed HDRE/wheat straw flour polymer composites showed improvement in the properties, which increased the fire resistance, physical properties, modulus and strength, heat resistance, and low gas permeability, even for small nanoclay scale (Babaei, Madanipour, Farsi, & Farajpoor, 2014). Liany, Tabei, Farsi, and Madanipour (2013) studied the wheat straw flour (WSF) and HDPE and found out that the exothermic chemical foaming agent (CFA) effect from azodicarbonamide (AZD) gave better enhanced properties on the physical, mechanical, and morphological. There were also increases in the water absorption tendency and swelling thickness of HDPE/WSF composites when AZD were added, while reduction on matrix was also witnessed by adding the NC. Adding NC and CFA also decreased the impact resistance. Adding the 2 phr of NC in the HDPE/WSF composites also improved the mechanical properties.

9.5.3 *Morphological properties*

Fig. 9.2 shows the $\times 2000$ magnification SEM Images, which showed no obvious sign of filler agglomeration (Redhwi et al., 2013). Significant agglomeration should be visible at this magnification. While, ascertained by scanning electron, the idea of some level of dispersion might be visible, even though the small sampled area does not yield an average picture of the samples. For studying nanoparticle dispersion, it's suggested that transmission electron microscopy (TEM) is better suited than SEM.

Fig. 9.3A–C showed the cross-section SEM micrograph of PI/10SiO_2, PI/20SiO_2, and PI/30SiO_2, respectively (Huang et al., 2007). Fig. 9.3D showed the comparison for polyimide blended with nanosilica powder (Degussa P25) SEM micrograph. Silica high loading in PI composite films is noted, especially without the surfactants and coupling agent are used. Homogeneous dispersion was seen for silica particles composites films in the micrograph. Fig. 9.3D showed poorer interfacial adhesion due to phase separation between the two phases. Fig. 9.3A and B, on the other hand, reveal a uniform dispersion and no obvious phase separation throughout the PI matrix. Fig. 9.3C showed aggregates various sizes of silica particles for PI/30SiO_2. Good interaction at the silica and PI interfaces at Fig. 9.3A–C does not exhibit holes and debonding (Huang et al., 2007).

9.6 Challenges and applications

By 2020, it is estimated around $3 trillion will be impacted due to advancement in nanotechnologies worldwide. When considering nano particles for nanotechnology industries production, at least six million workers will be needed to sustain this by the end of the decade globally (Roco, Mirkin, & Hersam, 2011). Compared with conventional composites, nanocomposites offer better performance in resistance, mechanical, and thermal properties (Gacitua, Ballerini, & Zhang, 2005). Polymer nanocomposites posed large diversity, which is variable in many applications in

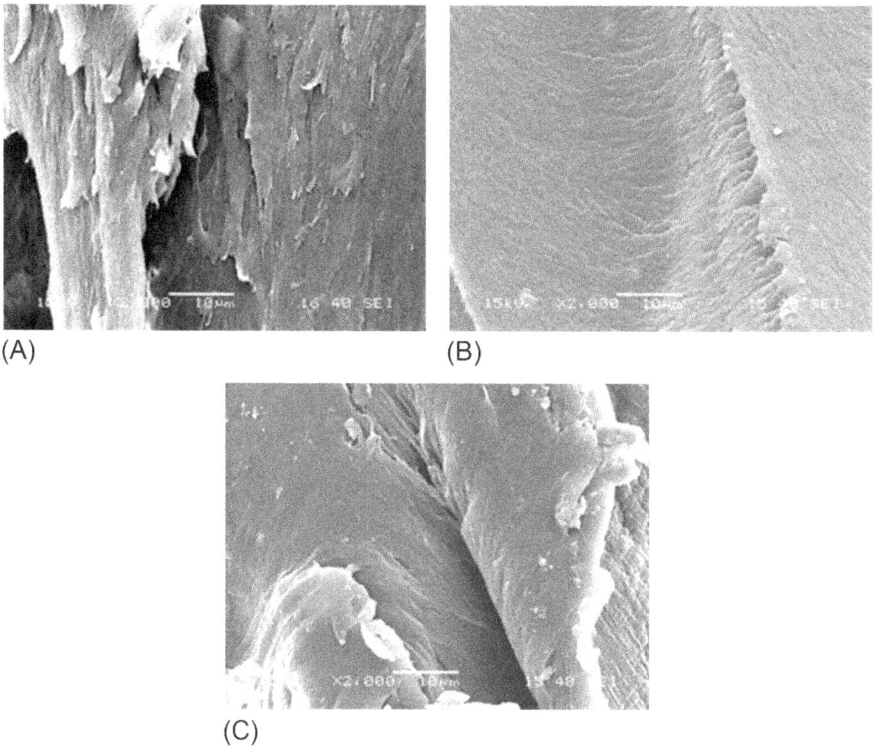

Fig. 9.2 The ×2000 magnification of (A) LDPE-5 wt%-ZnO sample, (B) LDPE-5 wt%-SiO_2 sample, and (C) LDPE-5 wt%-clay sample nanocomposites SEM.
Source taken from Redhwi, H. H., Siddiqui, M. N., Andrady, A. L., & Hussain, S. (2013). Durability of LDPE Nanocomposites with clay, silica, and zinc oxide Part I: mechanical properties of the Nanocomposite materials. *Journal of Nanomaterials, 1(1),* 1–6.

industry, such as automotive (bumpers, exterior and interior panels, insulator, gas tanks), aerospace (high performance components, flame, and water retardant panels), packaging (food containers, wrapping films, and bottled water) electrical and electronics (electrical components and printed circuit boards), construction (insulator, rooftop and structural panels, and certain building sections), and in medical (plastic surgery, biomedical plate and casts) industries (Wypych & Satyanarayana, 2005).

The oldest and traditional industries are construction. Significant changes in time with advanced technologies happened mainly with the building fast development (Chaturvedi & Dave, 2013) as most of nanotechnologies were used to enhance the functionality of wood, steel, glass and concreate, which is the primary area and materials construction. Nanoparticles in the micro-matrices embodiment, or surface coating on the areas of the materials improved the durability, tolerance, sustainability, and stress tolerance of the nanocomposites. Many routes in construction and environmental concerns were addressed. Nanotechnologies usage with the nature-based

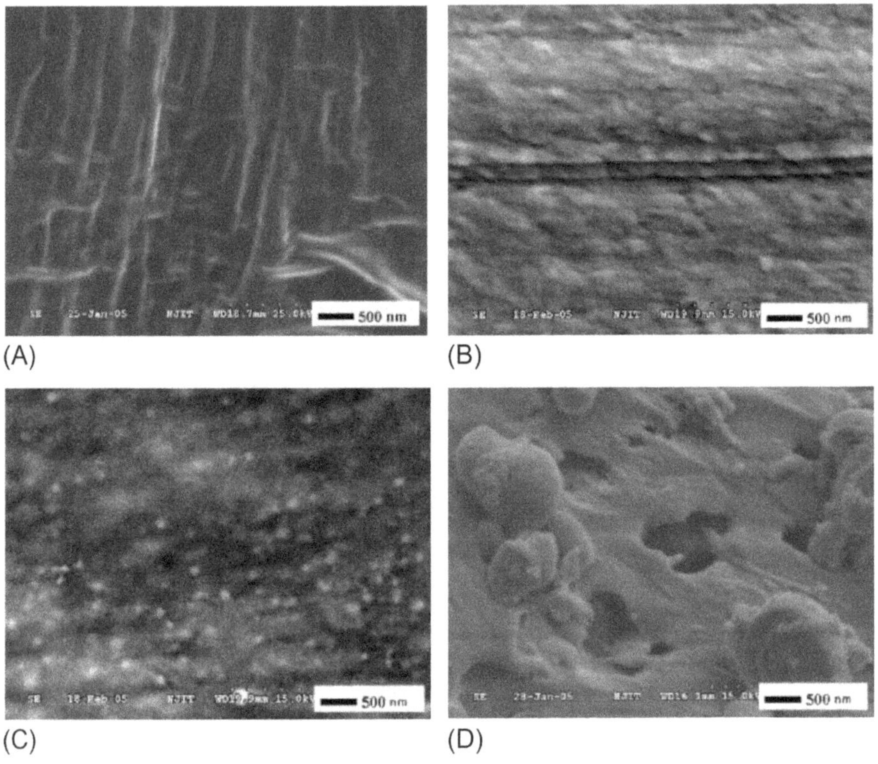

Fig. 9.3 PI/silica nanocomposite films SEM photographs: (A) PI/10SiO_2, (B) PI/20SiO_2, (C) PI/30SiO_2, (D) blend of PI and silica particle (Degussa P25).
Source taken from Huang, J.-W., Wen, Y.-L., Kang, C.-C., & Yeh, M.-Y. (2007). Preparation of polyimide-silica nanocomposites form nanoscale colloidal silica. *Polymer Journal, 39(7)*, 654–658.

nanotechnology can provide new and eco-efficient functionalities much more than the traditional construction materials. Some of the examples of nanotechnology applications are given below (Bartos et al., 2009):

- Nano-scaled coating for low maintenance self-cleaning windows
- Nano-enabled sensors to improved maintenance for building
- Nano enabled filter technologies for healthier indoor climates/air
- CNTs for super strong structural floors
- Nano-structured materials for long-lasting scratch resistant floors
- Nano scaled coatings for surfaces of antimicrobial steel
- Electrochromic "smart" windows for building with lower energy consumed

In automotive, the nanotechnology applications are manifold and diverse. Energy conversion, power train, pollution sensing and reduction, light-weight construction, wear reduction, surveillance control up to recycle potential, driving dynamics, and others.

(Presting & Konig, 2003). The development of quiet cars with the use of "nano in cars" concept, free CO_2 engines, choice mood-dependent color, self-forming and self-healing body, windscreens, and safe driving can be made using nanotechnology. The different nanoparticles forms such as dots, layers, fibers, tubes, and whisker, can be dispersed or distributed firmly within the matrix.

Due to increased awareness, an industry such as food safety and food processing has evolved in producing packaging materials effectively, using an antimicrobial agent which provides high-barrier packaging materials (Saba et al., 2014). Furthermore, with high technological sensors series available to trace contaminants, microbes, impurities, or gasses in packed foods that were made by polymer/clay nanocomposites, which are strong types of material packaging (Saba et al., 2014). Also reported in another study was silver nanoparticles have potential as nanomaterials, antimicrobial agents, and nanosensor-based assays, which are good at revealing analytes in food-relevant products, such as small organic molecules, food-borne pathogens, and gasses (Duncan, 2011). Virtually, every food industry segment is recognized and well-established with its prospective nanotechnology applications, either from the well-known food processing (e.g, food quality and textural improvement, gelation, odor enchanters, encapsulation of flavor, and viscosifying agents.) to nutrient supplements (e.g., higher nutraceuticals bioavailability and stability) to agriculture (e.g., targeted genetic engineering, fertilizer vaccine delivery or pesticide, and detection of plant and animal pathogens) and to food packaging (e.g., or abuse gas sensors, pathogen, UV-protection, anti-counterfeiting devices, and more stronger and impermeable polymer films) (Duncan, 2011).

In biomedical applications, most of the nanocomposites were used as dentistry cosmetic and dental apparatus and materials, which can be conveniently applied with a high quality of esthetical values. Most of the nanocomposites were used as a dental bracket, and used to fabricate the post and core systems, which can be applied in various restorations or refurbishment (crowns, veneers, onlays and inlays) (Saba et al., 2014). The novel nanocomposites which in the form of liquid crystalline epoxy, exhibited and reduced polymerization shrinkage effectively on the tooth structures, which were applied esthetically by dentistry, for indirect and direct restorations of the tooth and dental brackets for tooth alignment, which includes post and cores of the system (Tai, Hsu, Chen, Su, & Chen, 2014). Most of the nanotechnologies in the manufacturing of these applications of equipment are not recognizable by many people currently, but the profits of this fabrication process are remarkable in the future. Most of the cosmetic industries preferred materials which can gain benefits and applied with various manufacturing techniques that can easily be cleaned, especially on the surfaces, non-sticky materials either for its materials and machinery use, low details of abrasion and efficient energy consumption value-added products (Mihranyan, Ferraz, & Strømme, 2012). It is also aims to gave better stabilities, which actively infused the materials used, which enables production of more grateful products (Mihranyan et al., 2012).

In the field of polymer, nanofibers were also used to seal wounds and burns on human skin with hemostatic and anecdotic procedures, which are done exclusively and especially with features devices. With the use of electrical coated spray, finest fiber polymers, which are biodegradable, can be sprayed onto the injured and damaged skin area in the form of fibrous mat dressing (Saba et al., 2014). The use of these nanocomposites enable self-healing of wounds due to the elasticity of the nanocomposites, which boosts the skin growth development, reduces scar tissue formation, and eradicates wounds from diseases. Thus, the value of these nanofibers or nanocomposites is appealing and interesting for investment opportunities, especially in academia, government, and growing industries globally (Huang, Zhang, Kotaki, & Ramakrishna, 2003). Developing of nano-materials fiber composites related to paper and pulp industries might provide a good platform for developing new materials which give more value for barrier packaging, intelligent communication media, and printing (Kamel, 2007).

9.7 Summary

In a nutshell, there are a lot of potential improvements that can made by using clay and silica, providing improved versions of mechanical, thermal, spectral, and morphological properties, compared with non-addition types of nanocomposites. This advancement of technologies by using clay and silica also improves the weatherability and resistance on the nanocomposites when compared with conventional composites. The use of clay and silica in nanocomposites filled the gaps between the molecules, and by manipulating the amount of silica and clay contained, size, shape and chemical composition, it will create variable properties which can be used in various types of applications and environments, depending on the demands of usage. Manipulating the clay and silica also increases its durability and sustainability of the nanocomposites, which advances capabilities of resistance, which provides a layer of protection on the nanocomposites. Clay and silica are also widely available, and due to that, advances nanotechnological research. The use of clay and silica gives many ways and benefits to reduce the cost of the end-product of nanocomposites. If it is manipulated in such a way that it can reduce the amount of expensive polymer or through hybridization, it can be far more superior to extend synthetic and metallic nanocomposites successfully.

References

Abdullah, H., Ariyanto, N. P., Shaari, S., Yuliarto, B., & Junaidi, S. (2009). TEC and scintillation study of equatorial ionosphere: a month campaign over Sipitang and Parit Raja stations, Malaysia. *American Journal of Engineering and Applied Sciences*, *2*(1), 236–240.

Akbari, B., & Bagheri, R. (2007). Deformation mechanism of epoxy/clay nanocomposite. *European Polymer Journal*, *43*(3), 782–788.

Akbari, B., & Bagheri, R. (2013). Influence of nanoclay on morphology, mechanical properties and deformation mechanism of polystyrene. *Polymer-Plastics Technology and Engineering*, *53*(2), 156–161.

Alexandre, M., & Dubois, P. (2000). Polymer-layered silicate nanocomposites: preparation, properties and uses of a new class of materials. *Materials Science and Engineering: R: Reports*, *28*(1–2), 1–63.

Alexandre, M., Dubois, P., Sun, T., Garces, J. M., & Jerome, R. (2002). Polyethelene-layered silicate Nanocomposites prepared by the polymerization-filling technique: synthesis and mechanical properties. *Polymer*, *43*(8), 2123–2132.

Anadao, P. (2012). Polymer/clay nanocomposites: concepts, researches, applications and trends for the future. In F. Ebrahimi (Ed.), *Nanocomposites—New trends and development*: Intechopen.

Anthoulis, G. I., & Kontou, E. (2008). Micromechanical behaviour of particulate polymer Nanocomposites. *Polymer*, *49*(7), 1934–1942.

Arrakhiz, F. Z., Benmoussa, K., Bouhfid, R., & Qaiss, A. (2013). Pine cone fiber/clay hybrid composite: mechanical and thermal properties. *Materials & Design*, *50*(1), 376–381.

Azeez, A. A., Rhee, K. Y., Park, S. J., & Hui, D. (2013). Epoxy clay nanocomposites—processing, properties and applications: a review. *Composites Part B: Engineering*, *45* (1), 308–320.

Babaei, I., Madanipour, M., Farsi, M., & Farajpoor, A. (2014). Physical and mechanical properties of foamed HDPE/wheat straw flour/nanoclay hybrid composite. *Composites Part B: Engineering*, *56*(1), 163–170.

Boo, W. J., Sun, L. Y., Liu, J., Clearfield, A., Sue, H. J., Mullins, M. J., et al. (2007). Morphology and mechanical behavior of exfoliated epoxy/a-zirconium phosphate nanocomposites. *Composites Science and Technology*, *67*(2), 262–269.

Brune, D. A., & Bicerano, J. (2002). Micromechanics of nanocomposites: comparison of tensile and compressive elastic moduli, and prediction of effects of incomplete exfoliation and imperfect alignment on modulus. *Polymer*, *43*(2), 369–387.

Cannillo, V., Bondioli, F., Lusvarghi, L., Montorsi, M., Avella, M., Errico, M. E., et al. (2006). Modeling of ceramic particles filled polymer-matrix nanocomposites. *Composites Science and Technology*, *66*(7–8), 1030–1037.

Carteret, C. (2009). Mid- and near- infrared study of hydroxyl groups at a silica surface: H-bond effect. *The Journal of Physical Chemisty C*, *113*(20), 13300–13308.

Cech, V., Palesch, E., & Lukes, J. (2013). The glass fiber–polymer matrix interface/interphase characterized by nanoscale imaging techniques. *Composites Science and Technology*, *83* (1), 22–26.

Chan, C.-M., Wu, J., Li, J.-X., & Cheung, Y.-K. (2002). Polypropylene/calcium carbonate nanocomposites. *Polymer*, *43*(10), 2981–2992.

Chaturvedi, S., & Dave, P. N. (2013). Design process for nanomaterials. *Journal for Materials Science*, *48*, 3605–3622.

Chavarria, F., & Paul, D. R. (2004). Comparison of nanocomposites based on nylon 6 and nylon 66. *Polymer*, *45*(25), 8501–8515.

Chen, C., Justice, R. S., Schaefer, D. W., & Baur, J. W. (2008). Highly dispersed nanosilica-epoxy resins with enhanced mechanical properties. *Polymer*, *49*(17), 3805–3815.

Chen, J., & Taylor, A. (2012). Epoxy modified with triblock copolymers: morphology, mechanical properties and fracture mechanisms. *Journal of Materials Science*, *47*(11), 4546–4560.

Chen, X. D., Wang, Z., Liao, Z. F., Mai, Y. L., & Zhang, M. Q. (2007). Roles of Anatase and rutile TiO_2 nanoparticles in Photooxidation of polyurethane. *Polymer Testing*, *26*(2), 202–208.

Cheng, W., Miao, W., Peng, J., Zou, W., & Zhang, L. (2009). Synthesis of silica/polyolefin Nanocomposites via two-step method. *Iranian Polymer Journal*, *18*(5), 365–371.

Chiola, V., Ritsko, J. E., Vanderpool, C.D., 1971. Process for producing low-bulk density silica, Application No. US 3556725D A filed on 26-Feb-1969; Publication No. US 3556725 A published on 19-Jan-1971.

Choi, J., Yee, A. F., & Laine, R. M. (2004). Toughening of cubic silsesquioxane epoxy nanocomposites using core-shell rubber particles: a three-component hybrid system. *Macromolecules*, *37*(9), 3267–3276.

Ciprari, D., Jacob, K., & Tannenbaum, R. (2006). Characterization of polymer nanocomposite interphase and its impact on mechanical properties. *Macromolecules*, *39*(19), 6565–6573.

Dasari, A., Yu, Z.-Z., Yang, M., Zhang, Q.-X., Xie, X.-L., & Mai, Y. -W. (2006). Micro- and nano-scale deformation behavior of nylon 66-based binary and ternary nanocomposites. *Composites Science and Technology*, *66*(16), 3097–3114.

Deng, S., Ye, L., & Friedrich, K. (2007). Fracture behaviours of epoxy nanocomposites with nano-silica at low and elevated temperatures. *Journal of Materials Science*, *42*(8), 2766–2774.

Desai, A. V., & Haque, M. A. (2005). Mechanics of the interface for carbon nanotube–polymer composites. *Thin-Walled Structures*, *43*(11), 1787–1803.

Di Renzo, F., Cambon, H., & Dutartre, R. (1997). A 28-year-old synthesis of micelle-templated mesoporous silica. *Microporous Materials.*, *10*(4–6), 283–286.

Duncan, T. V. (2011). Applications of nanotechnology in food packaging and food safety: barrier materials, antimicrobials and sensors. *Journal of Colloid and Interface Science*, *363* (1), 1–24.

Esteves, A. C. C., Barros-Timmons, A., & Trindade, T. (2004). Nanocompósitos de matriz polimérica: estratégias de síntese de materiais híbridos. *Química Nova*, *27*(5), 798–806.

Fornes, T. D., & Paul, D. R. (2003). Modeling properties of nylon 6/clay nanocomposites using composite theories. *Polymer*, *44*, 4993–5013.

Frankland, S. J. V., Harik, V. M., Odegard, G. M., Brenner, D. W., & Gates, T. S. (2003). The stress–strain behavior of polymer–nanotube composites from molecular dynamics simulation. *Composites Science and Technology*, *63*(11), 1655–1661.

Gacitua, W., Ballerini, A., & Zhang, J. (2005). Polymer nanocomposites: synthetic and natural fillers a review. *Maderas Ciencia Y Tecnologia*, *7*(3), 159–178.

Gloaguen, J. M., & Lefebvre, J. M. (2001). Plastic deformation behaviour of thermoplastic/clay nanocomposites. *Polymer*, *42*(13), 5841–5847.

Gopakumar, T. G., Lee, J. A., Kontopoulou, M., & Parent, J. S. (2002). Influence of clay exfoliation on the physical properties of montmorillonite/polyethylene composites. *Polymer*, *43*(20), 5483–5491.

Hameed, N., Guo, Q., Xu, Z., Hanley, T. L., & Mai, Y.-W. (2010). Reactive block copolymer modified thermosets: highly ordered nanostructures and improved properties. *Soft Matter*, *6*(24), 6119–6129.

He, C., Liu, T., Tjiu, W. C., Sue, H.-J., & Yee, A. F. (2008). Microdeformation and fracture mechanisms in polyamide-6/organoclay nanocomposites. *Macromolecules*, *41*(1), 193–202.

Huang, J.-W., Wen, Y.-L., Kang, C.-C., & Yeh, M.-Y. (2007). Preparation of polyimide-silica nanocomposites form nanoscale colloidal silica. *Polymer Journal*, *39*(7), 654–658.

Huang, Z.-M., Zhang, Y.-Z., Kotaki, M., & Ramakrishna, S. (2003). A review on polymer nanofibers by electrospinning and their applications in nanocomposites. *Composites Science and Technology*, *63*(15), 2223–2253.

Hydro, R. M., & Pearson, R. A. (2007). Epoxies toughened with triblock copolymers. *Journal of Polymer Science Part B: Polymer Physics*, *45*(12), 1470–1481.

Jiang, L., Zhang, J., & Wolcott, M. P. (2007). Comparison of polylactide/nano-sized calcium carbonate and polylactide/montmorillonite composites: reinforcing effects and toughening mechanisms. *Polymer*, *48*(26), 7632–7644.

Kamel, S. (2007). Nanotechnology and its applications in lignoellulosic composites, a mini review. *Express Polymer Letter*, *1*(1), 546–575.

Katiyar, A., Yadav, S., Smirniotis, P. G., & Pinto, G. N. (2006). Synthesis of ordered large pore SBA-15 spherical particles for adsorption of biomolecules. *Journal of Chromatography*, *1122*(1–2), 13–20.

Kawasumi, M., Hasegawa, N., Kato, M., Usuki, A., & Okada, A. (1997). Preparation and mechanical properties of polypropylene-clay hybrids. *Macromolecules*, *30*(20), 6333–6338.

Ke, Y. C., & Stroeve, P. (2005). *Polymer-layered silicate and silica nanocomposites*. Amsterdam: Elsevier B.V.

Khosh, R. L., Bagheri, R., & Zokaei, S. (2008). Sequences of fracture toughness micromechanisms in PP/$CaCO_3$ nanocomposites. *Journal of Applied Polymer Science*, *110*(6), 4040–4048.

Kim, G. M., Qin, H., Fang, X., Sun, F. C., & Mather, P. T. (2003). Hybrid epoxy-based thermosets based on polyhedral oligosilsesquioxane: cure behavior and toughening mechanisms. *Journal of Polymer Science Part B: Polymer Physics*, *41*(15), 3299–3313.

Kishi, H., Kunimitsu, Y., Imade, J., Oshita, S., Morishita, Y., & Asada, M. (2011). Nano-phase structures and mechanical properties of epoxy/acryl triblock copolymer alloys. *Polymer*, *52*(3), 760–768.

Kord, B. (2011). Nanofiller reinforcement effects on the thermal, dynamic mechanical, and morphological behavior of HDPE/rice husk flour composites. *BioResources*, *6*(2), 1351–1358.

Kord, B. (2012). Effect of nanoparticles loading on properties of polymeric composite based on hemp Fiber/polypropylene. *Journal of Thermoplastic Composites Materials.*, *25*(7), 793–806.

Kord, B., & Kiakojouri, S. (2011). Effect of nanoclay dispersion on physical and mechanical properties of wood flour/polypropylene/glass fibre hybrid composites. *BioResources*, *6* (2), 1741–1751.

Koval'chuk, A. A., Shevchenko, V. G., Schegolikhin, A. N., Nedorezova, P. M., Klyamkina, A. N., & Aladyshev, A. M. (2008). Effect of carbon nanotube functionalization on the structure and mechanical properties of polypropylene/MWCNT composites. *Macromolecules*, *41*(20), 7536 7542.

Kumar, A. P., Depan, D., Tomer, N. S., & Singh, R. P. (2009). Nanoscale particles for polymer degradation and stabilization-trends and future perspectives. *Progress in Polymer Science*, *34*(6), 479–515.

Kumar, A. P., & Singh, R. P. (2007). Novel hybrid of clay, cellulose, and thermoplastics. I. Preparation and characterization of composites of ethylene–propylene copolymer. *Journal of Applied Polymer Science*, *104*(4), 2672–2682.

Li, X., Gao, H., Scrivens, W. A., Fei, D., Xu, X., Sutton, M. A., et al. (2004). Nanomechanical characterization of single-walled carbon nanotube reinforced epoxy composites. *Nanotechnology*, *15*(11), 1416–1423.

Liany, Y., Tabei, A., Farsi, M., & Madanipour, M. (2013). Effect of nanoclay and magnesium hydroxide on some properties of HDPE/wheat straw composites. *Fibers and Polymers*, *14* (2), 304–310.

Liu, J., Sue, H.-J., Thompson, Z. J., Bates, F. S., Dettloff, M., Jacob, G., et al. (2008). Nanocavitation in self-assembled amphiphilic block copolymer-modified epoxy. *Macromolecules*, *41*(20), 7616–7624.

Liu, J., Sue, H.-J., Thompson, Z. J., Bates, F. S., Dettloff, M., Jacob, G., et al. (2009). Effect of crosslink density on fracture behavior of model epoxies containing block copolymer nanoparticles. *Polymer*, *50*(19), 4683–4689.

Liu, X., & Wu, Q. (2001). PP/clay nanocomposites prepared by grafting-melt intercalation. *Polymer*, *42*(25), 10013–10019.

Luo, J.-J., & Daniel, I. M. (2003). Characterization and modeling of mechanical behavior of polymer/clay nanocomposites. *Composites Science and Technology*, *63*(11), 1607–1616.

Luo, Y., Feng, C., Wang, Q., Yi, Z., Qiu, Q., Lx, K., et al. (2013). Preparation and characterization of natural rubber/silica nanocomposites using rule of similarity in latex. *Journal of Wuhan University of Technology—Materials Science Edition*, *28*(5), 997–1002.

Ma, C. G., Mai, Y. L., Rong, M. Z., Ruan, W. H., & Zhang, M. Q. (2007). Phase structure and mechanical properties of ternary polypropylene/elastomer/nano-$CaCO_3$ composites. *Composites Science and Technology*, *67*(14), 2997–3005.

Maldas, D., & Kokta, B. V. (1993). Performance of hybrid reinforcements in PVC composites. II: use of surface-modified mica and different cellulosic materials as reinforcements. *Journal of Vinyl and Additive Technology*, *15*(1), 38–44.

Mark, J. E., Jiang, C. Y., & Tang, M. Y. (1984). Simultaneous curing and filling of elastomers. *Macromolecules*, *17*(12), 2613–2616.

Marouf, B.T., (2009) Effect of microstructure factors on fracture behavior of clay-rubber-epoxy hybrid nanocomposites, Ph.D. Dissertation, Sharif University of Technology: IR.

Mihranyan, A., Ferraz, N., & Strømme, M. (2012). Current status and future prospects of nanotechnologyb in cosmetics. *Progress in Materials Science*, *57*(5), 875–910.

Miltner, H. E., Assche, G. V., Pozsgay, A., Pukanszky, B., & Mele, B. V. (2006). Restricted chain segment mobility in poly(amide) 6/clay nanocomposites evidenced by quasi-isothermal crystallization. *Polymer*, *47*(3), 826–835.

Mitran, R.-A., Berger, D., Munteanu, C., & Matei, C. (2015). Evaluation of different mesoporous silica supports for energy storage in shape-stabilized phase change materials with dual thermal responses. *The Journal of Physical Chemistry C*, *119*(27), 15177–15184.

Najafi, A., Kord, B., Abdi, A., & Ranaee, S. (2012). The impact of the nature of nanoclay on physical and mechanical properties of polypropylene/reed flour nanocomposites. *Journal of Thermoplastic Composites Materials*, *25*(6), 717–727.

Nandiyanto, A. B. D., Iskandar, F., & Okuyama, K. (2008). Nano-sized polymer particle-facilitated preparation of mesoporous silica particles using a spray method. *Chemistry Letters*, *37*(10), 1040–1041.

Nandiyanto, A. B. D., Kim, S.-G., Iskandar, F., & Okuyama, K. (2009). Synthesis of silica nanoparticles with nanometer-size controllable mesopores and outer diameters. *Microporous and Mesoporous Materials*, *120*(3), 447–453.

Odegard, G. M., Clancy, T. C., & Gates, T. S. (2005). Modeling of the mechanical properties of nanoparticle/polymer composites. *Polymer*, *46*, 553–562.

Paul, D. R., & Robeson, L. M. (2008). Polymer nanotechnology: nanocomposites. *Polymer*, *49* (1), 3187–3204.

Pearson, R. A., Dittanet, P., Marouf, B. T., & Siotong, T. (2010). On the use of different nanofillers to toughen epoxies. *Polymer Nanocomposites 2010: a new decade of opportunities*, *1*(1), 1333.

Peng, D., & Qu, B. (2006). Synthesis of exfoliated PP/LDH nanocomposites via melt-intercalation: structure, thermal properties, and photo-oxidative behaviour in comparison with PP/MMT nanocomposites. *Polymer Engineering and Science*, *46*(9), 1153–1159.

Piggott, M. R., & Chua, P. S. (1987). Recent studies of the glass fiber-polymer interphase. *Industrial & Engineering Chemistry Research*, *26*(4), 672–677.

Presting, H., & Konig, U. (2003). Future nanotechnology developments for automotive applications. *Materials Science and Engineering: C*, *23*(6–8), 737–741.

Pukanszky, B. (2005). Interfaces and interphases in multicomponent materials: past, present, future. *European Polymer Journal*, *41*(4), 645–662.

Qin, H., Zhang, S., Liu, H., Xie, S., Yang, M., & Shen, D. (2005). Photo-oxidative degrataion of polypropylene/montmorillonite nanocomposites. *Polymer*, *46*(9), 3149–3156.

Rattanasom, N., Saowapark, T., & Deeprasertkul, C. (2007). Reinforcement of natural rubber with silica/carbon black hybrid filler. *Polymer Testing*, *26*(3), 369–377.

Ray, S. S., & Bousmina, M. (2005). Biodegradable polymers and their layered silicate nanocomposites: in greening the 21st century materials world. *Progress in Materials Science*, *50*(8), 962–1079.

Rebizant, V., Abetz, V., Tournilhac, F., Court, F., & Leibler, L. (2003). Reactive tetrablock copolymers containing glycidyl methacrylate. Synthesis and morphology control in epoxy-amine networks. *Macromolecules*, *36*(26), 9889–9896.

Rebizant, V., Venet, A.-S., Tournilhac, F., Girard-Reydet, E., Navarro, C., Pascault, J.-P., et al. (2004). Chemistry and mechanical properties of epoxy-based thermosets reinforced by reactive and nonreactive SBMX block copolymers. *Macromolecules*, *37*(21), 8017–8027.

Redhwi, H. H., Siddiqui, M. N., Andrady, A. L., & Hussain, S. (2013). Durability of LDPE Nanocomposites with clay, silica, and zinc oxide—Part I: mechanical properties of the Nanocomposite materials. *Journal of Nanomaterials*, *1*(1), 1–6.

Reichert, P., Nitz, H., Klinke, S., Brandsch, R., Thomann, R., & Mulhaupt, R. (2000). Poly (propylene)/organoclay nanocomposite formation: influence of compatibilizer functionality and organoclay modification. *Macromolecular Materials and Engineering*, *275*(1), 8–17.

Reynaud, E., Jouen, T., Gauthier, C., Vigier, G., & Varlet, J. (2001). Nanofillers in polymeric matrix: a study on silica reinforced PA6. *Polymer*, *42*(21), 8759–8768.

Roco, M. C., Mirkin, C. A., & Hersam, M. C. (2011). Nanotechnology research directions for societal needs in 2020: summary of international study. *Journal Nanoparticle Research*, *13* (3), 897–919.

Rothon, R. (2016). Nanofillers. In *Polymers and polymeric composites: A reference series* (2nd ed., pp. 1–21). Berlin Heindelberg: Springer.

Sanchez-Valdes, S., Lopez-Quintanilla, M. L., Ramirez-Vargas, E., Medellin-Rodriquez, F. J., & Gutierrez-Rodriguez, J. M. (2006). Effect of ionomeric compatibilizer on clay dispersion in polyethylene/clay nanocomposites. *Macromolecular Materials and Engineering*, *291*(2), 128–136.

Schadler, L., Brinson, L., & Sawyer, W. (2007). Polymer nanocomposites: a small part of the story. *JOM*, *59*(3), 53–60.

Seltzer, R., Kim, J.-K., & Mai, Y.-W. (2011). Elevated temperature nanoindentation behaviour of polyamide 6. *Polymer International*, *60*(12), 1753–1761.

Shen, L., Wang, L., Liu, T., & He, C. (2006). Nanoindentation and morphological studies of epoxy nanocomposites. *Macromolecular Materials and Engineering*, *291*(11), 1358–1366.

Sheng, N., Boyce, M. C., Parks, D. M., Rutledge, G. C., Abes, J. I., & Cohen, R. E. (2004). Multiscale micromechanical modeling of polymer/clay nanocomposites and the effective clay particle. *Polymer*, *45*(2), 487–506.

Shin, S. Y. A., Simon, L. C., Soares, J. B. P., & Scholz, G. (2003). Polyethelene-clay hybrid nanocomposites: in situ polymerization using bifunctional organic modifier. *Polymer*, *44*(18), 5317–5321.

Szazdi, L., Pozsgay, A., & Pukanszky, B. (2007). Factors and processes influencing the reinforcing effect of layered silicates in polymer nanocomposites. *European Polymer Journal*, *43*(2), 345–359.

Tai, Y.-Y., Hsu, S.-H., Chen, R.-S., Su, W.-F., & Chen, M.-H. (2014). Liquid crystalline epoxy nanocomposite material for dental application. *Journal of the Fomosan Medical Association, 114*(1), 46–51.

Tjong, S. C. (2006). Structural and mechanical properties of polymer nanocomposites. *Materials Science Engineering: R: Reports, 53*(3–4), 73–197.

Trewyn, B. G., Nieweg, J. A., Zhao, Y., & Lin, V. S.-Y. (2007). Biocompatible mesoporous silica nanoparticles with different morphologies for animal cell membrane penetration. *Chemical Engineering Journal, 137*(1), 23–29.

Trewyn, B. G., Slowing, I. I., Giri, S., Chen, H. T., & Lin, V. S.-Y. (2007). Synthesis and functionalization of a Mesoporous silica nanoparticle based on the sol–gel process and applications in controlled release. *Accounts of Chemical Research., 40*(9), 846–853.

Uddin, M. F., & Sun, C. T. (2010). Improved dispersion and mechanical properties of hybrid nanocomposites. *Composites Science and Technology, 70*(2), 223–230.

Vaia, R. A., & Wagner, H. D. (2004). Framework for nanocomposites. *Materials Todays, 7*(11), 32–37.

Wang, K. H., Choi, M. H., Koo, C. M., Xu, M. Z., Chung, I. J., & Jang, M. C. (2002). Morphology and physical properties of polyethylene/silicate nanocomposites prepared by melt intercalation. *Journal of Polymer Science B, 40*(14), 1454–1463.

Wang, K. H., Chung, I. J., Jang, M. C., Keum, J. K., & Song, H. H. (2002). Deformation behavior of polyethylene/silicate nanocomposites as studied by real-time wide-angle X-ray scattering. *Macromolecules, 35*(14), 5529–5535.

Wang, G.-T., Liu, H.-Y., Yu, Z.-Z., & Mai, Y.-W. (2009). Evaluation of methods for stiffness predictions of polymer/clay nanocomposites. *Journal of Reinforced Plastics and Composites, 28*(13), 1625–1649.

Wei, L., Tang, T., & Huang, B. (2004). Synthesis and characterization of polyethylene/clay-silica nanocomposites: a montmorillonite/silica-hybrid-supported catalyst and in situ polymerization. *Journal of Polymer Science A, 42*(4), 941–949.

Weichelt, F., Emmler, R., Flyunt, R., Beyer, B., Buchmeiser, M. R., & Beyer, M. (2009). ZnO-based UV Nanocomposites for wood coatings in outdoor applications. *Macromolecular Materials and Engineering, 295*(2), 130–136.

Wen, J., & Wilkes, G. L. (1996). Organic/inorganic hybrid network materials by the sol-gel approach. *Chemisty of Materials, 8*(8), 1667–1681.

Wilkes, G. L., Orler, B., & Huang, H. H. (1985). Ceramers-hybrid materials incorporating polymeric/Oligomeric species into inorganic glasses utilizing a sol-gel approach. *Polymer Preparation, 26*(1), 300–301.

Wu, S., Guo, Q., Kraska, M., Stuhn, B., & Mai, Y.-W. (2013). Toughening epoxy thermosets with block ionomers: the role of phase domain size. *Macromolecules, 46*(20), 8190–8202.

Wu, S., Guo, Q., Peng, S., Hameed, N., Kraska, M., Stuhn, B., et al. (2012). Toughening epoxy thermosets with block ionomer complexes: a nanostructure–mechanical property correlation. *Macromolecules, 45*(9), 3829–3840.

Wu, S., Guo, Q., Zhang, T., & Mai, Y.-W. (2013). Phase behavior and nanomechanical mapping of block ionomer complexes. *Soft Matter, 9*(1), 2662–2672.

Wu, J., Thio, Y. S., & Bates, F. S. (2005). Structure and properties of PBO–PEO diblock copolymer modified epoxy. *Journal of Polymer Science Part B: Polymer Physics, 43*(15), 1950–1965.

Wypych, F., & Satyanarayana, K. G. (2005). Functionalization of single layers and nanofibers: a new strategy to produce polymer nanocomposites with optimized properties. *Journal of Colloid and Interface Science, 285*(2), 532–543.

Xu, R., Pang, W., Yu, J., Huo, Q., & Chen, J. (2007). *Chemistry of zeolites and related porous materials: Synthesis and structure* (1st ed., pp. 1–616). Wiley-Interscience.

Young, R. J., Kinloch, I. A., Gong, L., & Novoselov, K. S. (2012). The mechanics of graphene nanocomposites: a review. *Composites Science and Technology*, *72*(12), 1459–1476.

Zan, L., Liu, Z., Fa, W., & Peng, T. (2006). Novel Polymethyl methacrylate (PMMA) TiO_2 Nanocomposites and its thermal and photic stability. *Wuhan University Journal of Natural Sciences*, *11*(2), 415–418.

Zao, H., & Li, R. K. Y. (2006). A study on the photo-degradation of zinc oxide (ZnO) filled polypropylene nanocomposites. *Polymer*, *47*(9), 3207–3217.

Zhang, M. Q., Rong, M. Z., Zhang, H. B., & Friedrich, K. (2003). Mechanical properties of low nano-silica filled high density polyethylene composites. *Polymer Engineering and Science*, *43*(2), 490–500.

Zhang, J., & Wilkie, C. A. (2003). Preparation and flammability Properites of polyethylene-clay nanocmoposites. *Polymer Degradation and Stability*, *80*(1), 163–169.

Zhao, D., Feng, J., Huo, Q., Melosh, N., Fredrickson, G. H., Chmelka, B. F., et al. (1998). Triblock copolymer syntheses of Mesoporous silica with periodic 50 to 300 angstrom pores. *Science*, *279*(5350), 548–552.

Zhong, Y., Poloso, T., Hetzer, M., & De Kee, D. (2007). Enhancement of wood/polyethylene composites via compatibilization and incorporation of organoclay particles. *Polymer Engineering and Science*, *47*(6), 797–803.

Further reading

Mittal, V. (2009). Polymer layered silicate nanocomposites: a review. *Materials*, *2*(1), 992–1057.

Nam, P. H., Maiti, P., Okamoto, M., Kotaka, T., Hasegawa, N., & Usuki, A. (2001). A hierarchical structure and properties of intercalated polypropylene/clay nanocomposites. *Polymer*, *42*(23), 9633–9640.

Low-density polyethylene/silica nanocomposites foams: Relationship between chemical composition, particle dispersion, cellular structure and physical, mechanical, dynamic mechanical, electrical, and thermal properties

10

Md. Rezaur Rahman, Md. Faruk Hossen, Sinin bin Hamdan
Department of Chemical Engineering and Energy Sustainability, Faculty of Engineering, University Malaysia Sarawak (UNIMAS), Kota Samarahan, Sarawak, Malaysia

10.1 Introduction

Nanoparticle-filled polymers show various benefits compared to micron-filled polymers, as these nanoparticle-filled polymers exhibit a better resistance to degradation, enhancement of thermal-mechanical properties, as well as improve the reduction in dielectric strength (Kozako et al., 2004). Fumed silica acquires characteristics like critical small particle size and large surface area, in which the surface of the fumed silica is categorized into three different chemical groups, including isolated hydroxy, hydrogen-bonded hydroxy, and siloxane groups (Parvinzadeh, Moradian, Rashidi, & Yazdanshenas, 2009). According to (Xanthos, 2005), the hydrophilic surface of fumed silica can be altered in order to change into hydrophobic by reaction to its surface hydroxyl groups with hydrophobic reagents like polydimethylsiloxane, dimethyldichlorosilane, and hexamethyldisilane. A great number of research has proven that nanosilica particles are able to promote better stiffness, tensile strength, modulus, impact strength, toughening, crystallinity, viscosity, creep resistance, as well as the interfacial adhesion in polyethylene, polypropylene, and thermoplastic elastomer nanocomposites (Garcia, Rooij, Winnubst, Zyl, & Verweij, 2004; Lin, Akil, & Ishak, 2009; Rottstegge et al., 2007; Xu, Li, Lu, Zhang, & Wang, 2008). It has been claimed that the coefficient of friction, wear resistance, and toughness of nylon 6 and nylon 6.6 nanocomposites are successfully enhanced by introducing different nanosilica particles into the polymers. Table 10.1 shows the specification of various nanosilica particles.

Silica and Clay Dispersed Polymer Nanocomposites. https://doi.org/10.1016/B978-0-08-102129-3.00010-5

Table 10.1 **Specification of various nanosilica particles (Parvinzadeh et al., 2009)**

Characteristic properties	Aerosol 200 hydrophilic (1)	Aerosol TT 600 hydrophilic (2)	Aerosol R 972 hydrophobic
BET surface area (m^2/g)	200±2.5	200±500	110±20
Primary particle size (nm)	12	40	16
Ignition loss (%)	≤1	≤2.5	
pH in 45 aqueous solution	3.7–4.7	3.6–4.7	3.6–5.5
Tapped density (g/cm^2)	Approx. 50	Approx. 60	Approx. 50
Moisture(%)	≤1.5	≤2.5	≤0.5
Purity (% of nanosilica)	≤99.8	≤99.8	≤99.8
C-content (%)	–	–	0.7–1.3
Al_2O_3-content (%)	<0.05	<0.05	<0.05
Fe_2O_3-content (%)	<0.003	<0.003	<0.003
TiO_2-content (%)	<0.03	<0.03	<0.03
HCl-content (%)	<0.025	<0.025	<0.025
Sieve residue-content (%)	<0.05	<0.05	<0.05

In this study, chemical composition, particle dispersion, cellular structure and physical, mechanical, dynamic mechanical, electrical, and thermal properties of low-density polyethylene (LDPE)/silica-based nanocomposites are discussed.

10.2 Chemical composition and physical properties

10.2.1 Fourier transform infrared spectroscopy (FTIR) analysis

FTIR analysis was conducted to analyze the chemical bonding formed between the nanofillers and the polymer matrices. There will be no significant changes in the FTIR testing if the materials used are not homogenously mixed (Qu, Gao, Wu, & Zhang, 2010). Fig. 10.1 showed the FTIR spectra of pure low-density polyethylene. The C—H region of both pure polyethylene and polypropylene are slightly different according to PerkinElmer (2011), by focusing on the C—H stretch region around 2950 cm^{-1}, the difference between polyethylene and polypropylene can be observed due to the differing ratios of both CH_2 and CH_3 groups. Thus, the pure polyethylene showed the peak intensity at 2846.93 cm^{-1} due to C—H stretching, and 3647.39 and

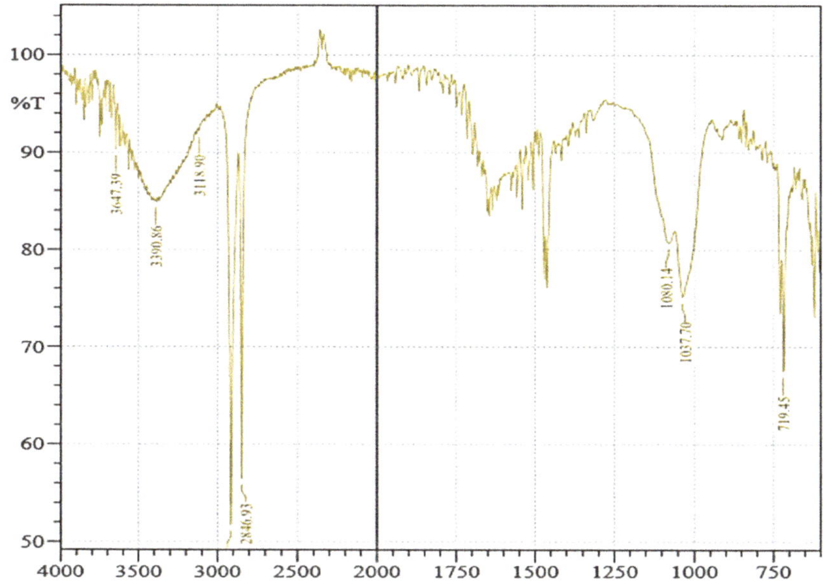

Fig. 10.1 FTIR spectra of pure low-density polyethylene.

3590.86 cm^{-1} due to —OH stretching. A nanofiller like fumed silica was introduced to improve the water uptake resistance of the polymer matrix. Hence, the analysis of —OH region for different filler contents of silica/polyethylene was made in order to analyze the improvement of the water uptake resistance.

The different percentages of silica loading to pure polyethylene are shown in Figs. 10.2–10.5. Based on Fig. 10.2, 0.5 wt.% silica/low-density polyethylene nanocomposite showed the similar peak intensity with pure polyethylene, which is at 2846.93 cm^{-1} due to C—H stretching. The peak intensity of —OH stretching at 3566.38 and 3383.14 cm^{-1} were observed, in which there was a slight decrease of —OH peak intensity as compared to the —OH peak intensity of pure polyethylene when 0.5% silica was added. The further decreasing of the peak intensity of —OH stretching was observed from 1.0 wt.% silica/low-density polyethylene nanocomposite as shown in Fig. 10.3, in which the peak intensity showed at 3749.62 and 3502.73 cm^{-1} due to —OH stretching. Slightly decreasing of —OH peak intensity was observed from 1.5 wt.% and 2.0 wt.% silica/low-density polyethylene nanocomposite, compared to 1.0 wt.% silica/low-density polyethylene nanocomposite as shown in Figs. 10.4 and 10.5. 1.5 wt.% silica/low-density polyethylene nanocomposite showed the peak intensity at 3647.39 and 3502.73 cm^{-1} due to —OH stretching, while 1.5 wt.% silica/low-density polyethylene nanocomposite showed the peak intensity at 3647.39 and 3566.38 cm^{-1}. Besides that, the peak intensity of C-H stretch region around 2950 cm^{-1} was decreased when the addition of silica increased up to 1.5 and 2.0 wt.%. Both 1.5 and 2.0 wt.% silica/low-density polyethylene nanocomposite showed the least peak intensity of C—H stretch region, which is at 2916.37 and 2848.86 cm^{-1}. A strong bending band at about 1000–1200 cm^{-1} and stretching bands

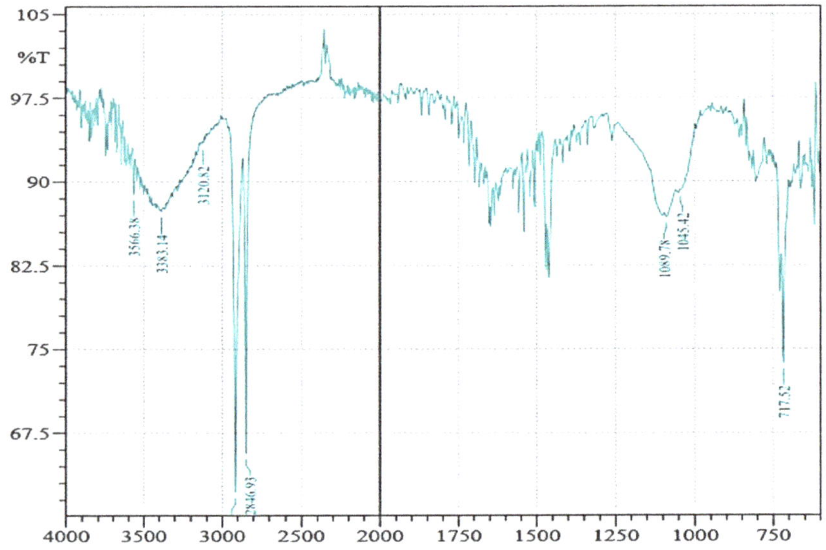

Fig. 10.2 FTIR spectra of 0.5 wt.% silica/low-density polyethylene nanocomposite.

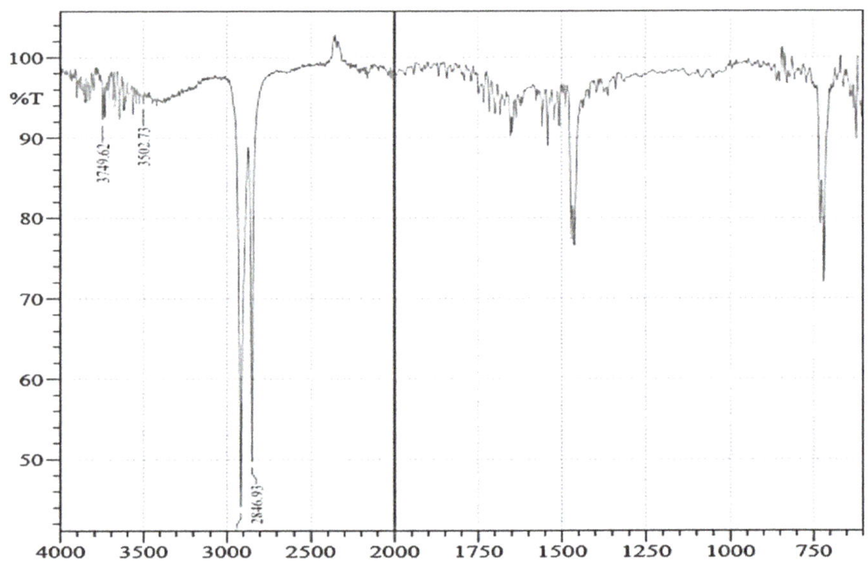

Fig. 10.3 FTIR spectra of 1.0 wt.% silica/low-density polyethylene nanocomposite.

at about 807–813 cm^{-1} indicated the presence of Si—O—Si (siloxane groups) by adding nanosilica particles to polyethylene (Parvinzadeh et al., 2009).

The FTIR spectra analysis of JCPEC, JCPECC, and JCPESC at 15 wt.% fibers loadings are shown in Fig. 10.6. The peak intensity in the absorption band at 2914 and 2846 cm^{-1} assigned to the C—H stretching in methyl and methylene groups in

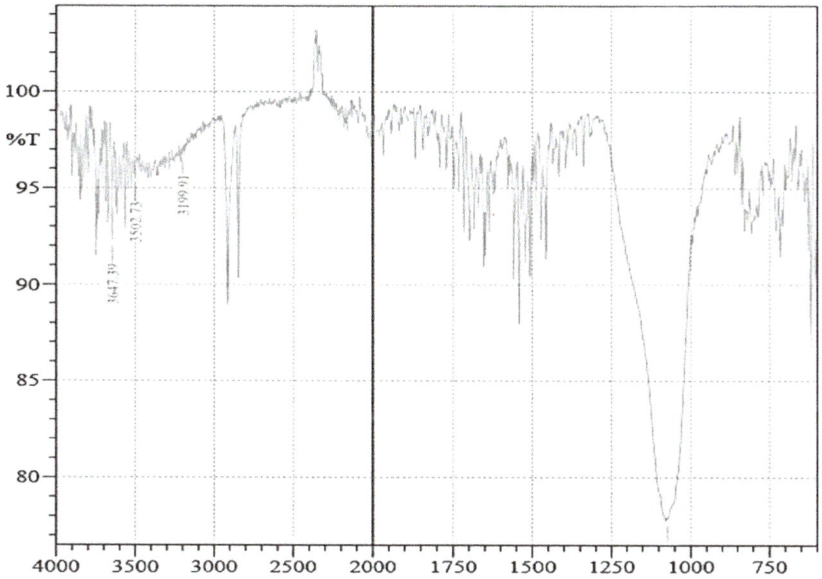

Fig. 10.4 FTIR spectra of 1.5 wt.% silica/low-density polyethylene nanocomposite.

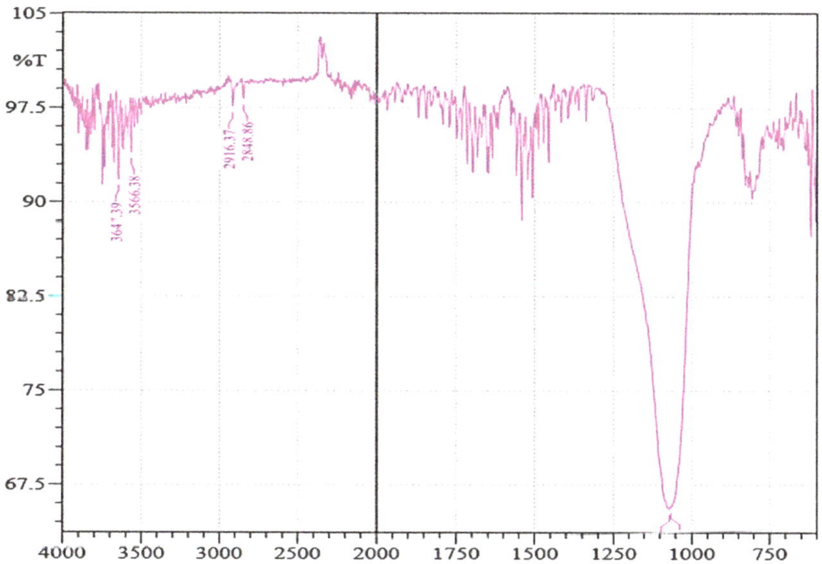

Fig. 10.5 FTIR spectra of 2.0 wt.% silica/low-density polyethylene nanocomposite.

cellulose and hemicelluloses, respectively. The peaks at 1718 and 1716 cm^{-1} of the JCPEC and JCPECC was assigned to the C=O stretching of the carboxylic acid and ester groups in hemicelluloses. The absorbance band near to 1471 cm^{-1} is assigned to the C—H deformation in lignin. The peaks intensity around at 1261 cm^{-1} showed the

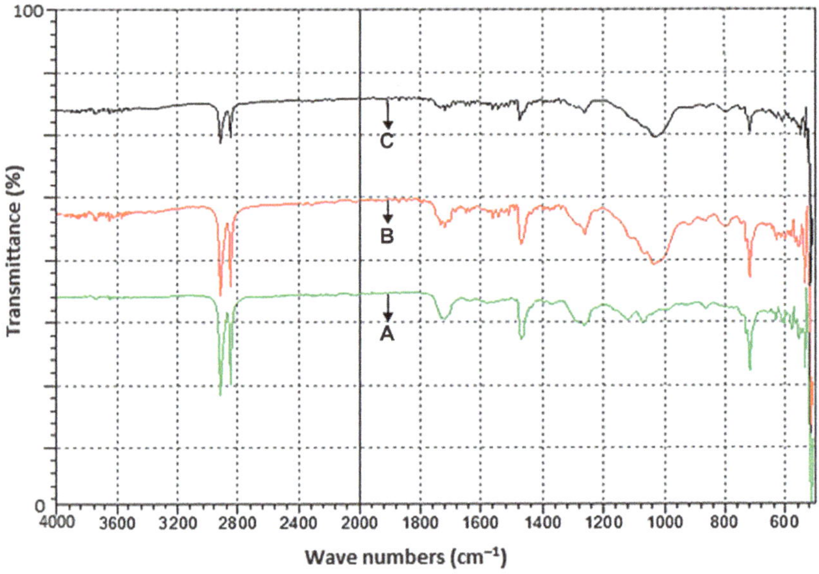

Fig. 10.6 FTIR spectra of (A) JCPEC, (B) JCPECC, and (C) JCPESC.

—C—O—C— bond in the cellulose molecules. The JCPECC had characteristic small peaks at 1109 and 1066 cm^{-1} is due to Si—O stretching (out-of-plane) for montmorillonite. The peak at 1035 cm^{-1} is assigned to Si—O stretching (in-plane) vibration for silicates layers. The Si—O—Si symmetric stretching vibration of JCPESC was attributed to 1033 cm^{-1}. However, the IR spectrum of silica filled composites showed peaks intensity at 1718 and 1716 cm^{-1} due to the addition of silica. It is indicated that silica and clay present in the composites system.

10.3 Mechanical properties

The calculated tensile strength and modulus of elasticity for silica-based nanocomposites are tabulated in Table 10.2. Fig. 10.7 shows the tensile strength for different percentage loading of silica to low-density polyethylene, while Fig. 10.8 shows the modulus of elasticity for different percentage loading of silica to low-density polyethylene.

The percentage of filler loading to both low-density polyethylene and polypropylene is chosen for a range of 0.5–2.0 wt.%. Based on the Fig. 10.7, the tensile properties of both composites showed similar trend result which silica-based nanocomposite and nanoclay-based nanocomposite showed significant improvement of mechanical properties by increasing their filler loading, as the tensile standard for silica-based composite and nanoclay-based nanocomposite are 12 and 40 MPa, respectively. The silica-based nanocomposite with 0.5 wt.% loading showed the lowest tensile properties and highest tensile properties are observed when the percentage loading of the

Table 10.2 Tensile strength and modulus of elasticity of silica/polyethylene nanocomposites

Composite	Silica loading (wt.%)	Tensile strength (MPa)	Modulus of elasticity (GPa)
Silica-based composite	0.5	25.9	0.82
	1.0	32.5	1.68
	1.5	38.6	2.75
	2.0	47.4	3.53

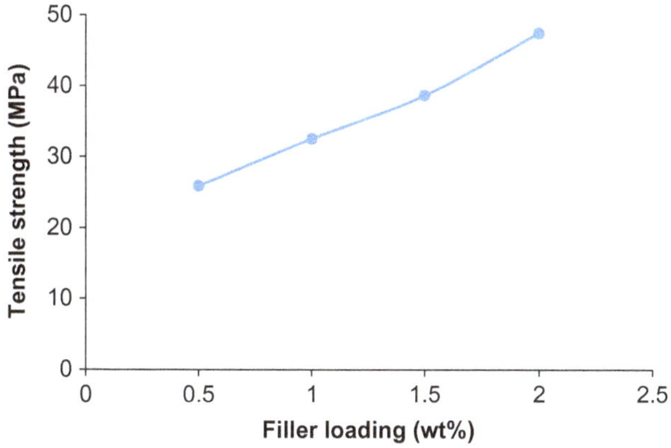

Fig. 10.7 Tensile strength of different percentage loading of silica/polyethylene

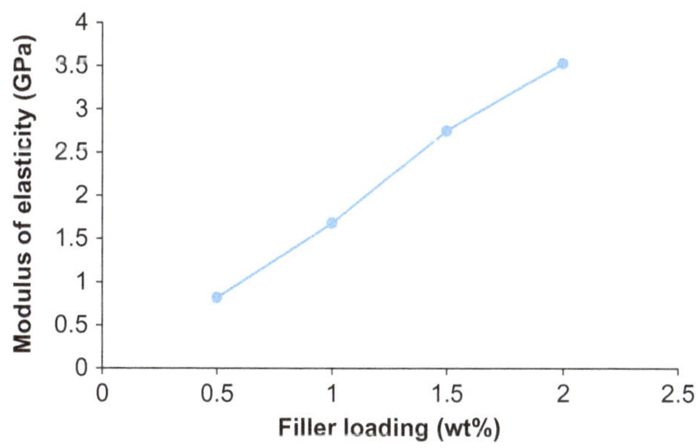

Fig. 10.8 Modulus of elasticity of different percentage loading of silica/polyethylene

silica to low-density polyethylene is increased to 2 wt.%. The addition of SiO_2 into the polymer matrix helps to improve both Young's modulus and impact strength of the polymer matrix. On the other hand, based on Fig. 10.8, the modulus of elasticity of both silica-based and nanoclay-based nanocomposites were shown as greatly improved as compared to their modulus of elasticity standard, which the standard Young's modulus of low-density polyethylene is 0.4 GPa. High modulus of elasticity indicates the resistance of both silica based nanocomposites against the elastic deformation when the load is applied on the composites. Thus, the modulus of elasticity of both silica based nanocomposites is improved as the percentage of filler loading increased.

Kontou and Niaounakis (2006) focused the study on the influence of filling on properties of LDPE/silica nanocomposites. They investigated two types of matrices such as: mLLDPE and zLLDPE prepared by metalocene and Ziegler-Natta catalysts, respectively. The filler silica was surface modified with silane coupling agent. The impact of concentration of particles of diameter on the tensile properties of the manufactures nanocomposites. Tensile strength and stress-strain curves were presented in Figs. 10.9, 10.10, and Table 10.3, respectively.

In most cases, nanosilica as nanofiller improved tensile and impact stress of the nanocomposites. As the spectrum of available polymers is broad, only several main types were described here. It was observed that less than several percent (mainly 0.5–5 wt.%) of nanosized particles are needed to incur the same modification effect as with the microsized ones which is reached with load about 20 wt.%. Therefore, the significant improvements in the mechanical behaviors can be achieved by surface modifications of silica grafted with PE matrix.

Hui, Chaki, and Chattopadhyay (2008) investigated the tensile properties of low-density polyethylene-ethylene vinyl acetate (LDPE-EVA) blend nanocomposites with silane (Si-69) modified silica manufactured at 1.5, 3, and 5 wt.% of silica loadings through 3 sequences (sequence 1, 2, and 3) according to Table 10.4 and the mechanical properties presented in Table 10.5. The silica loading is optimized by 3 wt.% in most

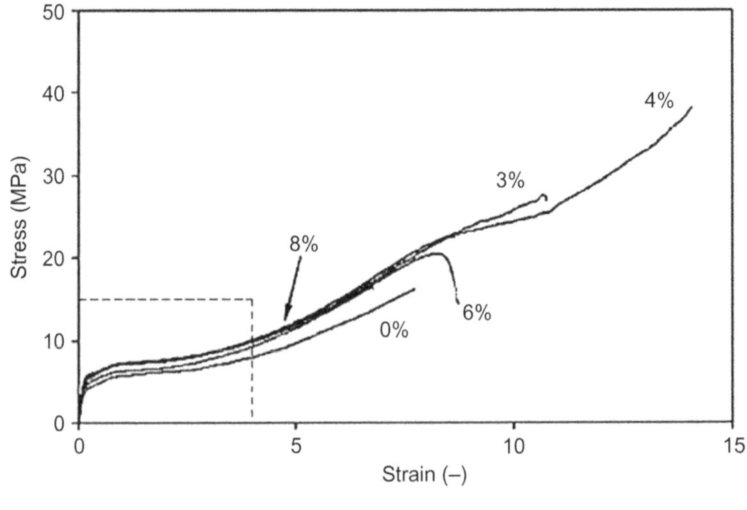

Fig. 10.9 Tensile stress–strain curves of the mLLDPE-SiO_2 nanocomposites. Polymer, Elsevier.

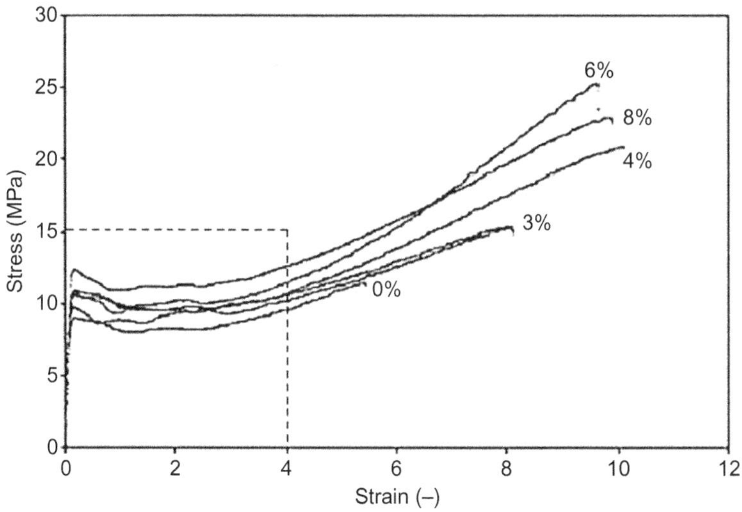

Fig. 10.10 Tensile stress–strain curves of the zLLDPE/SiO_2 nanocomposites
Polymer, Elsevier.

cases. In all the nanosilica-based films, a significant improvement in the tension set properties of the nanocomposites was observed. However, the tension set is improved in comparison with the control in all cases. Due to the increase in the interfacial interaction, the initial increase in the tensile strength up to a 3 wt.% loading of silica which eliminates the crack formation at the interface during stretching. The unmodified silica particles at a 3 wt.% loading are optimally dispersed in the elastomeric phases, providing a better reinforcement effect in comparison with other loadings. In contrast, filler-polymer interactions are not exhibited in increases in the elongation percentage and modulus because the surfaces of the silica particles are not modified. The tensile values are not affected greatly at higher loadings. This can be attributed to the aggregation of fillers leading to the occurrence of low-energy deformation modes, such as aggregate breakage, aggregate-polymer interface breakage, and breakage through the defect sites, which weakens the reinforcement effect. The maximum property enhancement is in sequence 3 compared to other sequences. The silica distribution is more uniform in this particular sequence, leading to the strengthening of the weaker phase of the blend system because silica is initially blended with the more polar EVA phase. However, in sequence 2, the chances of aggregate formation are obviously greater because silica is premixed with the relatively nonpolar LDPE matrix. Moreover, it may adversely affect the crystallization behavior of LDPE as such. On the other hand, an equivalent probability of the distribution of silica particles in both the LDPE and EVA phases exists in sequence 1. This observation explains the differences in the mechanical properties with respect to the variation of the sequence of mixing.

Table 10.3 **Tensile properties**

	mLLDPE/silica nanocomposites				zLLDPE/silica nanocomposites			
Nanosilica (wt.%)	**Elastic modulus (MPa)**	**Yield stress (MPa)**	**Strain at break (–)**	E_c/E_m[a] (–)	**Elastic modulus (MPa)**	**Yield stress (MPa)**	**Strain at break (–)**	E_c/E_m[a] (–)
0	51	3.7	7.8	1.00	230	10.5	5.5	1.00
3	55	4.2	10.8	1.23	237	11.0	8.1	1.03
4	70	5.0	14.0	1.36	240	9.0	10.0	1.04
6	82	5.4	8.3	1.61	245	11.0	9.5	1.07
8	103	5.5	11.0	2.03	280	12.9	9.6	1.22
10	65	7.2	9.0	1.28	153	10.8	9.0	0.66

[a] Fraction of modulus of composite (E_c) to modulus of pure matrix (E_m).
Polymer, Elsevier.

Table 10.4 Sample designations

Sample code	LDPE (wt.%)	EVA (wt.%)	Silica (wt.%)	Sequence of silica addition
EL 6/4	40	60	0	
ELS 6/4/1.5	40	60	1.5	
ELS 6/4/3-1	40	60	3	
ELS 6/4/3-2	40	60	3	
ELS 6/4/3-3	40	60	3	
ELS 6/4/3-3/ Si-69	40	60	3 phr + 10% Si-69	
ELS 6/4/5	40	60	5	
L 100	100	0	0	
E 100	0	100	0	

E, EVA; *L*, LDPE; *S*, silica.
Journal of Vinyl and Additive Technology, Wiley.

Upon the addition of a silane coupling agent, a considerable improvement in the tensile properties can be observed and the elongation at break is also slightly improved compared to that of the control sample. The addition of a coupling agent (Si-69) increases the interaction between the weak EVA network and silica, which reduces the filler-filler interaction by breaking the agglomerated structure. As a result, the interaction between the particles and polymer is increased, and also the interfacial adhesion between EVA and LDPE is possibly improved to a certain extent (as indicated by SEM images below). Thus, in the presence of silica fillers, the improvement of the set properties can be explained on the basis of an improvement of the recovery of the polar EVA matrix.

The nature of the solid filler and the processing conditions may affect the improvement in mechanical properties. Two methods: batch foaming (BF), using carbon dioxide and improved compression molding (ICM), using a chemical blowing agent (5% of azodicarbonamide) are performed to prepare LDPE foams shown in Fig. 10.11. In the BF method, due to the addition of nanosilica, it was seen to reduce the density with a slight improvement in modulus and collapse stress at the 3% and 6% silica loading. In the ICM method, the filler was seen to increase the modulus significantly up to 6% silica. It was learned from the two different processes that the difference in foam properties is not likely due to the state of dispersion of the nanosilica, but instead due to differences in the foam structure (in particular, the open cell content) obtained from the two different processes (Saiz-Arroyo, Rodríguez-Pérez, Velasco, et al., 2013). However, in the case of BF, the modulus increased while density reduced, which strongly suggests that the nanofiller does have a reinforcing effect. As Ogunsona, Ogbomo, Nar, and D'Souza (2011) discussed in such situations, the filler reinforcement effect is clearly evident even though samples are not compared at fixed foam density.

Table 10.5 **Mechanical properties of LDPE/silica nanocomposites with different silica loading**

Sample code	Tensile strength (MPa)	Elongation at break (%)	Modulus (MPa)				Set (%)
			100%	200%	300%	3%	
EL 6/4	3.4	480	2.63	2.90	3.13	6.4	16.8
ELS 6/4/1.5	3.8	478	2.62	2.93	3.26	5.9	15.0
ELS 6/4/3-1	3.9	479	2.48	2.87	3.23	6.1	14.0
ELS 6/4/3-2	3.4	465	2.24	2.62	2.97	5.3	26.0
ELS 6/4/3-3	4.0	520	2.56	2.85	3.13	6.2	13.0
ELS 6/4/3-3/Si-69	4.8	490	2.80	3.20	4.10	7.3	12.5
ELS 6/4/5	3.5	450	2.28	2.60	2.91	5.6	12.0

E, EVA; *L*, LDPE; *S*, silica
Journal of Vinyl and Additive Technology, Wiley.

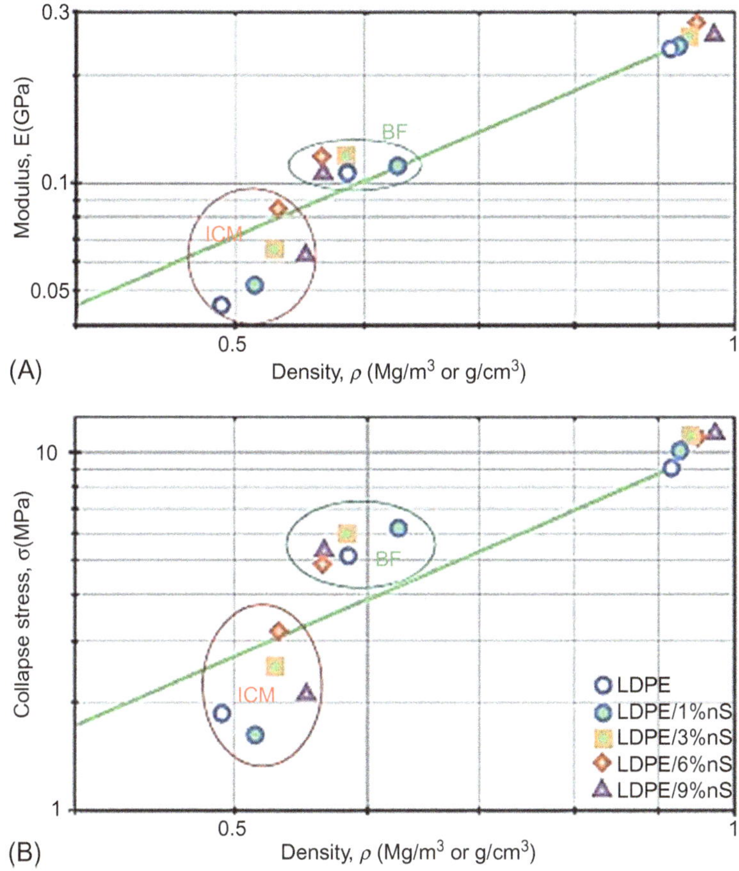

Fig. 10.11 Effect of processing conditions on the mechanical properties of LDPE foams with nanosilica (nS). (A) Modulus and (B) stress.
Journal of Cellular Plastics, Sage.

It is stated that the efficacy of nanofiller at improving the collapse strength and modulus of polymer foams can only be indicated after the effects of nanofiller on foam density. Besides, nanofiller can affect foam mechanical properties in many ways, for example mechanical reinforcement of the matrix, alignment of filler in the foam walls, and changes in the open cell content, etc., it becomes possible to identify the mechanism is active once the effects of foam density are accounted for.

Tensile strengths (TS) and Young's modulus (YM) at different fibers loaded for jute fiber/polyethylene (PE)/composite (JFPEC), jute fiber/PE/maleic anhydride composite (JFPEMAC), jute fiber/PE/maleic anhydride/clay nanocomposite (JFPEMACC), and jute fiber/PE/maleic anhydride/silica nanocomposite (JFPEMASC) are shown in Figs. 10.12 and 10.13, respectively. It was clearly seen that the tensile strengths and Young's modulus gradually increased up to 15% fiber loading and then decreased

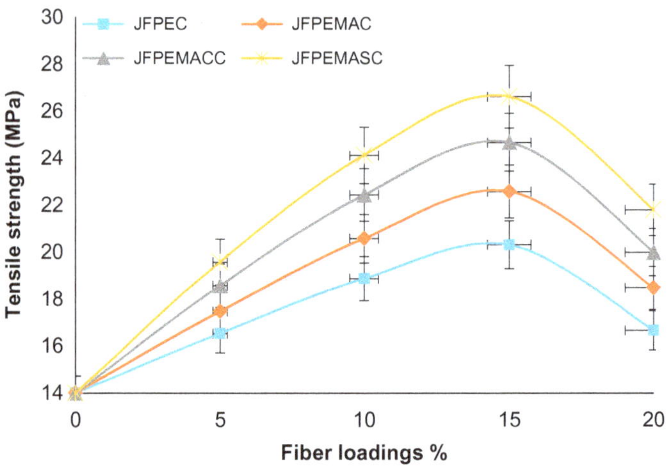

Fig. 10.12 Variation of Tensile strengths at different fiber loading for JFPEC, JFPEMAC, JFPEMACC and JFPEMASC.
Journal of Vinyl and Additive Technology, Wiley.

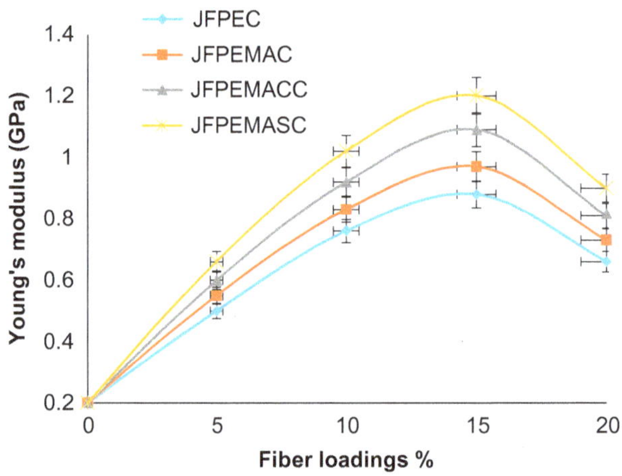

Fig. 10.13 Variation of Young's modulus at different fiber loading for JFPEC, JFPEMAC, JFPEMACC and JFPEMASC.
Journal of Vinyl and Additive Technology, Wiley.

for all composites with higher fiber loading. This is due to the weak interfacial bonding, lower compatibility, and higher agglomeration between fiber and matrixes. It has been reported that the composition at 15 wt.% jute fiber was then taken to be the optimum composition silica. Among the composites, JFPEMASC showed the maximum tensile strength compared with JFPEC, JFPEMAC, and JFPEMACC. The tensile strength increased significantly when maleic anhydride was added to the fiber and PE matrix. Clay and silica enhanced void spaces and improve the interfacial bonding between

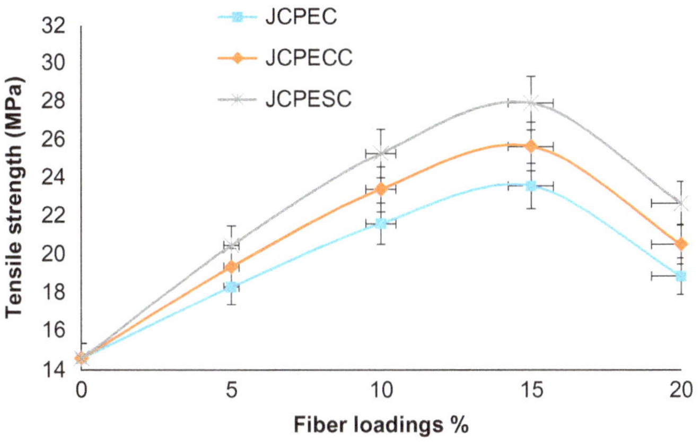

Fig. 10.14 Variation of tensile strengths at different fibers loading for JCPEC, JCPECC, and JCPESC.
Journal of Vinyl and Additive Technology, Wiley.

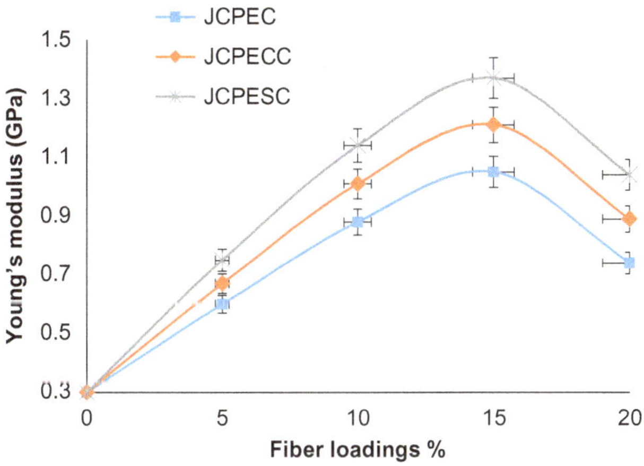

Fig. 10.15 Variation of Young's modulus at different fibers loading for JCPEC, JCPECC, and JCPESC.
Journal of Vinyl and Additive Technology, Wiley.

the fiber and PE matrix due to the surface modification of PE matrix with the maleic anhydride. On the other hand, JFPEMASC showed the highest Young's modulus compared with JFPEC, JFPEMAC, and JFPEMACC, respectively. This was due to better compatibility between fiber and matrix in the composites system. It was obvious that silica act as enforcement into fiber and matrix.

TS and YM at different fibers loaded for jute cellulose/PE composite (JCPEC), jute cellulose/PE/clay nanocomposite (JCPECC), and jute cellulose/PE/silica nanocomposite JCPESC are shown in Figs. 10.14 and 10.15, respectively. It was also observed that the 15 wt.% jute content proved to be the optimum composition. It has been seen that the tensile strength of JCPESC was higher than compared with JCPECC and JCPEC. The tensile strength improved significantly when silica was added to the PE matrix. Moreover, the exceptional phase morphology and enhanced interfacial properties are also responsible for the well performance properties of the silica composites compared to the others' composites. On the other hand, JCPESC showed the maximum Young's modulus compared with JCPECC and JCFPEC. This was due to well distribution of silica in the composites that enhance tensile strength and Young's modulus. It was noticeable that fumed silica acts as enforcement between fiber and matrix.

The effect of compatilizer on the mechanical properties of foams is plotted in Fig. 10.16. Generally, nanocomposite foams show higher mechanical values (elastic modulus and collapse stress) than the pure LDPE. The same type of analysis has been carried out for both the foamed and the solid. The elastic modulus of both foamed and solid composites, with and without compatilizer are presented in Fig. 10.17. Similar improvements in stiffness for foams and for solids are obtained up to a silica content of 3wt.% when the samples with a compatibilizer agent (Fig. 10.9A). Conversely, without a compatilizer, the improvements reached through the addition of silica are generally higher in foamed than in solid composites, 3 wt.% being the optimum silica content (Fig. 10.9B). In this case, the changes produced a positive effect in the cellular structure on the overall elastic modulus of the foamed composites. It is observed that the overall increments in elastic modulus are higher in the foams without compatibilizer when the improvements achieved in the foamed samples with and without compatibilizer are compared.

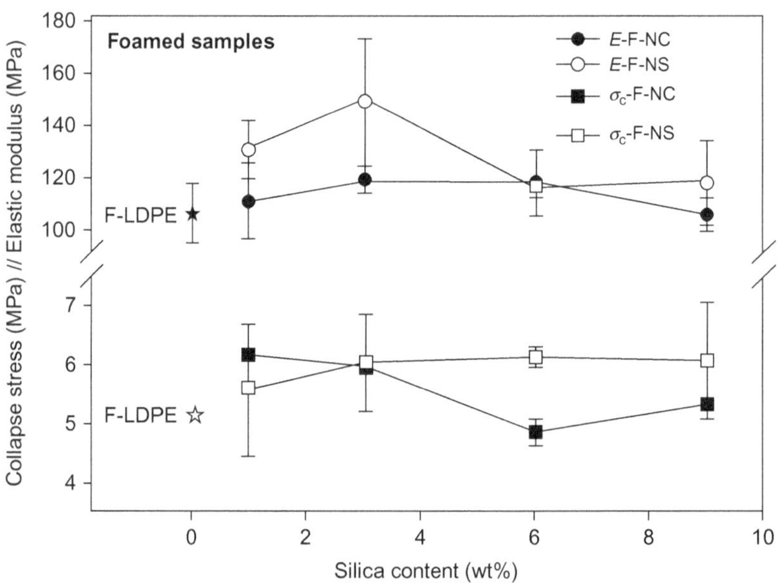

Fig. 10.16 Experimental results for elastic modulus and collapse stress for foamed nanocomposites with and without compatibilizer.
European Polymer Journal, Elsevier.

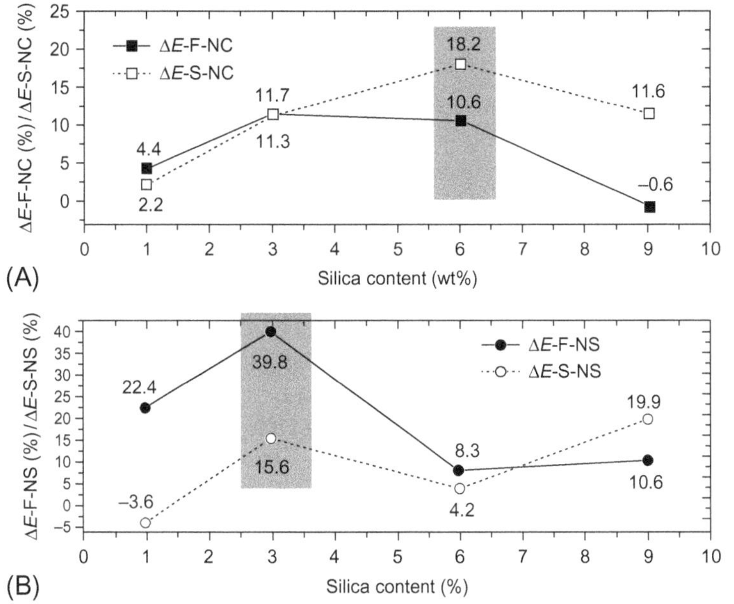

Fig. 10.17 Increments reached in the elastic modulus of both foamed and solid composites. Elastic modulus for foams (F) and solids (S) with compatibilizer (A) and elastic modulus for foams (F) and solids (S) without compatibilizer (B).
European Polymer Journal, Elsevier.

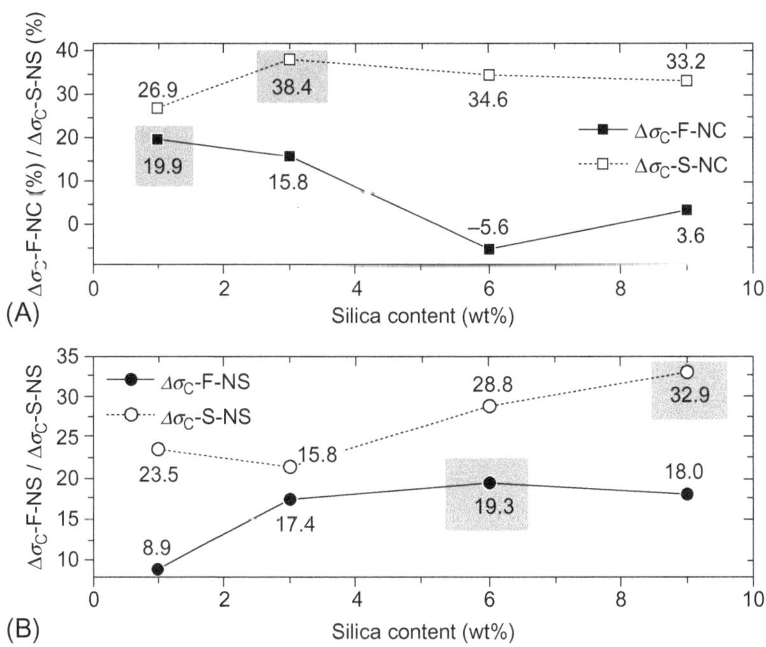

Fig. 10.18 Increments reached in the collapse stress of both foamed and solid composites. Collapse stress for foams (F) and solids (S) with compatibilizer (A) and collapse stress for foams (F) and solids (S) without compatibilizer (B).
European Polymer Journal, Elsevier.

The collapse stress for samples with and without compatibilizer is presented in Fig. 10.18. The strength gain reached in foam composites is always lower than that in the solid ones. An improvement in foam collapse stress was found at around 19%, a silica content of 1 wt.% and more than 3 wt.% are required in the samples with and without compatibilizer, respectively. Moreover, the overall increments for the solid composites with compatibilizer were higher than that without compatibilizer, while this behavior is completely the opposite for the foamed composites. The gain in strength is a consequence of the improved cellular structure in terms of cell size and cell density exhibited by the materials without compatibilizer. It is stated that silica particles lead to an overall increase in mechanical response of all the analyzed samples, with respect to the mechanical properties of the pure LDPE. The improvements in mechanical behavior for solids are always greater than that for foams, except in the case of elastic modulus of samples without compatibilizer, indicating the presence of a synergistic effect.

10.4 Thermal properties

10.4.1 Thermogravimetric analysis (TGA)

TGA was conducted to calculate the thermal constancy and decomposition process for silica/low-density polyethylene nanocomposites. The TGA curves of silica/low-density polyethylene nanocomposite loaded with different silica content were shown in Fig. 10.19. Table 10.6 also shows the weight loss for different silica loaded to polyethylene. Based on Fig. 10.19, there is no mass loss of both 1.5 wt.% and 2.0% silica polyethylene at around temperatures of 100°C and 200°C, but 1% of mass loss occurred for 0.5 and 1.0 wt.% silica/polyethylene. However, all the nanocomposites have the same amount of mass loss which is 1% at around 300–400°C. The variation

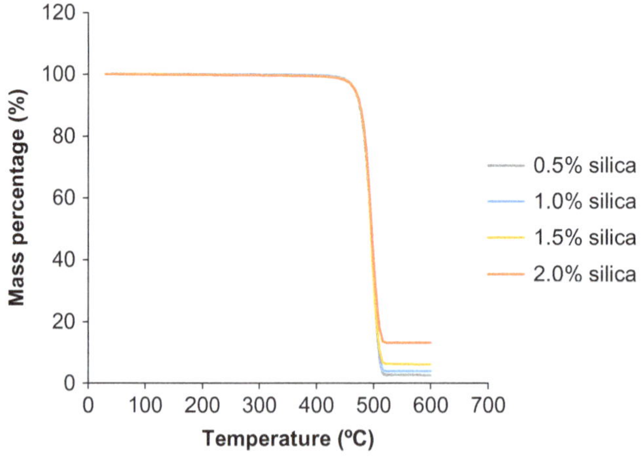

Fig. 10.19 TGA curves of LDPE/silica nanocomposite at different filler loading

Table 10.6 **Weight loss for different silica loading nanocomposites**

Nanocomposite	Weight loss (%)					
	100°C	200°C	300°C	400°C	500°C	600°C
0.5 wt.% silica/PE	1	1	1	1	78	97
1.0 wt.% silica/PE	1	1	1	1	77	96
1.5 wt.% silica/PE	0	0	1	1	76	94
2.0 wt.% silica/PE	0	0	1	1	70	87

Table 10.7 **Thermal stability of pure LDPE and LDPE/silica nanocomposites**

Sample	Thermal stability (°C)
NC0 (LDPE)	479.13
NC1(LDPE with 1 wt.% silica)	477.05
NC3 (LDPE with 3 wt.% silica)	489.12
NC6 (LDPE with 6 wt.% silica)	490.61
NC9 (LDPE with 9 wt.% silica)	493.30

Cellular Polymers, Smithers Rapra Technology.

that occurred in the thermal stability indicated that there is some increasing of the molecular weight of polyethylene by adding the silica nanofiller (Kaleel, Bahuleyan, Masihullah, & Al-Harthi, 2011). Silica/polyethylene nanocomposites showed enhanced thermal stability in which the peak degradation temperature of the nanocomposites was observed starting at 400°C. This is because of the addition of fillers promotes the varying degrees of high temperature stabilization in which the effect of adding fillers to polymer is proven by the delayed peak degradation temperature in the nanocomposites (Patwary & Mittal, 2015). According to the finding, the weight loss of nanocomposite was reduced when the loading of silica increased. At 600°C, 2.0 wt.% silica/polyethylene nanocomposite lost a total 87% of mass loss, while 0.5 wt.% silica/polyethylene nanocomposites have lost 97% of its mass. This is due to the reason that silicates can be used as nucleating agents, which helps to improve the crystallization kinetics, as well as provide better thermal stability imparted with the barrier properties toward oxygen and other volatiles (Prachum, Adam Strauss, & Kiatkamjornwong, 2011; Preschilla et al., 2010).

It was reported by researchers (Saiz-Arroyo, Escudero, Rodríguez-Pérez, & de Saja, 2011) that thermal stability of neat LDPE increased with the addition of silica nanoparticles (Table 10.7). The thermal decomposition temperature has been increased of 20°C by adding 9 wt.% of silica nanoparticles to LDPE.

The thermogravimetric analysis of JCPEC, JCPECC, and JCPESC are shown in Fig. 10.20. The TGA curve shows that the higher thermal stability of JCPESC was

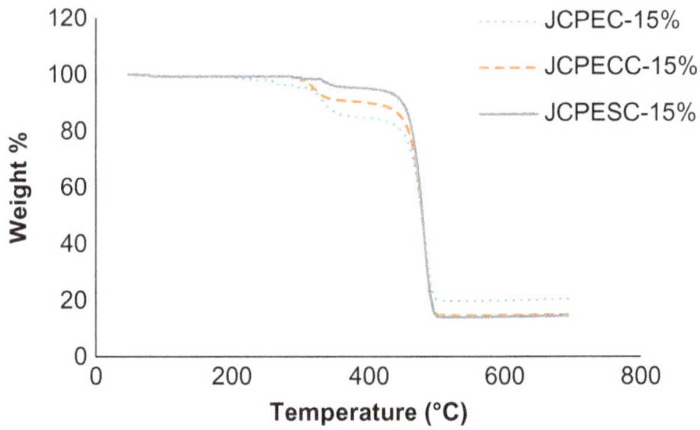

Fig. 10.20 TGA curve of JCPEC, JCPECC, and JCPESC.

Table 10.8 Thermal characteristics of JCPEC, JCPECC, and JCPESC

Name of the samples	T_i (°C)[a]	T_m (°C)[b]	T_f (°C)[c]	W_{T_i} (%)[d]	W_{T_m} (%)[e]	W_{T_f} (%)[f]	Activation energy, E_a (KJ/mol)
JCPEC	262	448	501	96.96	80.01	20.26	343.8754
JCPECC	308	438	507	97.94	87.28	14.83	375.718
JCPESC	319	432	511	98.41	93.46	14.04	466.9641

[a] Temperature corresponding to the beginning of decomposition.
[b] Temperature corresponding to the maximum rate of mass loss.
[c] Temperature corresponding to the end of decomposition.
[d] Mass loss of temperature corresponding to the beginning of decomposition.
[e] Mass loss of temperature corresponding to the maximum rate of mass loss.
[f] Mass loss of temperature corresponding to the end of decomposition.

significantly increased compared with JCPEC and JCPECC, respectively. The weight losses of the composites occurred at three stages, the first stage of weight loss was observed under 200°C, which correspondended to the release of moisture absorbed and water. At the second stage, weight loss of the composites due to the temperature between 200–380°C was decomposition of hemicellulose, lignin, and pectin. The last stage of weight loss cellulose and other cellulosic matters were decomposed at temperature range above 380°C. The TGA thermograph observed the weight loss about 3.04%, 2.06%, and 1.59% for JCPEC, JCPECC, and JCPESC, respectively. From Fig. 10.20, there is no degradation up to 160°C. Above this temperature, the thermal stability decreases step by step and degradation takes place. The initial decomposition started at 262, 308, and 319 for JCPEC, JCPECC and JCPESC, respectively.

The end of decomposition (T_f) of JCPESC was higher than those of JCPECC and JCPEC, which is presented in Table 10.8. Due to the accession of silica, the mass loss from JCPESC was lower than compared with JCPECC and JCPEC. The TGA thermograph shows that JCPESC was more thermally stable compared to the other composites. The activation energy could be helpful in reaching conclusions about the thermal stability of JCPESC. It was found that the activation energy of JCPESC was significantly higher than that of JCPECC or JCPEC. The larger activation energy showed greater stability of the composites.

10.4.2 Differential acanning calorimetry (DSC)

The DSC thermograms of silica/polyethylene nanocomposite with different filler loading were shown in Fig. 10.21. There are four heating curves of silica/low-density polyethylene nanocomposites with the silica loading of 0.5, 1.0, and 15 wt.% showed the almost similar melting temperature, $T_{m,}$ while the 2.0 wt.% silica/low-density polyethylene nanocomposite showed the highest melting temperature. According to the results, significant increment of the melting temperature started when the silica content added to the polymer is up to 2 wt.%. All four heating curves are conducted as to analyze the thermal properties of the silica/low-density nanocomposites. The melting properties of polymers are rely strongly on the strength of the chain interaction forces, in which the melting point of polymers becomes higher by rising the chain interaction force, as this shares the similar effect of adding the nanofiller to the polymers (El-Tonsy, Fouda, Oraby, Felfel, & El-Henawey, 2014). According to Chaudhry and Mittal (2015), the peak melting temperature of the silica/polyethylene nanocomposites usually is higher than the pure polyethylene, which indicates the presence of silica nanofillers successfully enhanced the thermal resistance of the crystal structures formed in the nanocomposites. Hence, with the addition of silica to low-density polyethylene, it promotes the formation of crystal structures within the nanocomposites, as these crystal structures greatly improved the heat or flame resistance of the nanocomposites.

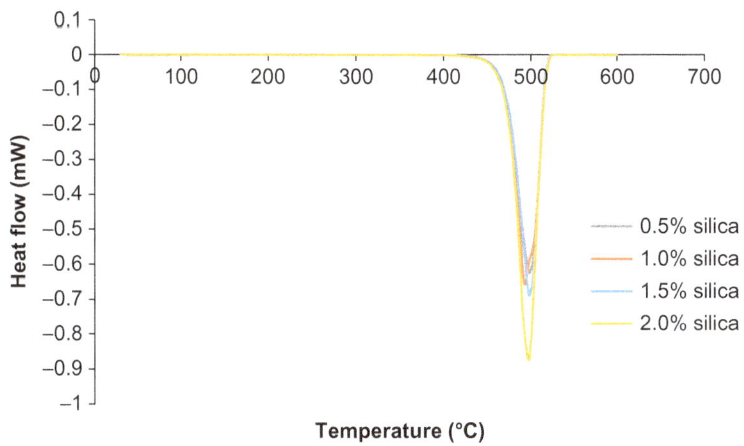

Fig. 10.21 DSC thermograms of LDPE/silica nanocomposite at different filler loading.

The effect of filling on the dynamic mechanical properties of LDPE/silica nanocomposite foams has been reported by Kontou and Niaounakis (2006). In this case, silane modified silica particles and two types of polyethylenes (mLLDPE and zLLDPE) were used. Obtained DSC data for LDPE/Silica nanocomposites are presented in Table 10.9. Silica content was optimized as 4 wt.%, whereas 20 wt.% filling with microsized silica is needed in order to achieve similar performance. It was also seen that more than 8 wt.% silica content is disadvantageous to the properties of the LLDPE composites. Finally, it is stated that nanofillers can increase the stiffness of the LLDPE matrix and also can modify morphology, and introduce new energy dissipation mechanisms.

The DSC thermogram for EVA-LDPE/silica nanocomposites is shown in Fig. 10.22 and the results are summarized in Table 10.10. A β transition (glass-rubber relaxation) was observed at −30°C for pure EVA, which is related to the branch points containing vinyl acetate group. It was seen that when the concentration of the side groups exceeds a certain limit, the β peak becomes prominent (Oakes & Robinson, 1954). Moreover, an endothermic decomposition peak also was observed at 41°C for EVA, indicating a very small number of crystalline features. On the other hand, one broad endothermic relaxation peak around 107°C found for pure LDPE due to the crystalline melting, which may also be related to the α relaxation (the segmental motion of the crystalline phase). The γ relaxation occurs in the range of −150°C to −100°C associated with the crankshaft motion of the polymethylene groups of the main-chain backbone (glass transition of LDPE).

Due to interference from the crystalline zone, the γ transition of pure LDPE has not been detected clearly in this study. However, in a control blend sample, one additional endothermic peak appeared, demonstrating a possibility of cocrystallization. It was observed that there is no significant change in the main melting peak for LDPE phase upon the addition of nanosilica particles to the blend. The melting enthalpy of the filled matrix was lower than that of the control. It is noted that the melting enthalpy remains more or less constant when 3 wt.% silica is added in sequence 3. On the other hand, the silica content is more in LDPE for the other two sequences and their effect on crystallinity is expected to be greater. The area belonging to main melting endothermic peak diminished significantly upon the use of a silane coupling agent.

The effect of silica particles in LDPE was obtained from the DSC results (Table 10.11). Is has been observed that the melting temperature is not affected by the addition of silica, whereas the degree of crystallinity increases as nanosilica content. Thus, it can be stated that silica nanoparticles show their activity on the crystallization of the polymeric matrix [Saiz-Arroyo et al., 2011]. Finally, it was observed that the crystallinity of LDPE increased by adding nanosilica up to a 6%, (for 9 wt.% of filler).

10.4.3 Dynamic mechanical thermal analysis (DMTA)

DMTA is used to investigate the interfacial interactions between fillers and polymer matrix into the composite system. Usually, strong interfacial interactions between the nanofillers and matrix can restrict the movement of polymer segments near

Table 10.9 DSC data for LDPE/silica nanocomposites

	mLLDPE/silica				zLLDPE/silica			
Nanosilica (wt.%)	T_m (°C)	ΔH (J/g)	X_c (%)	T_g (°C)	T_m (°C)	ΔH (J/g)	X_c (%)	T_g (°C)
0	103.0	100.8	34.8	−40	127.1	113.7	39.2	−50
3	101.7	114.5	39.5	−30	125.0	119.0	41.0	−40
4	105.0	116.3	40.1	−30	126.1	116.3	40.1	−30
6	104.0	103.5	36.7	−30	125.8	110.3	38.0	−25
8	103.3	97.2	33.5	−40	123.8	108.8	37.5	−25
10	103.3	84.6	29.2	−15	126.4	103.6	35.7	−20

Polymer, Elsevier.

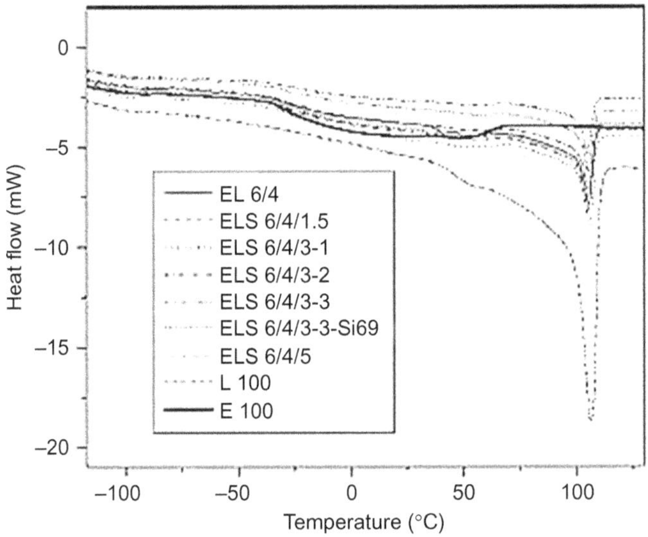

Fig. 10.22 DSC thermogram of 60:40 blends of EVA and LDPE with different loadings of nanosilica particles.
Journal of Applied Polymer Science, Wiley.

Table 10.10 DSC data for EVA-LDPE/silica nanocomposites with different silica loading

Sample code	Third cycle β relaxation (°C)	First cycle			
		T_1 (°C)	T_2 (°C)	Peak area (J/g)	T_m (°C)
E 100	−30	38	73	9.65	–
L 100	–	65	117	79.81	107
EL 6/4	−30	71	114	30.13	105
ELS 6/4/1.5	−31	74	113	28.86	105
ELS 6/4/3-1	−30	72	115	27.71	106
ELS 6/4/3-2	−30	72	112	26.33	106
ELS 6/4/3-3	−30	71	113	30.05	105
ELS 6/4/3-3/Si-69	−30	72	113	26.67	105
ELS 6/4/5	−30	70	113	30.01	105

E, EVA; *L*, LDPE; *S*, silica; T_1, temperature for onset of melting; T_2, temperature for completion of melting; T_m, melting temperature.
Journal of Vinyl and Additive Technology, Wiley.

the particles, which is reflected on the modulus-temperature response of the matrix. The effect of filling on the dynamic mechanical properties of LDPE/silica nanocomposite foams has been reported by Kontou and Niaounakis (2006). In this case, silane modified silica particles and two types of polyethylenes (mLLDPE and zLLDPE) were used. The DMA results were obtained in terms of storage modulus and loss modulus—plotted in Figs. 10.23–10.26 and summarized in Table 10.12. It was seen that a small amount of silica can rapidly change matrix properties and optimum value was established as 4 wt.%. It has been mentioned that in order to achieve similar performance, 20 wt.% of microsized silica is needed, and more than 8% is detrimental to the properties of the LLDPE composites. Finally, it is stated that nanofillers can increase the stiffness of the LLDPE matrix and can improve the morphology also.

The presence of silica nanofiller increases the glass transition temperature or melt temperature of LDPE matrix, whereas it shows negligible effect on that of polypropylene (PP) and propionic anhydride (PA) matrices. It was observed that silica can increase the rate of nucleation, whereas it has no influence on the rate of spherulite growth. It had been reported that about 20 wt.% of microsized silica have the same modification effect as with the less than several percent (0.5–5 wt.%) of nanosized ones. The improved stability and good dispersion of silica in polymer matrix are due to surface chemical modifications of primarily particles and significant improvements can be achieved by grafting macromolecular chains of the same nature as the matrix.

The storage modulus versus the temperature and tan δ versus the temperature for various samples was performed by dynamic mechanical thermal analysis (DMTA) are plotted in Figs. 10.27 and 10.28, respectively. The three major relaxations (α, β, and γ) exhibited by control LDPE. The α and β relaxations occurred in the temperature ranges of 30–100°C and −30°C to 10°C, which are associated with the crystalline phase and the amorphous phase, respectively. The β relaxation occurs in the range of −30°C to 10°C and is attributed to the amorphous phase. The γ relaxation occurred in −150°C to −100°C, conventionally considered the glass transition of LDPE, which is not detectable at this time. The vibration–rotational motion of the polymer chain within the crystalline zone is responsible for the α transition (Lee et al., 1997). Furthermore, a broad peak at 62°C was found for LDPE. If the concentration of the branch

Table 10.11 DSC results for melting point, crystallinity degree, and percentage of crystallinity increase for prepared nanocomposites

Sample	T_m (°C)	X_c (°C)	ΔX_c (%)
NC0 (LDPE)	112.25	44.55	–
NC1(LDPE with 1 wt.% Silica)	112.37	45.39	1.89
NC3 (LDPE with 3 wt.% Silica)	113.03	46.21	3.72
NC6 (LDPE with 6 wt.% Silica)	110.76	46.89	5.25
NC9 (LDPE with 9 wt.% Silica)	112.55	47.28	6.13

Cellular Polymers, Smithers Rapra Technology.

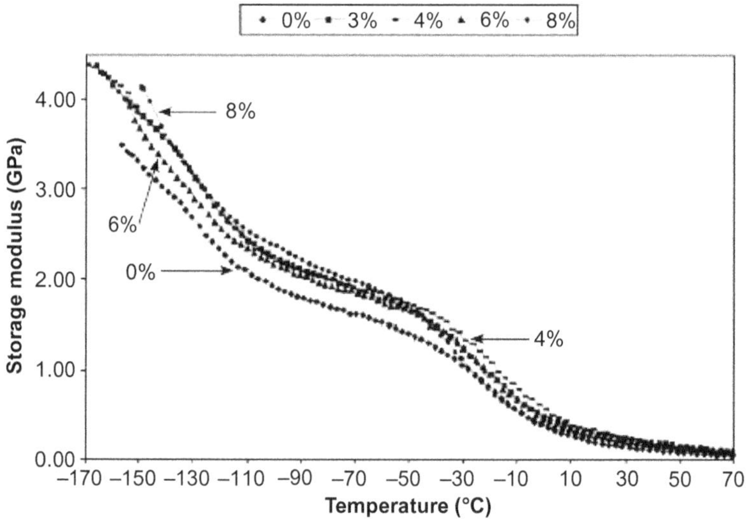

Fig. 10.23 Storage modulus (E′) dependence on temperature of the mLLDPE/Silica nanocomposites at frequency of 1 Hz.
Polymer, Elsevier.

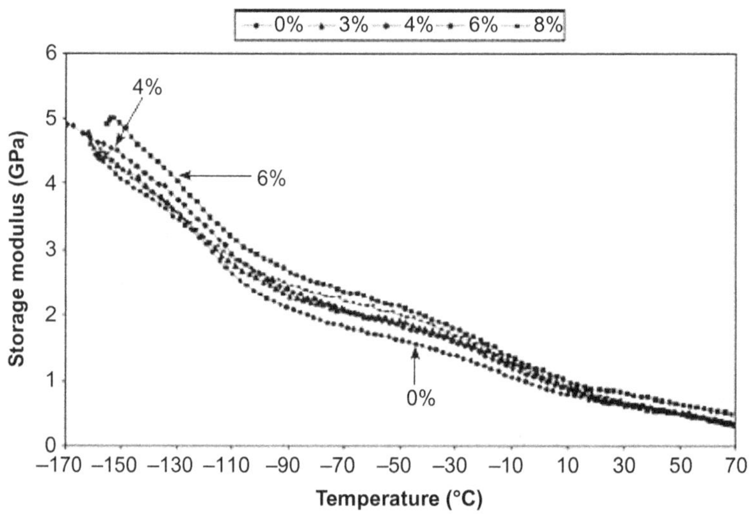

Fig. 10.24 Storage modulus (E′) dependence on temperature of the zLLDPE/Silica nanocomposites at frequency of 1 Hz.
Polymer, Elsevier.

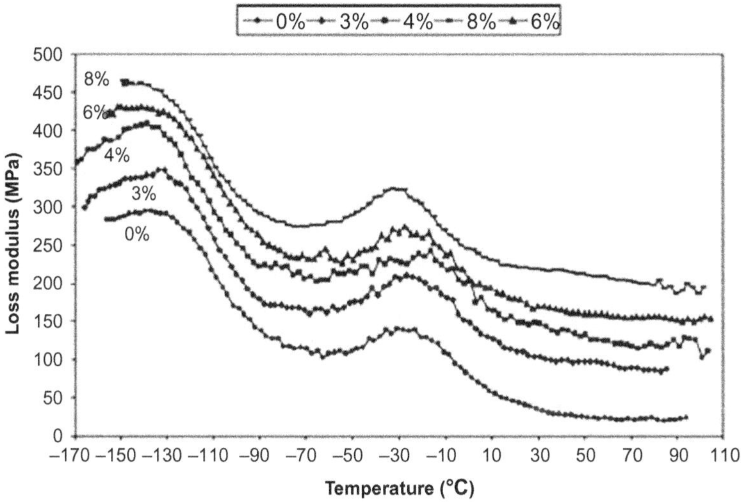

Fig. 10.25 Loss modulus (E″) dependence on temperature of the mLLDPE/SiO_2 nanocomposites at frequency of 1 Hz.
Polymer, Elsevier.

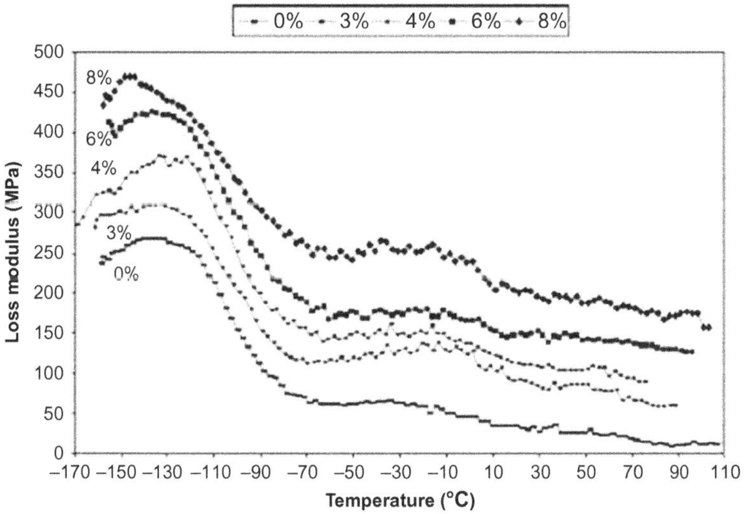

Fig. 10.26 Loss modulus (E″) dependence on temperature of the zLLDPE/silica nanocomposites at frequency of 1 Hz.
Polymer, Elsevier.

Table 10.12 **DMTA results of LLDPE/silica nanocomposites with different filler loading**

Nanosilica (wt.%)	mLLDPE/silica loss modulus				zLLDPE/silica loss modulus			
	Peak position (°C)		Peak position (MPa)		Peak position (°C)		Peak position (MPa)	
	T_γ	T_β	T_γ	T_β	T_γ	T_β	T_γ	T_β
0	−136	−30	0.29	0.14	−139	−40	0.32	0.12
3	−135	−26	0.32	0.21	−136	−28	0.33	0.16
4	−135	−20	0.39	0.23	−135	−29	0.27	0.17
6	−137	−22	0.34	0.18	−137	−31	0.30	0.11
8	−145	−32	0.31	0.17	−145	−37	0.39	0.17
10	−147	−31	0.28	0.18	−142	−22	0.40	0.16

Polymer, Elsevier.

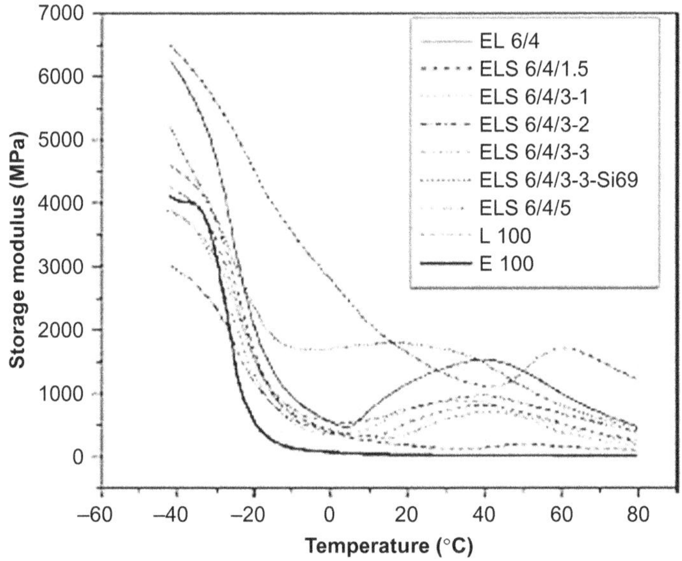

Fig. 10.27 Variation of the storage modulus with the temperature for 60:40 EVA-LDPE blends with various loadings of nanosilica and with variations in the sequence.
Journal of Vinyl and Additive Technology, Wiley.

points containing side groups exceeds a certain limit, the β relaxation becomes prominent. On the other hand, two main relaxations at −19°C and 57°C occurred for pristine EVA are assigned to the β relaxation (considered to be the glass transition of EVA) and the α relaxation, respectively (Arsac, Carrot, & Guillet, 1999).

For control blend, the relaxations at 57°C and 62°C corresponding to EVA and LDPE, respectively, are merged and shifted toward a higher temperature. This trend also observed for silica containing blend and the polymer–nanosilica interaction is evidenced by the shifting of the β-relaxation peak (corresponding to the maximum tan δ) of EVA toward a higher value. Three peaks found at −19, 4 (interface between the EVA and LDPE), and 67°C for the tan δ plot of the control blend (Fig. 10.28) are shifted to higher values (−16°C, 7°C, and 71°C) due to the addition of 1.5 wt.% pristine silica. After the addition of 3 wt.% silica, two peaks can only be clearly observed at −18°C and 70°C. It is noted that an intermediate relaxation temperature is not prominent there. The following order: 1.5 wt.% > 5 wt.% > 3 wt.% > control observed for tan δ value corresponding to β relaxation, which indicates that an optimum dispersion of silica in the EVA phase is obtained by 3 wt.% silica loading. Two transitions at −17°C and 71°C, along with a feeble relaxation at 16°C were observed after the addition of 5 wt.% silica. The storage modulus increased with the temperature up to 40°C indicating cocrystallization of all the blends which was absent in pristine polyethylene. Because of the onset of the α relaxation, the modulus essentially starts falling off.

However, a relaxation at an intermediate temperature is not observed in the case of other two sequences. Thus, it can be stated that the silica dispersion and distribution have an important role on the relaxation behavior of the blends. The changes of dynamic storage modulus-temperature behavior in sequence 1 are similar to those in sequence 3. Finally, it was seen that the tan δ (maximum) value is highest among

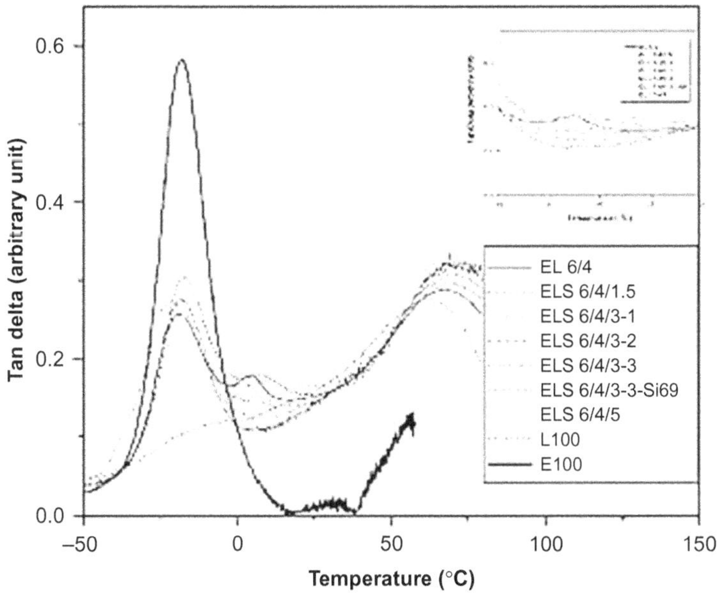

Fig. 10.28 Variation of tan δ with the temperature for 60:40 EVA-LDPE blends with different loadings of nanosilica and with variations in the sequence.
Journal of Vinyl and Additive Technology, Wiley.

all the blends. Nanosilica dispersed in LDPE-EVA composite systems alters the microscale morphology, as well as their crystalline parameters which is a strong function of the sequence and extent of nanosilica addition. It is stated that the dynamic mechanical properties result for sequences 1 and 3 was found to be greater compared to those of sequence 2. The optimum silica content was established to be 3 wt.% in most cases. The associations of silica is found to be more in the EVA phase than with the LDPE even in the case of sequence 1. The significant increases in the dynamic properties were found for modified (with Si-69) silica filled polymer blends. Therefore, it can be concluded that due to crystallization, along with a finer nanoscale dispersion of silica to reinforce weaker elastomeric EVA domains, a greater degree of reinforcement can be achieved by facilitation of reinforcement.

10.5 Scanning electron microscopy (SEM) study

The interfacial bonding of different percentages of filler loadings for silica-based nanocomposites was investigated through SEM analysis. Fig. 10.29 shows the SEM images for the pure low-density polyethylene and pure polypropylene. Based on Fig. 10.29, uneven surfaces were observed from the surface morphologies of pure low-density polyethylene. Fig. 10.30 shows the SEM images of fractured surfaces of silica-based nanocomposites (filler loading from 0.5 to 2.0 wt.%). The improvement of the surface roughness of different percentages of silica loaded to pure low-density polyethylene is shown in Fig. 10.30(A–D). Fig. 10.30(A) shows surface morphologies of 0.5 wt.% silica/low-density polyethylene nanocomposite. With 0.5 wt.% of silica loaded to low-density polyethylene, the uneven and hard surface displayed by the pure low-density polyethylene has been slightly improved. However, some gaps were observed between the matrix and silica due to the insufficient amount of silica was loaded to the pure polyethylene. This supports that 0.5 wt.% silica/low-density

Fig. 10.29 Surface morphologies of pure low-density polyethylene.

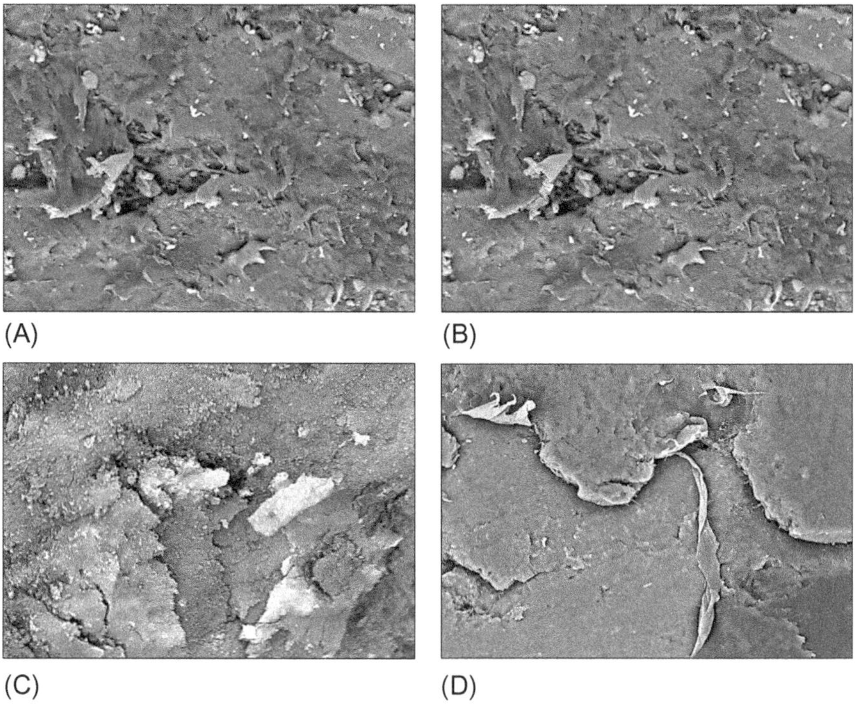

Fig. 10.30 (A) Surface morphologies of 0.5 wt.% silica/low-density polyethylene nanocomposite. (B) Surface morphologies of 1.0 wt.% silica/low-density polyethylene nanocomposite. (C) Surface morphologies of 1.5 wt.% silica/low-density polyethylene nanocomposite. (D) Surface morphologies of 2.0 wt.% silica/low-density polyethylene nanocomposite.

polyethylene nanocomposite showed the weakest tensile properties as compared to other different percentage of silica like 1.0, 1.5, and 2.0 wt.%. With the addition of 1.0 and 1.5 wt.% of silica to polyethylene, as shown in Fig. 10.30(B) and (C), the surface roughness of the fracture surfaces was further enhanced and showed a smoother surface as compared to 0.5 wt.% silica/low-density polyethylene nanocomposite. Large agglomeration of micron sized of silica was found at the addition of 1.0 wt.% silica composite, which might be due to the poor silica dispersion. The smoothest fracture surface of silica-based composite was observed with the addition of 2.0 wt.% of silica to the pure low-density polyethylene, which is shown in Fig. 10.30(D). Besides enhancing the surface roughness, the uneven surface displayed by the pure polyethylene also was improved as well. The smooth and flat surface was achieved by increasing the loading of silica to the pure low-density polyethylene, as the addition of higher percentage of silica helps to reduce the presence of gaps existing between the matrix. Thus, the polyethylene nanocomposite with the highest addition of silica, which is 2.0 wt.%, exhibited the highest tensile properties.

The study of the influence of filling on properties of linear low density polyethylene (LLDPE) has been reported by Kontou and Niaounakis (2006). Two matrices (metalocene; mLLDPE and Ziegler-Natta; zLLDPE) and silica with surface modified by dimethyldichlorosilane content from 2 wt.% up to 10 wt.% have been used for the manufacturing of LDPE/Silica nanocomposite foams. Surface morphology was obtained by scanning electron microscope. It has been seen from Fig. 10.31 that silica in mLLDPE creates larger agglomerates than in zLLDPE. It can be stated that nanofillers do not only increase the stiffness of the LLDPE matrix, but also modify morphology and introduce new energy dissipation mechanisms.

SEM micrographs of methyl ethyl ketone etched samples of low-density polyethylene-ethylene vinyl acetate (LDPE-EVA) blend nanocomposites with various loadings of silica are shown in Fig. 10.32. The black domains indicate the positions of the extracted (EVA) phase. LDPE is found as the continuous matrix in control sample EL 6/4 with 40 wt.% LDPE and the positions of the extracted EVA phase indicated by the black domains. It was seen that except in sequence 2, the silica particles are not clearly visible on the LDPE matrix. Upon increasing the silica loading, the average size of the EVA phase becomes larger in sequence 3, which is due to the decrease in the flowability of EVA. The average size of the EVA phase is smaller in the blend prepared via sequence 1 compared to that in sequence 3. Whereas, a co-continuous morphology was observed for sequence 2 only, which is due to the silica particles

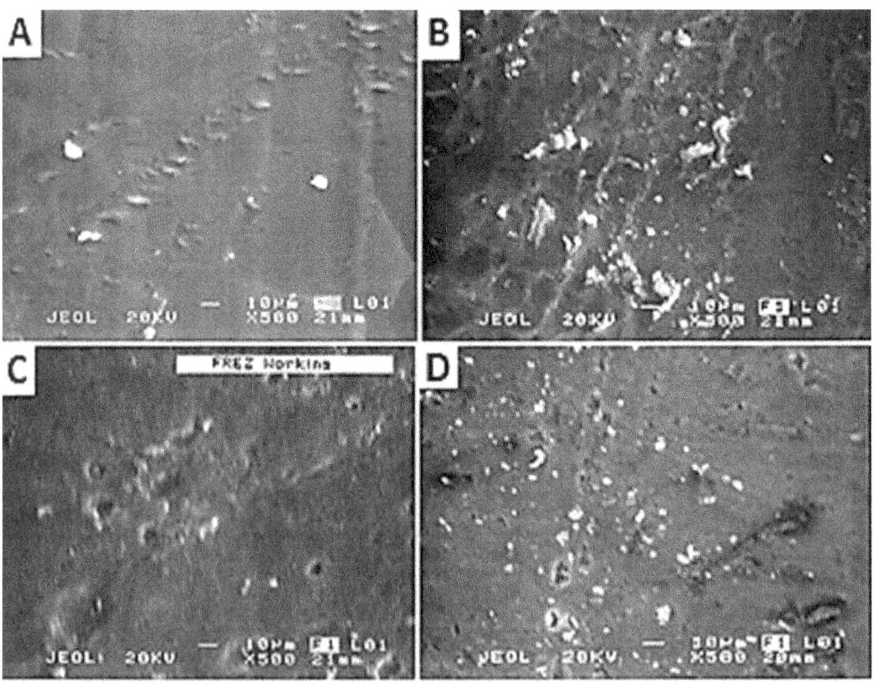

Fig. 10.31 SEM micrograph of the mLDPE-SiO_2 nanocomposites with 4% (A) and 8% (B) silica content and zLDPE-SiO_2 nanocomposites with 4% (C) and 8% (D) silica content. Polymer, Elsevier.

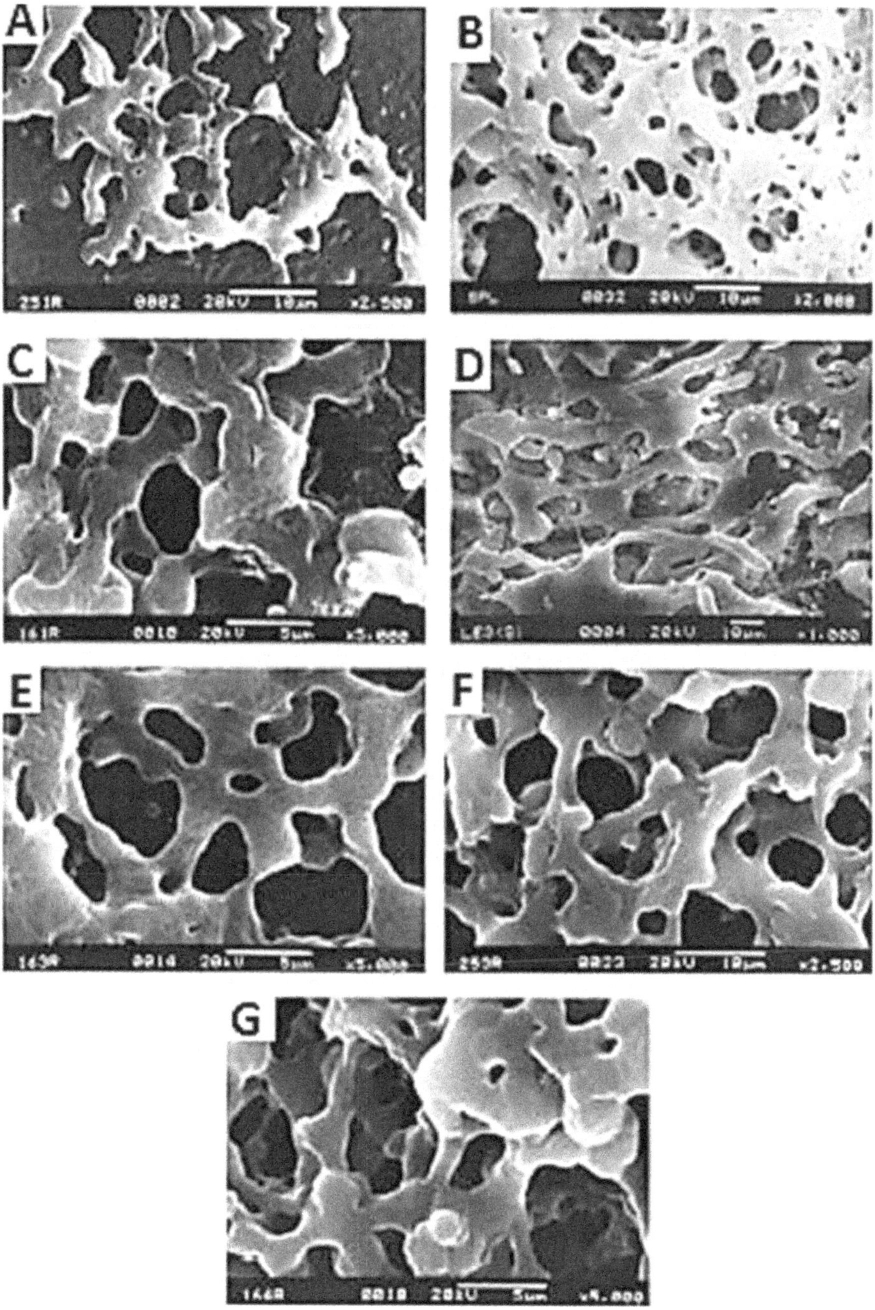

Fig. 10.32 SEM micrographs of 60:40 EVA-LDPE blends with various loadings of silica: (A) EL 6/4, (B) ELS 6/4/1.5, (C) ELS 6/4/3-1, (D) ELS 6/4/3-2, (E) ELS 6/4/3-3, (F) ELS 6/4/3-3/Si-69, and (G) ELS 6/4/5.
Journal of Vinyl and Additive Technology, Wiley.

aggregates appearing clearly on LDPE phases. The average size-scale distribution of EVA domains becomes narrower due to the addition of silane coupling agent, although the overall domain size increases (Hui et al., 2008).

SEM micrographs of the tensile fracture surfaces of the jute cellulose/PE composite (JCPEC), jute cellulose/PE/clay nanocomposite (JCPECC), and jute cellulose/PE/silica nanocomposite (JCPESC) at 15% fiber loadings are shown in Fig. 10.33(A-C). The figure clearly shows that there was a significant difference in the interfacial interaction among the three composites. The Fig. 10.33(A) clearly shows the number of pull-out traces of fiber with rough surfaces, microvoids, and agglomeration in the composites due to the weak interfacial bonding and poor dispersion between fibers and matrix. On the other hand, Fig. 10.33(B) shows better dispersion and considerable decreasing in fiber pull-out of the fiber into the matrix. This indicates that improvement of the interfacial bonding between the fiber and matrix in addition of clay. The better dispersion, strong interfacial bonding, smoother surface, less void and agglomeration were observed in the case of JCPESC, which are seen in Fig. 10.33(C). The silica particles were integrated between the fiber and matrix to reduce the agglomeration and increased strong bonding between fiber and matrix in the composite.

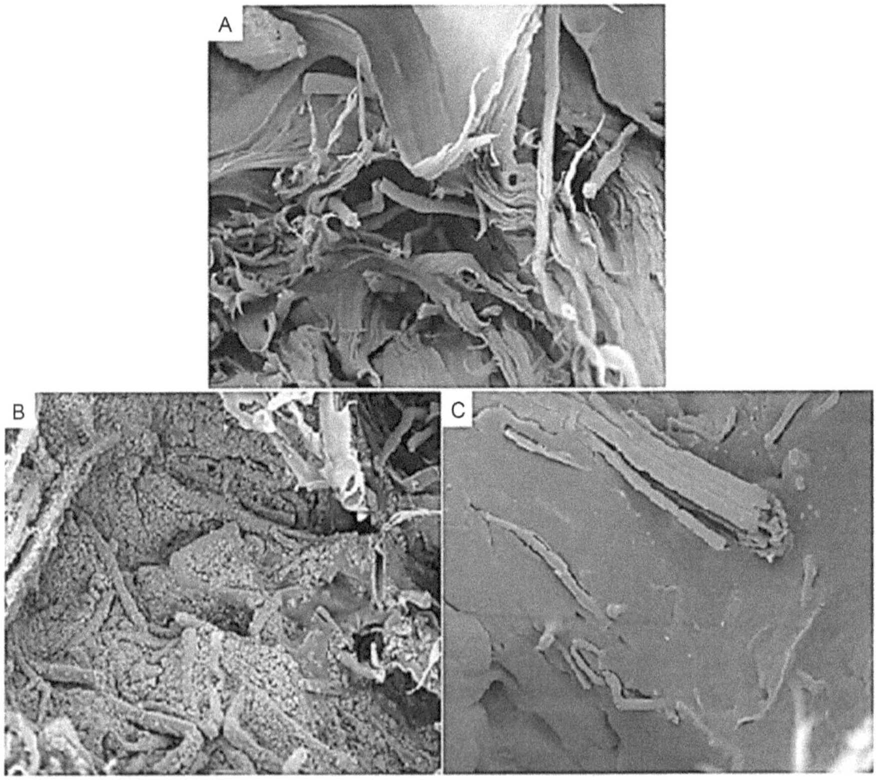

Fig. 10.33 SEM images of the fracture surfaces of (A) JCPEC, (B) JCPECC, and (C) JCPESC.

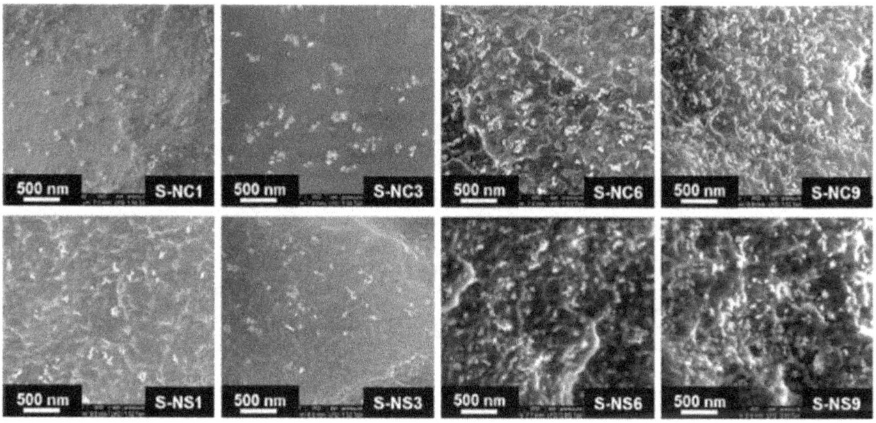

Fig. 10.34 SEM micrographs corresponding to solid nanocomposites with compatibilizer (superior row) and without compatibilizer (inferior row).
European Polymer Journal, Elsevier.

According to the above discussed results, the enhanced morphological properties also reflect the improvement of the mechanical and thermal properties.

SEM micrographs of the solid LDPE/different amounts of silica nanocomposites are shown in Fig. 10.34. The first and second row corresponds to samples with and without compatibilizer, respectively. The bright white spots indicated silica particles, which are not individually dispersed, but are forming well-scattered aggregates with a size between 100 nm and 200 nm. Similar studies have been reported using LDPE as matrix and silica particles with similar diameters (Barus, Zanetti, Lazzari, & Costa, 2009; Garcia, Hoyos, Guzman, & Tiemblo, 2009). Generally, when the silica content is increased, as it could be expected, the number of particle aggregates increases. In addition, when the silica content is increased, the real number of particles per unit volume increases, regardless of the presence of the compatibilizer. It is seen that the larger aggregates are accomplished as increasing the silica content with the addition of compatibilizer to the LDPE/silica nanocomposites. Conversely, when only the surface modification of the silica is acting, and no compatibilizer is added to the polymer matrix, aggregates with similar sizes are achieved independently on the silica content. While the similar dispersion levels are achieved independently on the clay content in the samples with and without compatibilizer, and the best dispersion degrees are achieved with the lowest contents of particles.

SEM micrographs of foamed LDPE and LDPE/silica nanocomposites are shown in Fig. 10.35. The micrographs show the cellular structures of both pure LDPE and LDPE nanocomposite foams and also stated how the addition of small amounts of silica leads to a significant reduction in cell size; moreover, this effect is more pronounced for samples without compatibilizer. Due to the high density of the cellular materials, the effects of nanoparticles on the mechanisms of degeneration of the cellular structure are not considered. It is known that the higher the dispersion level, the greater the nucleation effect (Famili, Janani, & Enayati, 2011; Ibeh & Bubacz, 2008;

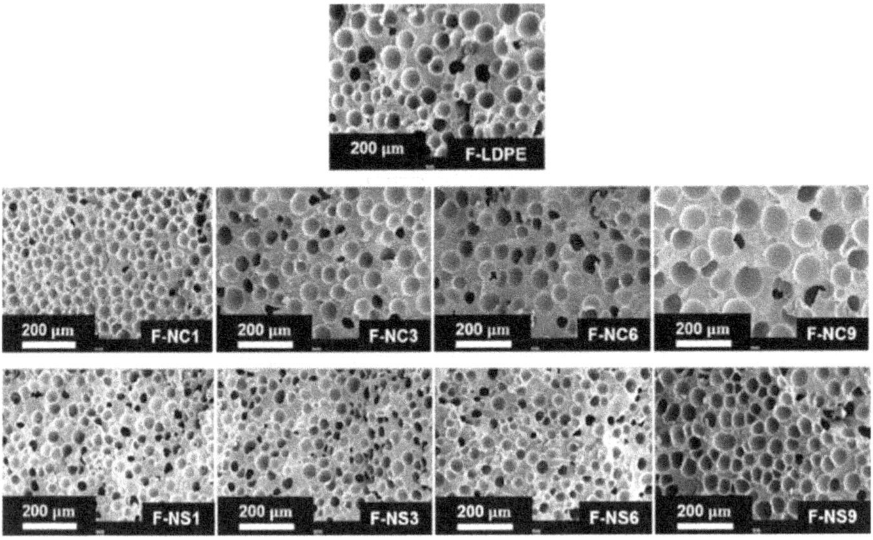

Fig. 10.35 SEM micrographs showing the cellular structure of foamed samples. European Polymer Journal, Elsevier.

Lee et al., 2005). In contrast, a high energy is required to force the interface apart if there is a high degree of bonding between the nanoparticles and matrix. However, due to the poor compatibilization, a lower surface tension is found for the interface between the matrix and the filler which enhancing cell nucleation (Famili et al., 2011).

10.6 Cellular structure of nanocomposite foams

Due to the nucleating effect of nanofillers in the nanocomposite foams, an improved cellular structure is achieved. Fig. 10.36 is corresponding to the analysis of the cellular structure of the foamed samples, which has been observed by SEM micrographs in a previous section shown in Fig. 10.35. The obtained results are presented in Fig. 10.36A (average cell size) and Fig. 10.36B (cell density). The samples for more than 1 wt.% silica contents with and without compatibilizer behave in a very different way. The cell size increases as the silica content increases for foam nanocomposites with compatibilizer (F-NC), reaching values even higher than that of the pure LDPE. Whereas, the average cell size does not show any significant variation as 3 and 6 wt.% silica contents for foam nanocomposites without compatibilizer (F-NS), obtaining the values are always smaller than that of the neat polymer. The cell densities are smaller than that achieved for samples containing only surface-modified silica nanoparticles and even smaller than that achieved for the control one. Due to the poor adhesion that exists between the polymer and nanofiller, it can be stated that the

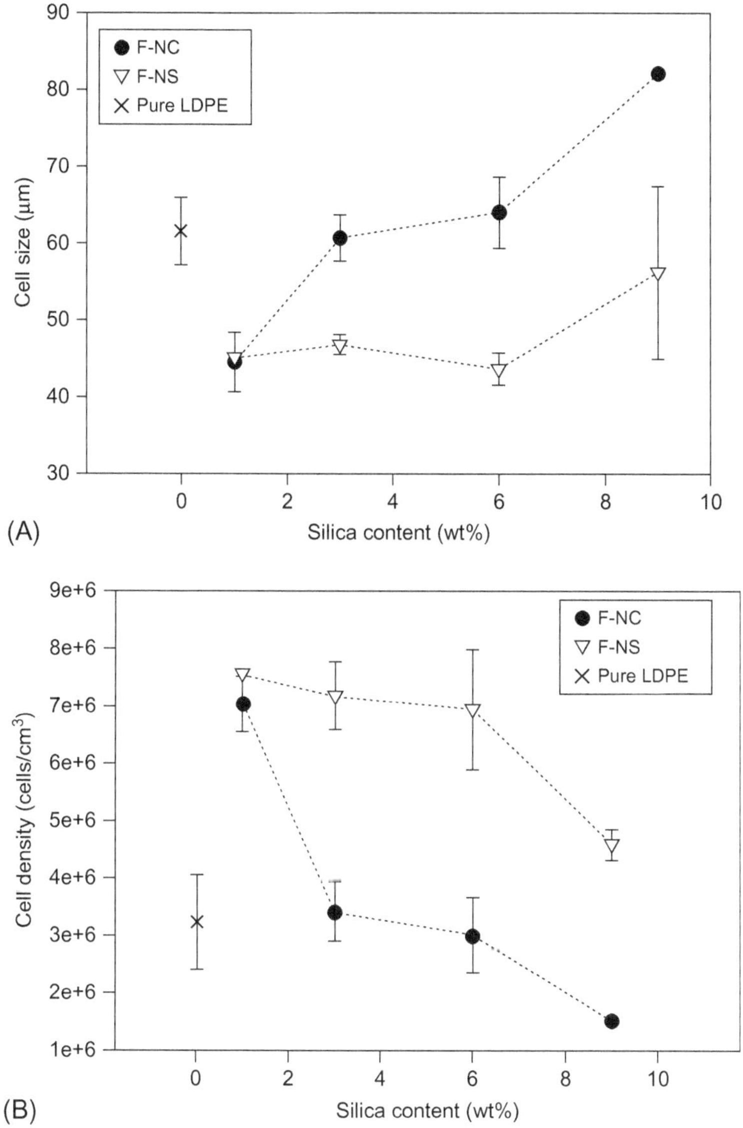

Fig. 10.36 Cellular structure parameters of LDPE and LDPE/silica nanocomposites. Average cell size (A) and average cell density (B).
European Polymer Journal, Elsevier.

effectiveness of silica particles as nucleating agents is greater in the samples without compatibilizer (Famili et al., 2011; Ibeh & Bubacz, 2008; Lee et al., 2005). The compatibilizer enhances the bonding degree between the two phases and hence, the particles nucleating effect is mitigated.

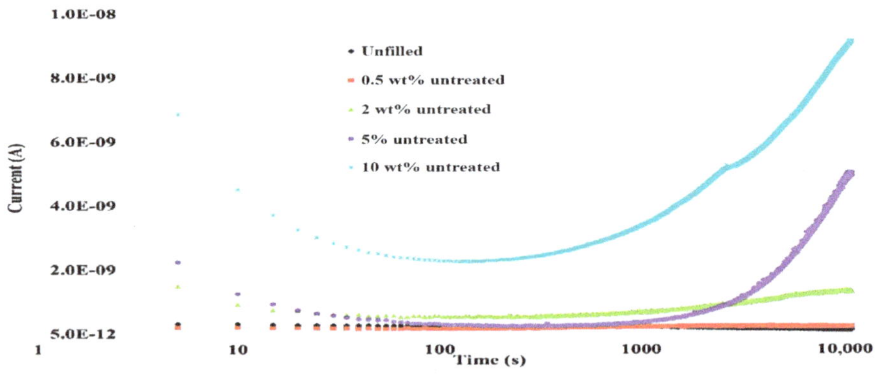

Fig. 10.37 Currents versus time for all nanocomposite samples at an applied electric field of 30kV/mm (A) untreated (B) C-3 treated.

10.7 Electrical properties (DC current)

Fig. 10.37 shows the conventional current versus time at an applied electric field of 30 kV/mm for different samples of nanocomposites, which also includes the polymer without reinforcing agent. According to Wang, Xu, Chen, and Vaughan (2015), the current versus time for a classic material is expected to exhibit a decreasing trend initially, which is also known as absorption current, then followed by a steady current, also known as conduction current. Much research also claimed that the latter process will have a dynamic period before becoming stable when it is at a high electric field (Chen, Banford, Fourance, & Tedford, 1989; Patsch, 1990). Based on Fig. 10.37, the absorption current of all nanocomposites samples is observed to be decreased by following the time, while the phenomenon of the increasing of absorption current with the nanofillers loading ratio in the nanocomposites samples is observed for the untreated nanofillers. Thus, the magnitude of the current is considered greater as compared to the unfilled samples, except the nanocomposite sample, which consists of the lowest nanosilica reinforcing ratio (0.5%). The dynamic second process will only happen in some of the untreated nanocomposite samples, consisting of 2%, 5%, and 10% of nanosilica loading ratio. The increasing trend of the current versus time is observed instead of a steady conduction current, thus the surface treatment done on the nanofillers are considered as a vital procedure in the observed dynamic process, which shows that the increase in current is not as high as the one observed with the untreated nanofillers (Wang et al., 2015).

References

Arsac, A., Carrot, C., & Guillet, J. (1999). Rheological characterization of ethylene vinyl acetate copolymers. *Journal of Applied Polymer Science*, *74*(11), 2625–2630.

Barus, S., Zanetti, M., Lazzari, M., & Costa, L. (2009). Preparation of polymeric hybrid nanocomposites based on PE and nanosilica. *Polymer*, *20*, 2595–2600.

Chaudhry, A. U., & Mittal, V. (2015). High-density polyethylene nanocomposites using masterbatches of chlorinated polyethylene/graphene oxide. *Polymer Engineering & Science*, 78–88. https://doi.org/10.1002/pen.23241.

Chen, G., Banford, H. M., Fourance, R. A., & Tedford, D. J. (1989). Electrical conduction in low density polyethylene. In *3rd International Conference on Conduction and Breakdown in Solid Dielectrics* (pp. 277–281).

El-Tonsy, M. M., Fouda, I. M., Oraby, A. H., Felfel, R. M., & El-Henawey, M. I. (2014). Dependence of physical properties of linear low-density polyethylene on the silicon dioxide filler size. *Journal of Thermoplastic Composite Materials*, 1–14. https://doi.org/10.1177/0892705714533376.

Famili, M. H. N., Janani, H., & Enayati, M. S. (2011). Foaming of a polymer–nanoparticle system: Effect of the particle properties. *Journal of Applied Polymer Science, 119*, 2847–2856.

Garcia, N., Hoyos, M., Guzman, J., & Tiemblo, P. (2009). Comparing the effect of nanofillers as thermal stabilizers in low density polyethylene. *Polymer Degradation and Stability, 54*, 39–48.

Garcia, M., Rooij, M. D., Winnubst, L., Zyl, W. E. V., & Verweij, H. (2004). Friction and wear sudies on nylon-6/SiO_2 nanocomposites. *Journal of Applied Polymer Science, 92*, 1855–1862.

Hui, S., Chaki, T. K., & Chattopadhyay, S. (2008). Effect of silica-based nanofillers on the properties of a low-density polyethylene/ethylene vinyl acetate copolymer based thermoplastic elastomer. *Journal of Applied Polymer Science*, *110*, 825–836.

Ibeh, C. C., & Bubacz, M. (2008). Current trends in nanocomposite foams. *Journal of Cellular Plastics*, *44*, 493–515.

Kaleel, S. H., Bahuleyan, B. K., Masihullah, J., & Al-Harthi, M. (2011). Thermal and mechanical properties of polyethylene/doped-TiO_2 nanocomposites synthesized using in situ polymerization. *Journal of Nanomaterials*, 1–6. https://doi.org/10.1155/2011/964353.

Kontou, E., & Niaounakis, M. (2006). Thermo-mechanical properties of LLDPE/SiO_2 nanocomposites. *Polymer*, *47*, 1267–1280.

Kozako, M., Kido, R., Fuse, N., Ohki, Y., Okamoto, T., & Tanaka, T. (2004). In *Difference in surface degradation due to partial discharges between polyamide nanocomposites and microcomposites. IEEE Conference on Electrical Insulation and Dielectric Phenomenon* (pp. 398–401).

Lee, H., Cho, K., Ahn, T. K., Choe, S., Kim, I. J., Park, I., et al. (1997). Solid-state relaxations in linear low-density (1-octene comonomer), low-density, and high-density polyethylene blends. *Journal of Polymer Science Part B: Polymer Physics*, *35*(10), 1633–1642.

Lee, L. J., Zeng, C., Cao, X., Han, X., Shen, J., & Xu, G. (2005). Polymer nanocomposite foams. *Composites Science and Technology*, *65*, 2344–2363.

Lin, O. H., Akil, H. M., & Ishak, Z. A. M. (2009). Characterization and properties of activated nanosilica/polypropylene composites with coupling agents. *Polymer Composites, 30*, 1693–1700.

Oakes, W. G., & Robinson, D. W. (1954). Dynamic electrical and mechanical properties of polythene over a wide temperature range. *Journal of Polymer Science, Part A: Polymer Chemistry*, *14*(77), 505–507.

Ogunsona, E., Ogbomo, S., Nar, M., & D'Souza, N. A. (2011). Thermal and mechanical effects in polystyrene montmorinonite nanocomposite foams. *Cellular Polymers*, *30*, 79–94.

Parvinzadeh, M., Moradian, S., Rashidi, A., & Yazdanshenas, M. E. (2009). Surface characterization of polyethylene terephthalate/silica nanocomposites. *Applied Surface Science*, *256*, 2792–2802. https://doi.org/10.1016/j.apsusc.2009.11.030.

Patsch, R. (1990). Space charge phenomena in polyethylene at high electric fields. *Journal of Physics D: Applied Physics*, *23*, 1497–1505.

Patwary, F., & Mittal, V. (2015). Degradable polyethylene nanocomposites with silica, silicate and thermally reduced graphene using oxo-degradabale pro-oxidant. *Heliyon*. https://doi.org/10.1016/j.heliyon.2015.e00050.

PerkinElmer (2011). *Polymer idenfication using mid infrared spectroscopy. Available from: https://www.perkinelmer.com/labsolutions/resources/docs/APP_PolymerIdentificationMidInfaredSpectroscopy.pdf.*

Prachum, Y., Adam Strauss, R. H., & Kiatkamjornwong, S. (2011). The physical and mechanical properties of beta-nucleated polypropylene/montmorillonite nanocomposites. *Journal of Applied Polymer Science*, *122*(2), 1066–1076.

Preschilla, N., Rasheed, A. S. A., Sahadevan, S., Biswas, A., Bellare, J. R., & Shyamroy, S. (2010). Study of layered silicate clays as synergistic nucleating agent for polypropylene. *Journal of Polymer Science Part B: Polymer Physics*, *48*(16), 1786–1794.

Qu, P., Gao, Y., Wu, G., & Zhang, L. (2010). PLA/cellulose nanocomposites. *BioResources*, *5*(3), 1811–1823.

Rottstegge, J., Qiao, Y. K., Zhang, X., Zhou, Y., Xu, D., Han, C. C., et al. (2007). Polymer nanocomposites powders and melt spun fibers filled with silica nanoparticles. *Journal of Applied Polymer Science*, *103*, 218–277.

Saiz-Arroyo, C., Escudero, J., Rodríguez-Pérez, M. A., & de Saja, J. A. (2011). Improving the structure and physical properties of LDPE foams using silica nanoparticles as an additive. *Cellular Polymers*, *30*(2), 63–78.

Saiz-Arroyo, C., Rodríguez-Pérez, M. A., Velasco, J. I., et al. (2013). Influence of foaming process on the structure–properties relationship of foamed LDPE/silica nanocomposites. *Composites Part B: Engineering*, *48*, 40–50.

Wang, Y., Xu, Z. Q., Chen, G., & Vaughan, A. (2015). DC current in nanosilica-based polyethylene nanocomposites. *IEEE*, 515–518.

Xanthos, M. (2005). *Functional fillers for plastics.* Weinheim: Wiley.

Xu, X., Li, B., Lu, H., Zhang, Z., & Wang, H. (2008). The effect of the interface structure of different surface-modified nano-SiO_2 on the mechanical properties of Nylon 66 composites. *Journal of Applied Polymer Science*, *107*, 2007–2014.

Further reading

Al-Samhan, M., Samuel, J., Al-Attar, F., & Abraham, G. (2017). Comparative effects of MMT clay modified with two different cationic surfactants on the thermal and rheological properties of polypropylene nanocomposites. *International Journal of Polymer Science*. https://doi.org/10.1155/9484.

Chen, L., Wong, S. C., & Pisharath, S. (2002). Fracture properties of nanoclay-filled polypropylene. *Journal of Applied Polymer Science*, *88*, 3298–3305.

Lei, Y., & Cebe, P. (2009). crystal polymorphism in electrospun composite nanofibers of poly(vinylindene fluoride) with nanoclay. *Polymer*, *50*, 2133–2141.

Leszczynska, A., Njuguna, J., Pielichowski, K., & Banerjee, J. R. (2007). Polymer montmorillonite nanocomposites with improved thermal properties. Part I. Factors influecing thermal stability and mechanisms of thermal stability improvement. *Thermochimica Acta*, *453*, 75–96.

Li, Y., Wang, S., Zhang, Y., & Zhang, Y. (2006). crystalization behavior of carbon black-filled polypropylene and polypropylene/epoxy composites. *Journal of Applied Polymer Science*, *102*, 104–118.

Nafchi, H. R., Abdouss, M., Najafi, S. K., Gargari, R. M., & Mazhar, M. (2015). Effects of nanoclay particles and oxidied polypropylene polymers on improvements of the thermal properties of wood plastic composite. *Maderas. Ciencia y tecnología*, *17*(1), 45–54. https://doi.org/10.4067/S0718-221X2015005000005.

Raka, L., Bogoeva-Gaceva, G., Lu, K., & Loos, J. (2009). Characterization of latex-based isotactic polypropylene/clay nanocomposites. *Journal of Polymer Science Part B: Polymer Physics*, *50*, 3739–3746.

Rodriguez-Llamazares, S., Rivas, B. L., Perez, M., Perrin-Sarazin, F., Maldonado, A., & Venegas, C. (2011). The effect of clay type and of clay-masterbatch product in the preparation of polypropylene/clay nanocomposites. *Journal of Applied Polymer Science*, *122* (3), 2013–2025.

Zhang, J., Manias, E., & Wilkie, C. A. (2008). Polymerically modified layered silicates: An effective route to nanocomposites. *Journal of Nanoscience and Nanotechnology*, *8*(4), 1597–1615.

The effect of clay dispersion on polypropylene nanocomposites: Physico-mechanical, thermal, morphological, and optical properties

11

Md. Rezaur Rahman, Sinin bin Hamdan, Md. Faruk Hossen
Department of Chemical Engineering and Energy Sustainability, Faculty of Engineering, University Malaysia Sarawak (UNIMAS), Kota Samarahan, Sarawak, Malaysia

11.1 Overview of the nanocomposites

Nanocomposites are defined if one of the components are in the nanometer range ($1\,nm = 10^{-9}\,m$) (Roy, Roy, & Roy, 1986). Due to the limitation of microcomposites and monolithics, nanocomposite materials have emerged as appropriate replacements to overcome limitations and preparation challenges which areassociated with the mechanism of fundamental composition and their nanocluster stoichiometry phase. Conventional composites are unable to meet the uniqueprocessing design for their property combinations, which are reported to be the materials of 21st century. Though the interface of the composites was reported early, general properties of the nanocomposites are yet to be investigated (Gleiter, 1992; Schmidt, Shah, & Giannelis, 2002).

In the late 1990s, use of the words nanoscience, nanotechnology, and nanomaterials, doubled in 1.6 years according to the published paper (Braun, Schubert, & Sindelys, 1997). Based on the literature review, it was summarized that about 13,420 papers (of which 4028 contain the keywords nanocomposite and polymer in Web of Science-ISI: updated on 10 February 2009) have been published on nanocomposites, whereas about 4663 patents with comprehensive document on nanocomposites were accessible throughout the same period per Scirus (www.scirus.com). The developing science and technology of nanomaterials have been dedicated and presented in specific conferences and special issues of some journals. The critical size of the particle is a special particle which particle size is less than a particular level that has been reported in literature (Kamigaito, 1991). Moreover, the materials properties can be enhanced as dimensions reach the nanometer level, which largely improved interactions at materials interfaces and the surface area/volume ratio of reinforcement materials engaged in the preparation of nanocomposites is crucial to the understanding of their structure–property relationships. Further, in 1991, carbon

Silica and Clay Dispersed Polymer Nanocomposites. https://doi.org/10.1016/B978-0-08-102129-3.00011-7

nanotubes (CNTs) opened a new avenue due to their unique properties and subsequent use to fabricate composites, exhibiting CNT related mechanical, thermal, and electrical properties (Biercuk, Llaguno, & Radosvljevicm, 2002; Iijima, 1991; Ounaies, Park, Wise, Siochi, & Harrison, 2003; Wiesenberger, Grulke, Jacques, Ramtell, & Andrews, 2003). The processing and applications of CNT-containing nanomaterials will go from the spinning CNTs into composite products and textiles, whereas the nanocomposites offer new technology and business opportunities for all sectors of industry which have been environmentally friendly (Choa et al., 2003; Dalton et al., 2003).

According to their matrix materials, nanocomposite materials can be classified in three different categories, such as: (1) Ceramic Matrix Nanocomposites (CMNC); (2) Metal Matrix Nanocomposites (MMNC), and (3) Polymer Matrix Nanocomposites (PMNC).

There has been continuous research on nanocomposites since the 1990s, whereas CNTs-reinforced nanocomposite systems are comprehensively studied and have had uninterrupted growth in the number of publications on the subject (Alexandre & Dubois, 2000; Andrews & Weisenberger, 2004; Aruna & Rajam, 2003; Choi & Awaji, 2005; Fischer, 2003; Gall et al., 2002; Gangopadhyay & Amitabha, 2000; Giannelis, 1996; Jordan, Jacob, Tannenbaum, Sharaf, & Jasiuk, 2005; Kickelbick, 2003; Pandey, Reddy, Kumar, Singh, 2005; Pandey, Kumar, et al., 2005; Peigney et al., 2000; Pokropivnyi, 2002a; Pokropivnyi, 2002b; Ray & Bousmina, 2005; Ray & Okamoto, 2003; Schmidt et al., 2002; Sternitzke, 1997; Thostenson, Li, & Chou, 2005; Thostenson, Ren, & Chou, 2001; Wang, Guo, Fu, Wu, Zhu, 2004; Wang, Padture, Tanaka, 2004; Wypych, Adad, & Grothe, 1998; Wypych, Seefeld, & Denicolo, 1997; Xie, Mai, & Zhou, 2005). In spite of this development, this chapter describes the current status of various types of nanocomposites. In the case of PMNC, some researchers deal with processing aspects, especially those on layered silicates (Alexandre & Dubois, 2000; Ray & Okamoto, 2003). However, conducting, fiber reinforced and biodegradable polymer-based systems show the improvement structure/morphology/property, as well as applications and perspectives. It also counts significant prospects and challenges in the development of structural and functional fiber nanocomposites (Alexandre & Dubois, 2000; Gangopadhyay & Amitabha, 2000; Pandey, Reddy, et al., 2005; Pandey, Kumar, et al., 2005; Peigney et al., 2000; Ray & Bousmina, 2005; Ray & Okamoto, 2003; Sternitzke, 1997; Thostenson et al., 2001). It is recent history that conducting polymer-based composites are novel materials. However, conducting polymer-based composite systems have achieved electrical conductivity and colloidal stability, and which optimization of their physical properties is expected with their appropriate utilization in the near future, and a big leap forward for materials science (Gangopadhyay & Amitabha, 2000). Preparation, characterization, and properties of recently developed biodegradable polymer-based nanocomposites have been discussed, including crystallization behavior and melt rheology of both the matrix and the layered (montmorillonite) (Pandey, Reddy, et al., 2005; Pandey, Kumar, et al., 2005; Ray & Bousmina, 2005). Likewise, CNTs and polymer matrices are discussed to highlight the stress transfer from the matrix and the potential of these composites for possible macro scale CNT-polymer production, which emphasizes toughness and interfacial bonding between CNT and polymer (Andrews & Weisenberger, 2004). It was observed that microcomposites encounteredd problems regarding a critical volume fraction of CNTs, which is appropriate to strengthening microcomposites. According to the review, there was very limited work on metal-based nanocomposites, including the ones with CNT reinforcements.

This book chapter gives an overview of metal, ceramic, and polymer-based nanocomposites including the incorporation of CNTs, whereas the three different kinds of nanocomposites consider these facts and also including the absence of a more general review (Schmidt et al., 2002; Braun et al., 1997; Biercuk et al., 2002; Ounaies et al., 2003; Wiesenberger et al., 2003; Choa et al., 2003; Giannelis, 1996; Wypych et al., 1998; Sternitzke, 1997; Peigney et al., 2000; Alexandre & Dubois, 2000; Gangopadhyay & Amitabha, 2000; Thostenson et al., 2001; Pokropivnyi, 2002a, 2002b; Gall et al., 2002; Kickelbick, 2003; Fischer, 2003; Ray & Okamoto, 2003; Andrews & Weisenberger, 2004; Wang, Guo, et al., 2004; Wang, Padture, et al., 2004; Pandey, Reddy, et al., 2005; Pandey, Kumar, et al., 2005; Thostenson et al., 2005; Jordan et al., 2005; Choi & Awaji, 2005; Xie et al., 2005; Ray & Bousmina, 2005; Awaji, Choi, & Yagi, 2002; Niihara, 1991; Nakahira & Niihara, 1992; Ferroni, Pezzotti, Isshiki, & Kleebe, 2001; Tjong & Wang, 2004; Athawale, Bhagwat, Katre, Chandwadkar, & Karandikar, 2003; Akita & Hattori, 1999; Akita & Kobayashi, 1999; Akita, Kobayashi, Hattori, & Kagawa, 1999; Chang & An, 2002; Zavyalov, Pivkina, & Schoonman, 2002; Thompson, Herring, Gates, & Connel, 2003; Liu, Phang, Shen, Chow, & Zhange, 2004; Theng, 1974; Ogawa & Kuroda, 1997a, 1997b; Kojima et al., 1993a, 1993b; Kojima, Okada, Fukushima, et al., 1993; Stearns, Zhao, & Harmer, 1992; Borsa & Brook, 1995; Riedel, Strecker, & Petzow, 1989; Riedel, Seher, & Becker, 1989; Riedel, Seher, Mayer, & Szabo, 1995; Livage, 1997; Vorotilov, Yanovskaya, Turevskaya, & Sigov, 1999; Hench & West, 1990; Ennas et al., 1998; Sen, Choudharya, & Pramanik, 2004; Viart, Richard-Plouet, Muller, & Pourroy, 2003, Kundu, Mukherjee, Chakravorty, & Sinha, 1998; Baiju et al., 2005; Ananthakumar, Prabhakaran, Hareesh, Manoharan, & Warrier, 2004; Sivakumar, Sibu, Mukundan, Pillai, & Warrier, 2004; Warrier & Anilkumar, 2001; Wunderlich, Padmaja, & Warrier, 2004; Ghosh & Pramanik, 2001; Camargo et al., 2003; Camargo et al., 2004; Mathur, Veith, Shen, Hufner, & Jilavi, 2002; Ning, Zhang, Pan, & Guo, 2003; Xia, Riester, et al., 2004; Xia, Wang, Li, Hu, 2004; An, You, & Lima, 2003; Ruhle, 2003; Jiang & Gao, 2003; Chaisan, Yimnirun, & Ananta, 2009; Cha, Kim, Lee, Mo, & Hong, 2005; Balázsi, Kónya, Wéber, Biró, & Arató, 2003; Chung, 2004; Lim, You, Choi, Lim, & Jang, 2005; Peigney, Flahaut, Lautent, Chastel, & Rousset, 2002; Siegel et al., 2001; Lee et al., 2005; Laurent, Peigney, Dumortier, & Rousset, 1998; Sinnott et al., 1999; Lim, An, & Lee, 2002; Bajwa, Rainfoth, & Lee, 2005; Sealy, 2004; Tai, Kim, & Kim, 2003; Li, Huang, & Guo, 2003; Goujon & Goeuriot, 2001; Chen, Zhang, Cai, 2003; Chen, Lee, Liu, 2003; Sawaguchi, Toda, & Niihara, 1991; Suzuki, Sekino, & Niihara, 1995; Sasaki, Suga, Yanai, Suganuma, & Niihara, 1994; Sakka, Bidinger, & Aksay, 1995; Ge, Lei, & Zhou, 1991; Nawa, Sekino, & Niihara, 1994; Anya, 1999; Baron, Kumar, Le Gonidec, & Hampshire, 2002; Timms, Ponton, & Strangwood, 2002; Mabuchi,

Tsuda, Ohtsuka, Matsui, & Morii, 1999; Weimer & Bordia, 1999; Ma, Wu, Wei, Liang, & Wu, 1998; Zhan, Kuntz, Wan, & Mukherjee, 2003; Zhan, Kuntz, Wan, Garay, & Mukherjee, 2003; Natile & Glisenti, 2003; Balázsi et al., 2005; Baker, Ismat Shah, & Hasanain, 2004; Yoon, Lee, Oh, & Kim, 2002; Provenzano, Louat, Imam, & Sadananda, 1992; Contreras, López, & Bedolla, 2004; Bhattacharya & Chattopadhyay, 2004; Bhattacharya & Chattopadhyay, 2001; Srinivasan & Chattopadhyay, 2004; Branagan, 2000; Branagan & Tang, 2002; Xiaochun, Yang, & Cheng, 2004; Ying & Zhang, 2000; Choy, 2003; Joseph et al., 2005; Chow et al., 1990; Haubold & Gertsman, 1992; Holtzt & Provenzano, 1994; Cushing, Kolesnichenko, & O'Connor, 2004; West, Wang, & Goodson, 2003; Kamat, Flumiani, & Dawson, 2002; Carpenter et al., 2000; Chen, Xia, Peng, Li, & Xie, 2000; Chen, Wong, Pisharath, 2002; Chen, Lee, Liu, 2002; Xu et al., 1999; Noguchi, Magario, Fuzukawa, & Shimizu, 2004; Kuzumaki, Miyazawa, Ichinose, & Ito, 1998; Yang & Schaller, 2004; Marchal, Delannay, & Froyen, 1996; Jang, Kim, Lee, Kim, & Um, 2005; Liu et al., 1997; Venkatraman & Sundararajan, 1996; El-Eskandarany, 1998; Pathak, Tiwari, & Malhotra, 1986; Quin, Dickey, Andrews, & Rantell, 2000; Schadler, Giannaris, & Ajayan, 1998; Jimenez, Ogata, Kawai, & Ogihara, 1997; Ogata, Jimenez, Kawai, Ogihara, & Ogata, 1997; Jeon, Jung, Lee, & Hudson, 1998; Aranda & Ruiz-Hitzky, 1992; Greenland, 1963; Francis, 1973; Zhao, Urano, & Ogasawara, 1989; Usuki, Kojima, et al., 1993; Usuki, Kawasumi, et al., 1993; Okamoto et al., 2000; Okamoto, Morita, & Kotaka, 2001; Yao, Song, Hourston, & Luo, 2002; Messersmith & Giannelis, 1994; Vaia & Giannelis, 1997a, 1997b; Gilmann, 1999; Vaia, Vasudevan, Krawiec, Scanlon, & Giannelis, 1995; Kawasumi, Hasegawa, Kato, Usuki, & Okada, 1997; Tomasko, Han, Liu, & Gao, 2003; Watkins & Mccarthy, 1994; Watkins & Mccarthy, 1995a; Watkins & Mccarthy, 1995b; Watkins & Mccarthy, 1995c; Carrado & Xu, 1998; Fernando & Satyanarayana, 2005; Park, Kwon, Woo, & Kim, 2005; Mbhele et al., 2003; Aymonier, Bortzmeyer, Thomann, & Lhaupt, 2003; Evora & Shukla, 2003; Di Lorenzo, Errico, & Avella, 2002; Park, Bernet, De La Roche, Hanh, 2003; Park, Cho, Lim, Choi, Jhon, 2003; Xu, Yin, Ge, Wu, & Zhang, 1998; Liu, Gao, Wang, Li, & Xu, 2002; Jackson, Bauer, Nakatani, & Barnes, 1996; Avadhani & Chujo, 1997; Sandler et al., 1999; Qian, Dickey, Andrews, & Rantell, 2000; Ding et al., 2003; Lin et al., 2003; Wong, Paramsothy, Xu, Ren, & Liao, 2003; Koerner, Price, Pearce, Alexer, & Vaia, 2004; Kubayashi & Hayashi, 1992; Tang, Santare, & Advani, 2003; Andrews, Jacques, Minot, & Rantell, 2002; Maser et al., 2003; Park et al., 2002; Philip, Xie, Abraham, & Varadan, 2005; Azioune et al., 1999; Xia, Riester, et al., 2004; Xia, Wang, et al., 2004; Kim, Kim, & Choi, 2003; Valentini, Biagiotti, Kenny, & Santucci, 2003; Bharadwaj, 2001; Kojima et al., 1993a; Kojima, et al., 1993; Bourbigot, LeBras, Dabrowski, Gilman, & Kashiwagi, 2000; Okada et al., 1990; Dabrowskii, Bourbigot, Delobel, & Lebras, 2000; Lee, Baljon, Loring, & Panagiopoulos, 1998; Fredrickson & Bicerano, 1999; Kuznetsov & Balazs, 2000; Manias et al., 2000; Herron & Thorn, 1998; Favier, Canova, Shrivastava, & Cavaille, 1997; Chazeau, Cavaille, Canova, Dendievel, & Boutherin, 1999; Ogawa & Kuroda, 1997a, 1997b; Chen, Yang, & Yang, 1997; Lu & Huang, 2002;

Liu et al., 2003; Trindade & O'Brien, 1996; Trindade, Neves, & Barros, 2000; Vaia & Wagner, 2004; Curtin & Sheldon, 2004; Thostenson & Chou, 2003; Thostenson & Chou, 2002; Tai, Yeh, & Liu, 2004; Gojny, Wichmann, Kopke, Fiedler, & Schulte, 2004; Ogasawara, Ishida, Ishikawa, & Yokota, 2004; Giannelis, Krishnamoorti, & Manias, 1999; LeBaron, Wang, & Pinnavaia, 1999; Vaia, Price, Ruth, Nguyen, & Lichtenhan, 1999; Biswas & Ray, 2001; Giannelis, 1998; Xu, Manias, Snyder, & Runt, 2001; Yano, Usuki, Okada, Kurauchi, & Kamigaito, 1993; Gilman et al., 2000; Hori et al., 1999; Blumstein, 1965; Theng, 1979; Vaia, Ishii, & Giannelis, 1993; Passaglia, Bertoldo, Ciardelli, Prevosto, & Lucchesi, 2008; Krishnamoorti, Vaia, & Giannelis, 1996; Hasegawa, Kawasumi, Kato, Usuki, & Okada, 1998; Lee & Jang, 1996; Noh & Lee, 1999; Wang & Pinnavaia, 1998a; Wang & Pinnavaia, 1998b; Yano, Usuki, & Okada, 1997; Liu, Qi, & Zhu, 1999; Fornes, Yoon, Keskkula, & Paul, 2001; Oakada & Usuki, 1995; Laus, Francesangeli, & Sandrolini, 1997; Becker, Varley, & Simon, 2002; Wang, Long, Wang, Li, & Qi, 1998; Zhu, Morgan, Lamelas, & Wilkie, 2001; Ray & Okamoto, 2003; Nielsen, 1967; Okada, Kawasumi, Kurauchi, & Kamigaito, 1987; Lwan, 2002; Oriakhi, Farr, & Lerner, 1997; Wilson et al., 1999; Do Nascimento, Constantino, & Temperini, 2002; Yang, Zhu, Yin, Wang, & Qi, 1999; Lee, Takekoshi, & Giannelis, 1997; Ray & Bousmina, 2005; Morgan, Harris, Kashiwagi, Chyall, & Gilman, 2002; Scherer, 1999; Kim, Kim, Choi, & Jhon, 1999; Lan, Kavitatna, & Pinnavia, 1994; Mohanty, Drzal, & Misra, 2003; Dutta et al., 2003; Ghose et al., 2006; Lijie & Jinbo, 2006; Xu, Ray, & Abdel-Magid, 2006; Lucas & Young, 2004; Shofner, Rodríguez, Vaidyanathan, & Barrera, 2003; Bai & Allaoui, 2003; Hu, Jang, & Sinnott, 2003; Hussain & Simon, 2003; Frankl, Harik, Odegard, Brenner, & Gates, 2003; Haque & Ramasetty, 2005; Karger-Kocsis & Zhang, 2004; Merkulov, Lowndes, Wei, Eres, & Voelkl, 2000; Endo et al., 2002; Endo et al., 2003; Finegan, Tibbetts, & Glasgow, 2003; Finegan, Tibbetts, & Gibson, 2003; Ma, Zeng, Realff, Kumar, & Schiraldi, 2003; Sandler et al., 2003; Valentini et al., 2004; Qian, Wagner, Liu, Yu, & Ruoff, 2002; Srivastava, Wei, & Cho, 2003; Veith et al., 2000; Veith et al., 2001; Voevodin & Zabinski, 2005; Meda et al., 2005; Bafna, Beaucage, & Mirabella, 2003; Bafna, Beaucage, Mirabella, & Mehta, 2002). The area of clay nanocomposites covered are the main features, current status, and recent developments which focuses on the preparation methods, structure, properties, and applications of these nanocomposites. It also presented the opportunities and potential uses of nanocomposites, along with perspectives for the future market and safety aspects.

11.2 Possibilities and prospects in nanocomposites

The overall opportunities and the potential of three types of nanocomposites systems deal with their processing, structure, properties, and applications. Ceramics have good wear resistance, brittle and high thermal and chemical stability, whereas it shows the low toughness of the uncertain block for their broader use in industry. Due to the significant improvement in mechanical properties, ceramic-matrix nanocomposites have been receiving attention in research and the need for further research to overcome this limitation. The ceramic matrix may lead to increased fracture toughness for its incorporation of energy-dissipating components such as whiskers, fibers, platelets, or particles (Lange, 1973; Becher, 1991; Harmer, Chan, & Miller, 1992, pp. 309–311). Moverover, the stress field of a transmitting crack, contributing to the toughening and strengthening processes combined phase transition in conjunction with the volume expansion, whereas reinforcements deflect the crack and/or provide bridging elements, hindering further opening of the crack (Awaji et al., 2002, p. 36).

It was revealed by the pioneering researcher Niihara that Al_2O_3/SiC system is the main component of ceramic matrix nanocomposites (CMNC) (Niihara, 1991). According to the literature, most studies reported that addition of a low (i.e., ~10%) volume fraction of SiC particles or suitable size confirmed the noticeable strengthening of the Al_2O_3 during hot pressing of the resulting mixture. The preparation of advanced nanocomposites with high toughness and superior failure characteristics due to the incorporation of high strength nanofibers into ceramic matrices and the toughening mechanism based on the crack-bridging role of the nanosized reinforcements. (Ferroni et al., 2001, p. 39).

Metal matrix nanocomposites (MMNC) is a composite which consists of a ductile metal or alloy matrix and implanted some nanosized reinforcement which combines metal and ceramic features and has a high strength and modulus with ductility and toughness. It has high shear/compression strength and high service temperature capabilities which are suitable for production of materials. It shows a surprising potential for various application areas, such as aerospace and automotive industries and structural materials (Tjong & Wang, 2004). Not only do CNT with CMNC and MMNC give us the assurance, but also pose challenges for real success.

Due to their lightweight and often ductile nature and ease of production, polymer materials are broadly used in industry. However, they have some disadvantages, such as low modulus and strength compared to metals and ceramics. To overcome the abovementioned disadvantages, fibers, whiskers, platelets, or particles are widely used as reinforcements in the polymer matrix to enhance mechanical properties. Polymers have been filled with several inorganic compounds, either synthetic or natural to manage the increment of heat and impact resistance, flame retardancy, and mechanical strength, and to decrease electrical conductivity and gas permeability with respect to oxygen and water vapor. An addition of inorganic nanoparticles into the metal and ceramic reinforcements offers striking routes to positive single magnetic, electronic, optical, or catalytic properties, such as processability and film forming capability (Athawale et al., 2003). Keeping their lightweight and ductile nature, polymers can be improved with nanoscale reinforcements which have an exceptional potential to generate new phenomena and lead to special properties (Akita & Hattori, 1999; Akita & Kobayashi, 1999; Chang & An, 2002; Jordan et al., 2005). For microcomposites, 40%–50% fibers are added as a reinforcement, which may be point out the reinforcing efficiency of these composites even at low volume fractions (Ray & Bousmina, 2005).

The dramatic improvement in their biodegradability is due to the addition of reinforcements to a wide variety of polymer resins and emphasizes a good example of polymer matrix nanocomposites [PMNC] as favorable systems for ecofriendly applications (Kickelbick, 2003). Further, future space mission concepts involve large, ultra-lightweight spacecraft which are require to possess and continue a specific combination of properties over a long period (10–30 years) in comparatively harsh environments such as 173–373 K for satellites and cycling temperatures of 1273 K for re-entry vehicles, exposure to atomic O_2, and solar radiation (Thompson et al., 2003). It is described that those available in conventional launch vehicles should possess many of the common mission concepts that these parts will have to be fabricated from flexible, appropriate materials and which can be folded or packaged into small volume, similarly the Gossamer spacecraft devices are movable mechanical parts such as gears and gyroscopes, and others including solar arrays/sails, antennae and drives, sunshields, rovers, radars, solar concentrators, and reflector arrays. The required orbit will be achieved if the structure consists of ultra-lightweight parts and could be deployed mechanically the large ultra-lightweight functioning spacecraft because the inflation is needed. It is imperative that the above mentioned characteristics should be available in one single material. Metal oxide-incorporated polymer nanocomposites fulfill these requirements and spacecrafts offer a significant cost advantage compared to on-orbit construction and the large size can enable some unique missions, whereas rocket propellants are prepared from a polymer-Al/Al_2O_3 nanocomposite to develop ballistic performance (Meda et al., 2005). This chapter discusses polymer nanocomposites including nanomaterials, nanoindustries, and a host of possible A to Z applications.

On the other hand, CNTs have not had their potential recognized as nanoscopic reinforcements in polymer matrices, but unlimited challenges and opportunities are still predictable for the system (Dresselhaus, Dresslhaus, & Eklund, 1996; Nalwa, 2000; Dresselhaus, Dresslhaus, & Avouris, 2001; Ajayan, Schadler, & Braun, 2003; Alpha Gary, 2003; Agnihotri, Rostam-Abadi, & Rood, 2004; Dresselhaus, Dresselhaus, Saito, & Jorio, 2005; Serp, Corrias, & Kalck, 2003; Penn, He, & Natan, 2003; Darkrim, Malbrunot, & Tartaglia, 2002; Kupfer, 2001).

These are based on the following:

(a) On the basis of equivalent volume fraction of a typical carbon fiber, high aspect ratio (~1000), very high tensile properties and electrical and thermal conductivities will be expected from CNTs with a small number of defects per unit length, which possess 500 times more surface area per gram. (Andrews & Weisenberger, 2004).

(b) From 2001 to June 2006, CNTs-related research has been most active with publications doubling within six months and about 3000 in patenting activity in this area has been impressive 335.

(c) C) The use of CNTs in composites is more challenging in obtaining homogeneous dispersions and strong interfacial interactions because of their hollow nature and CNTs can be opened and filled with a variety of materials, including biological molecules which generate technological opportunities. Added to this, the surface grafting/functionalization, making the use of CNTs in composites more interesting.

(d) Various applications of CNTs in composites have been reported extensively (Thostenson et al., 2005; Xie et al., 2005).

(e) In the electronic and thermal management sectors, the possibility of spinning polymers to obtain certain textiles constitutes which have great promise for their extended use in a variety of applications.

(f) The environmentally friendly "green materials" for future applications will be based on nano reinforcements with biodegradable polymers which have a high potential for the design, opportunities, and rewards to be great with nanocomposites and hence there is a tremendous worldwide interest in these materials.

11.3 Preparation of nanocomposites

11.3.1 Raw materials

For the microcomposites, Al_2O_3, SiC, SiN, etc. matrix materials were used to produce the CMNC. However, Al, Mg, Pb, Sn, W are employed to produce the metal matrices in MMNC, while PMNC is produced by use of the entire range of polymers, vinyl polymers, condensation polymers, polyolefins, speciality polymers which are biodegradable. The nanorange size in these materials are used for the reinforcement whereas Fe and other metal powders, clays, silica, TiO_2 and other metal oxides, clays and layered silicates have been used as synthetic and natural crystalline reinforcements (Fernando & Satyanarayana, 2005). The reinforcement materials improved properties even when they are used at very low concentrations due to their availability with very low particle sizes and well-known intercalation chemistry and most of these reinforcements are prepared by known techniques such as chemical, mechanical (e.-g., ball milling), vapor deposition, etc. (Alexandre & Dubois, 2000; Theng, 1974).

The preparation of CNTs is commonly done by chemical/vapor deposition methods and analysis of CNTs revealed that about 49% of the patents filed related to the processing of CNTs and about 14% to their structure, properties, and models (Belin & Epron, 2005; Dresselhaus et al., 2005). Graphene cylinders are the components which are the major part of CNTs and there are two varieties such as; single walled (SWCNT) and multi walled (MWCNT), with about 70% yield in the case of SWCNT, whereas SWCNTs are single graphene cylinders and MWCNTs contain two or more concentric cylindrical central hollow core sheets of graphene (Gary, 2003). The physical characteristics both types CNTs exhibit are metallic or semiconducting in nature, which was very high aspect ratios and micro crystalline, whereas improved interfacial bonding between the matrix and the nanosized reinforcements due to the surface modifications of reinforcements which reduce the agglomeration and improve the homogeneous distribution (Anya, 1999).

In situ polymerization is used for improving dispersion in the case of CNT-reinforced polymer composites, while alignment of CNTs could be achieved by techniques such as filtration, template and plasma-enhanced chemical vapor deposition, force field-inducements, whereas chemical methods may be more effective especially for polymer and ceramic matrices and application of the CNTs use for surfactants (Xie et al., 2005).

11.3.2 Preparation methods for ceramic nanocomposites

(a) Powder Process

Al_2O_3/SiC were used as raw materials to prepare the nanocomposites. The following steps need to be followed to prepare the nanocomposites:

(i) The raw materials were selected as a small average size powder, uniformity, and high purity.
(ii) Mixture was made by mixing wet ball milling or attrition milling techniques in organic or aqueous media.
(iii) Sample was dryed by using lamps or ovens or by freeze-drying.
(iv) Hot pressing or gas pressure sintering or slip casting or injection molding and pressure filtration techniques were used to consolidate the solid material.

There are sample advantages of this method. However, there were numerous limitations such as low formation rate, high temperature, agglomeration, poor phase dispersion, and formation of secondary phases in the product (Nakahira & Niihara, 1992).

(b) Polymer Precursor Process

Al_2O_3/SiC, SiN/SiC were used as raw materials to prepare the nanocomposites. Microwave oven was used to pyrolysis of mixing a Si-polymeric precursor to produce the reinforcing particles. The advantages of this method are opportunity to prepare finer particles and superior reinforcement dispersion. The limitations of this process are inhomogeneous and phase-segregated materials due to agglomeration and dispersion of ultra-fine particles (Riedel, Strecker, et al., 1989; Riedel, Seher, et al., 1989).

(c) Sol-Gel Process

SiO_2/Ni, ZnO/Co, TiO_2/Fe_2O_3, La_2O_3/TiO_2, Al_2O_3/SiC, TiO_2/Al_2O_3, Al_2O_3/SiO_2, $Al_2O_3/SiO_2/ZrO_2$, TiO_2/Fe_2TiO_5, $NdAlO_3/Al_2O_3$ were used as a raw materials to prepare the nanocomposites. Organic molecular precursor reactions dissolved in organic media, which is called the hydrolysis and polycondensation, whereas the formation of three-dimensional polymers containing metal-oxygen bonds (sol or gel) which is solid material and dried by thermal treatment (Livage, 1997; Vorotilov et al., 1999; Hench & West, 1990; Ennas et al., 1998; Sen et al., 2004; Viart et al., 2003; Kundu et al., 1998).

The advantages of this method are simple, low processing temperature; versatile; high chemical homogeneity; rigorous stoichiometry control; high purity products; formation of three dimensional polymers containing metal-oxygen bonds. Single or multiple matrices are applicable for the production of composite materials with liquids or with viscous fluids that are derived from alkoxides. The limitations of this process are greater shrinkage and lower amount of voids, compared to the mixing method.

(d) Hot pressing

SiO_2/CNT, SiC/CNT were used as raw materials to prepare the nanocomposites. SiO_2 glass powders were dispersed into CNTs in ethanol medium with stirring and ultrasonic treatment conducted to dry and hot pressure sintering in pure N_2 atmosphere. Mixing of nanoparticles of SiC and carbon nanotubes (Ning et al., 2003).

(e) CVD or Spray pyrolysis:

Al_2O_3/CNT were used as a raw materials to prepare the nanocomposites. The alumina matrix was used for the preparation of CNTs into its porous walls which show the anodizing growth where it grows into hexagonal array of straight pores extending from the substrate to the matrix surface (Ruhle, 2003; Xia, Riester, et al., 2004; Xia, Wang, et al., 2004).

(f) Catalytic decomposition

Al_2O_3/CNT were used as a raw materials to prepare the nanocomposites. In the presence of iron catalysts, Al_2O_3 powder was impregnated with acetylene (An et al., 2003)

(g) Solvothermal process

Fe_3O_4/CNT were used as a raw materials to prepare the nanocomposites. CNTs were dispersed in EDA (ethylenediamine) using ultrasonic treatment in the presence of iron (III)-urea complex and heating in a Teflon-lined autoclave maintained at 200°C for 50 hours which was followed by cooling to room temperature (Jiang & Gao, 2003).

11.3.3 Ceramic matrix nanocomposites (CMNC)

The preparation of ceramic matrix nanocomposites has been pronounced for various methods (An et al., 2003; Ananthakumar et al., 2004; Baiju et al., 2005; Borsa & Brook, 1995; Camargo et al., 2003; Camargo et al., 2004; Ennas et al., 1998; Ghosh & Pramanik, 2001; Hench & West, 1990; Jiang & Gao, 2003; Kundu et al., 1998; Livage, 1997; Mathur et al., 2002; Ning et al., 2003; Riedel et al., 1995; Riedel, Strecker, et al., 1989; Riedel, Seher, et al., 1989; Ruhle, 2003; Sen et al., 2004; Sivakumar et al., 2004; Stearns et al., 1992; Viart et al., 2003; Vorotilov et al., 1999; Warrier & Anilkumar, 2001; Wunderlich et al., 2004; Xia, Riester, et al., 2004; Xia, Wang, et al., 2004). There are many common methods used for microcomposites, namely conventional powder method, Polymer precursor route, Spray pyrolysis, Vapor techniques (CVD and PVD), and chemical methods, which include the sol-gel process, colloidal and precipitation approaches, and the template synthesis. There are some advantages and limitations for the abovementioned method, but the polymer precursor route used in the synthesis of an Al_2O_3/SiC nanocomposite.

The sol-gel process is affected by various type of parameters such as type of solvent, timing, pH, precursor, water/metal ratio, etc., which allows a versatile control of structural and chemical properties of the final oxide materials (Ennas et al., 1998). Many methodologies have been pronounced regarding the processing of carbon nanotubes (CNT)-reinforced ceramic nanocomposites (An et al., 2003; Ananthakumar et al., 2004; Baiju et al., 2005; Borsa & Brook, 1995; Camargo et al., 2003; Camargo et al., 2004; Ennas et al., 1998; Ghosh & Pramanik, 2001; Hench & West, 1990; Jiang & Gao, 2003; Kundu et al., 1998; Livage, 1997; Mathur et al.,

2002; Ning et al., 2003; Riedel et al., 1995; Riedel, Strecker, et al., 1989; Riedel, Seher, et al., 1989; Ruhle, 2003; Sen et al., 2004; Sivakumar et al., 2004; Stearns et al., 1992; Viart et al., 2003; Vorotilov et al., 1999; Warrier & Anilkumar, 2001; Wunderlich et al., 2004; Xia, Riester, et al., 2004; Xia, Wang, et al., 2004).

11.3.4 Metal matrix nanocomposites (MMNC)

Spray pyrolysis; liquid metal infiltration; rapid solidification; vapor techniques (PVD, CVD); electrodeposition and chemical methods are the most common techniques for the processing of metal matrix nanocomposites (Baker et al., 2004; Yoon et al., 2002; Provenzano et al., 1992; Contreras et al., 2004; Bhattacharya & Chattopadhyay, 2001; Bhattacharya & Chattopadhyay, 2004; Srinivasan & Chattopadhyay, 2004; Branagan, 2000; Branagan & Tang, 2002; Xiaochun et al., 2004; Ying & Zhang, 2000; Choy, 2003; Joseph et al., 2005; Chow et al., 1990; Haubold & Gertsman, 1992; Holtzt & Provenzano, 1994; Cushing et al., 2004; West et al., 2003; Kamat et al., 2002; Carpenter et al., 2000; Chen et al., 2000; Chen, Wong, et al., 2002; Chen, Lee, et al., 2002; Chen, Zhang, et al., 2003; Chen, Lee, et al., 2003; Xu et al., 1999; Noguchi et al., 2004; Kuzumaki et al., 1998; Yang & Schaller, 2004). There are advantages and limitations of the mentioned methods, such as Fe-based nanocomposites prepared by solidification techniques. Steel nanocomposite was obtained by quenching and the metallic glass obtained from a Fe-based alloy which followed by devitrifying the glass precursor through heat treatment above its crystallization temperature where it showed a crystalline multiphase microstructure. According to the Branagan and Tang, novel nanostructures were obtained in bulk Fe alloys by designing alloy compositions with different amounts of W and C to get maximum solubility and it shows the very high hardness of these nanocomposites, whereas the formation of nanophases was the high nucleation frequency within the limited time for growth of grains before impingement. However, due to the induced agglomeration and nonhomogeneous distribution have been encountered in preparing composites with very fine particles, while the wettability between the matrix and the particles improved by ultrasound.

11.3.5 Preparation methods for metal matrix nanocomposites

(a) Spray Pyrolysis

Fe/MgO, W/Cu were used as raw materials to prepare the nanocomposites. The following steps need to be followed to prepare the nanocomposites:

i. Prepared the suitable solvent to get the liquid source dissolution of the inorganic precursors (starting materials)
ii. Generation of a mist from this liquid source using an ultrasonic atomizer
iii. Use of a carrier gas to carry the mist into a preheated chamber
iv. Promoting their decomposition to give the respective oxide materials with vaporization of the droplets in the chamber and trapping with a filter.

v. The metal oxides to produce the corresponding metallic materials along with selective reduction

The advantages of this method are the effective preparation of ultra-fine, spherical, and homogeneous powders in multicomponent systems, reproductive size and quality, while the limitation of this method is high cost associated with producing large quantities of uniform, nanosized particles. (Choa et al., 2003).

(b) Liquid Infiltration

Pb/Cu, Pb/Fe, W/Cu/Nb/Cu, Nb/Fe, Al-C60 were used as a raw materials to prepare the nanocomposites. The following steps need to be followed to prepare the nanocomposites:

i. Introduce the fine reinforcement particles with the matrix metal material.
ii. The liquid infiltration was conducted by thermal treatment which influences reinforcements in the matrix melts.
iii. The internal porosity was associated and eliminated by the thermal treatment, which is below the matrix melting point. The advantage of this method is short contact times between matrix and reinforcements; molding into different and near net shapes of different stiffness and enhanced wear resistance; rapid solidification; both lab scale and industrial scale production. Limitations of this method are use of high temperature; segregation of reinforcements; formation of undesired products during processing. (Choa et al., 2003; Yoon et al., 2002).

(c) Rapid Solidification Process (RSP)

Al/Pb, Al/X/Zr (X = Si, Cu, Ni), Fe alloy were used as raw materials to prepare the nanocomposites. The following steps need to be followed to prepare the nanocomposites:

i. The metal components melting together
ii. To ensure homogeneity, the miscibility gap between the different components is keeping the melt above the critical line.
iii. Melt spinning process used to rapid solidification of the melt

The advantages of this method are simple and effective, while the limitations of this method are only metal-metal nanocomposites, which induced agglomeration and non-homogeneous distribution of fine particles. (Bhattacharya & Chattopadhyay, 2001; Bhattacharya & Chattopadhyay, 2004; Srinivasan & Chattopadhyay, 2004)

(d) RSP with ultrasonics

Al/SiC were used as raw materials to prepare the nanocomposites. To improve wettability between the matrix and the reinforcements applied the ultrasonics mixing and the advantage of this method is good distribution without agglomeration, even with fine particles.

(e) High Energy/Ball Milling

Cu-Al_2O_3 were used as raw materials to prepare the nanocomposites. Ball milling process is applied until the required nanosized alloy is obtained to produce nanocomposite. The advantages of this method are homogeneous mixing and uniform distribution (Ying & Zhang, 2000).

(f) CVD/PVD

Al/Mo, Cu/W, Cu/Pb were used as raw materials to prepare the nanocomposites. The following steps need to be followed to prepare the nanocomposites:

PVD:

i. The vapor-phase is produced to sputtering/evaporation of different components.
ii. To promote the condensation of metal nanoparticles by supersaturation of the vapor phase in an inert atmosphere.
iii. Consolidation of the nanocomposite by thermal treatment under inert atmosphere

CVD: Use of chemical reactions to get vapors of materials, followed by consolidation

The advantage of this method is to have the capability to produce highly dense and pure materials, uniform thick films, adhesion at high deposition rates, and good reproducibility. Whereas the limitation of this method is optimization of many parameters, such as cost and relative complexity (Choy, 2003; Joseph et al., 2005).

(g) Chemical Processes (Sol-gel, Colloidal)

Ag/Au, Fe/SiO_2, Au/Fe/Au were used as raw materials to prepare the nanocomposites. The following steps need to be followed to prepare the nanocomposites:

Colloidal Method:

i. To synthesize, metal particles are used to cause chemical reduction of inorganic salts in solution.
ii. This process needs to use the consolidation of the dry material.
iii. In order to promote selective oxide reduction and generate the metal component, drying and thermal treatment are needed to solidify in reducing atmosphere.

Sol gel process:

i. Preparation of two micelle solutions using mesoporous silica containing 0.1 M $HAuCl_4$ (aq.) and 0.6 M $NaBH_4$ (aq.);
ii. Mixing under ultraviolet light until complete reduction of the gold.

For Fe/Au-containing nanocomposites the following steps need to be followed to prepare the nanocomposites:

i. Synthesis of the iron shell.
ii. Preparation of the second shell and drying of the powders after second gold coating.
iii. Pressing of the mixture to get the final material.

The advantages of this method are simple, low processing temperature, versatile, high chemical homogeneity, rigorous stoichiometry control, and high purity products, while limitations of this method are weak bonding, low wear-resistance, high permeability, and difficult control of porosity (Cushing et al., 2004; Kamat et al., 2002; West et al., 2003).

(h) Electroless Coating

Co-CNT were used as raw materials to prepare the nanocomposites. The following steps need to be followed to prepare the nanocomposites:

i. Electrolysis plating bath used to contain the activated CNTs with the cobalt precursor and the reducing agent $CoSO_4{\cdot}7H_2O$, the complexing agent and a buffer are used in this system. CNT with deposit of Co coating results Co-CNT.
ii. Thermal treatment at 873°K, 200 Torr, under a 10% H_2/N_2 flow gas.

(i) Electroless Coating

Sn/CNTs, SnSb0.5/CNT and Sn_2Sb/CNT were used as raw materials to prepare the nanocomposites. Reduction of $SnCl_2$ and $SbCl_3$ precursors by KBH_4 in the presence of CNTs.

(j) Hot Pressing

Al/CNT was used as a raw material to prepare the nanocomposites. The nanocomposites are prepared by mixing of powders through grinding for 30 minutes and hot pressing at 793°K under a pressure of 25 MPa

(k) Nanoscale Dispersion

Al/CNT was used as a raw material to prepare the nanocomposites. This was done by grounding of the precursor of MWCNT (13 nm dia and 10–50 μm long) with natural rubber and ethyl propylene; mixing with Al powder and rolling into sheets by compression molding at 353°K and placing of this precursor on an Al (99.85%) plate of 28 μm grain size and heating to 1073°K in N_2 atmosphere for 1 hour and finally cooling.

(l) PM/Infiltration

Mg-Al_2O_3-CNT was used as a raw material to prepare the nanocomposites. The following steps need to be followed to prepare the nanocomposites:

i. In a medium of alcohol and acid mechanical mixing of Mg powders with MWCNT (1 vol.%) sinterization at 550°C under 25 MPa pressure.
ii. Infiltration of molten Mg through performs of Al_2O_3 fibers (25 vol.%; 40–100 μm long) covered with MWCNTs under gas pressure.

11.3.6 Polymer matrix nanocomposites (PMNC)

There are numerous methods that have been designated for the preparation of polymer nanocomposites which includes layered materials and those containing CNTs (Schadler et al., 1998; Jimenez et al., 1997; Ogata et al., 1997; Jeon et al., 1998; Aranda & Ruiz-Hitzky, 1992; Greenland, 1963; Francis, 1973; Zhao et al., 1989; Usuki, Kojima, et al., 1993; Usuki, Kawasumi, et al., 1993; Okamoto et al., 2000; Okamoto et al., 2001; Yao et al., 2002; Messersmith & Giannelis, 1994; Vaia & Giannelis, 1997a, 1997b; Gilmann, 1999; Vaia et al., 1995; Kawasumi et al.,

1997; Tomasko et al., 2003; Watkins & Mccarthy, 1994; Watkins & Mccarthy, 1995a; Watkins & Mccarthy, 1995a; Watkins & Mccarthy, 1995a; Carrado & Xu, 1998; Fernando & Satyanarayana, 2005; Park et al., 2005; Mbhele et al., 2003; Aymonier et al., 2003; Evora & Shukla, 2003; Di Lorenzo et al., 2002; Park, Bernet, et al., 2003; Park, Cho, et al., 2003; Xu et al., 1998; Liu et al., 2002; Jackson et al., 1996; Avadhani & Chujo, 1997; Sandler et al., 1999; Qian et al., 2000; Ding et al., 2003; Lin et al., 2003; Wong et al., 2003; Koerner et al., 2004; Kubayashi & Hayashi, 1992; Tang et al., 2003; Andrews et al., 2002; Maser et al., 2003; Park et al., 2002; Philip et al., 2005; Azioune et al., 1999; Xia, Riester, et al., 2004; Xia, Wang, et al., 2004; Kim et al., 2003; Valentini et al., 2003; Bharadwaj, 2001; Kojima et al., 1993a; Kojima, et al., 1993; Bourbigot et al., 2000; Okada et al., 1990; Dabrowskii et al., 2000; Lee et al., 1998; Fredrickson & Bicerano, 1999; Kuznetsov & Balazs, 2000; Manias et al., 2000; Herron & Thorn, 1998; Favier et al., 1997; Chazeau et al., 1999; Ogawa & Kuroda, 1997a, 1997b; Chen et al., 1997; Lu & Huang, 2002; Liu et al., 2003; Trindade & O'Brien, 1996; Trindade et al., 2000; Vaia & Wagner, 2004; Curtin & Sheldon, 2004; Thostenson & Chou, 2002; Thostenson & Chou, 2003; Tai et al., 2004; Gojny et al., 2004; Ogasawara et al., 2004; Giannelis et al., 1999; LeBaron et al., 1999; Vaia et al., 1999; Biswas & Ray, 2001; Giannelis, 1998; Xu et al., 2001; Yano et al., 1993; Gilman et al., 2000; Hori et al., 1999; Blumstein, 1965; Theng, 1979; Vaia et al., 1993; Passaglia et al., 2008; Krishnamoorti et al., 1996; Hasegawa et al., 1998; Lee & Jang, 1996; Noh & Lee, 1999; Wang & Pinnavaia, 1998a; Wang & Pinnavaia, 1998b; Yano et al., 1997; Liu et al., 1999; Fornes et al., 2001). The most popular methods are namely:

i. Intercalation of the polymer or prepolymer from solution
ii. In situ intercalative polymerization
iii. Melt intercalation
iv. Direct mixture of polymer and particulates
v. Template synthesis
vi. In situ polymerization and
vii. Sol-gel process.

The most prominent one is probably the incorporation of inorganic building blocks in organic polymers, whereas various methods for the incorporation of nanodispersoids into conducting polymers are also available (Azioune et al., 1999; Ogasawara et al., 2004).

Scheme 11.1 is the intercalative processes which is employed for the preparation of polymer-based nanocomposites, counting those comprising layered silicates and this method's range of nanocomposites with structures from intercalated to exfoliated, which depends on the degree of penetration of the polymer chains into the silicate galleries. As a result, this procedure has become standard for the preparation of polymer-layered silicate combinations.

The various methods, namely direct mixing, solution mixing, melt-mixing and in situ polymerization are used to prepare of CNT-reinforced polymer nanocomposites,

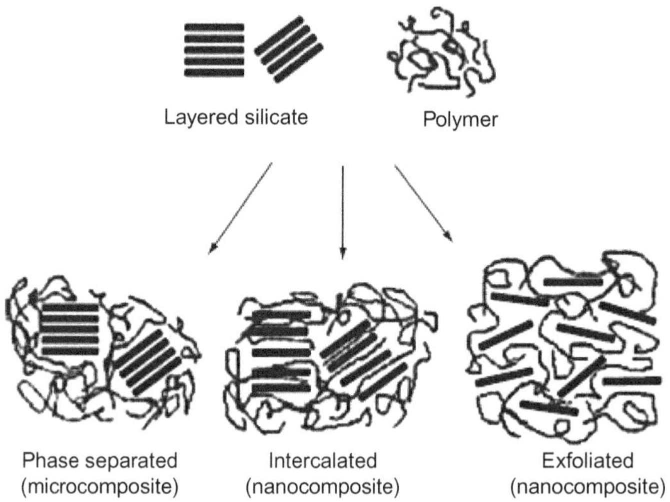

Scheme 11.1 Propose interaction between Polymer and Layered silicate.

while different processing techniques, mostly chemical and electrochemical methods, have been applied for the preparation of conducting polymer nanocomposites (Gangopadhyay & Amitabha, 2000). If the reinforcements contain layered silicate, organic cation, and polymer matrix, then the method of preparation and the strength of interfacial interactions between the polymer matrix and the layered silicate which are obtained are three different types of PLS nanocomposites.

A phase-separated composite is obtained if the polymer is unable to intercalate between the silicate sheets, whereas the properties stay in the same range as that of traditional microcomposites. However, due to the clay to polymer ratio the insertion of a polymer matrix into the layered silicate structure occurs in a crystallographically regular fashion, which is called the intercalated nanocomposites and it will build the multilayer morphology, which is built up with alternating polymeric and inorganic layers in the composites system. Normally, only a few molecular layers of polymer can be intercalated in these materials.

The in situ method can be used with mineral/vegetal fibers which improves the mechanical properties with the possibility of attaching the polymer to the grafted surface through coupling agents which optimize the interface bonding (Fernando & Satyanarayana, 2005) Various methods are successfully used for preparation of polymer-based nanocomposites, which give information on various factors still lacking, such as:

i) The use of an appropriate method for a specific matrix-reinforcement combination.
ii) The maximum amount of reinforcements to give optimum property combinations and lower the cost of the processes, etc.

Therefore, it is still necessary to look into these aspects, including use of simulation and modeling techniques.

11.3.7 Preparation methods for polymer-based nanocomposite systems

(a) Intercalation/Prepolymer from Solution

Clay with PCL, PLA, HDPE, PEO, PVA, PVP, PVA, etc. were used as raw materials to prepare the nanocomposites. Introduction of the layered reinforcing material in which the polymer may intercalate while the layered silicates produce the solution from intercalation of the polymer or prepolymer. The solvent that was used in the system in which the polymer or prepolymer is soluble and the silicate layers are swellable.

The advantages of this method is synthesis of intercalated nanocomposites based on polymers with low, or even no polarity, and preparation of homogeneous dispersions of the filler. Limitation of this method is industrial use of large amounts of solvents (Jimenez et al., 1997; Ogata et al., 1997).

(b) In situ Intercalative Polymerization

Montmorillonate with N6/PCL/PMMA/PU/Epoxy were used as raw materials to prepare the nanocomposites. The formation of polymer between the intercalated sheets with encasing of the layered silicate within the liquid monomer or a monomer solution that are polymerization by heat or radiation or by the diffusion of a suitable initiator or a catalyst fixed through cation exchange inside the interlayer. The advantage of this method is the easy procedure, based on the dispersion of the filler in the polymer precursors, while the limitation of this method is difficult control of intragallery polymerization (Usuki, Kojima, et al., 1993; Usuki, Kawasumi, et al., 1993).

(c) Melt Intercalation

Montmorillonate with PS/PEO/PP/PVP, Clay-PVPH were used as raw materials to prepare the nanocomposites. The result is the layered host above the softening point of the polymer which is statically or under shear by annealing of a mixture of the polymer, whereas the polymer chains from the bulk polymer melt into the galleries between the host layers during annealing. The advantage of this method is that it is environmentally gentle and compatible with industrial polymer processes, while the limitation of this method is limited applications to polyolefins which represent the majority of polymers used (An et al., 2003; Ananthakumar et al., 2004; Baiju et al., 2005; Bajwa et al., 2005; Balázsi et al., 2003; Camargo et al., 2003; Camargo et al., 2004; Cha et al., 2005; Chaisan et al., 2009; Chen, Zhang, et al., 2003; Chen, Lee, et al., 2003; Chung, 2004; Ghosh & Pramanik, 2001; Goujon & Goeuriot, 2001; Jiang & Gao, 2003; Kundu et al., 1998; Laurent et al., 1998; Lee et al., 2005; Li et al., 2003; Lim et al., 2002; Lim et al., 2005; Mathur et al., 2002; Ning et al., 2003; Peigney et al., 2002; Ruhle, 2003; Sawaguchi et al., 1991; Sealy, 2004; Siegel et al., 2001; Sinnott et al., 1999; Sivakumar et al., 2004; Tai et al., 2003; Wang, Guo, et al., 2004; Wang, Padture, et al., 2004; Warrier & Anilkumar, 2001; Wunderlich et al., 2004; Xia, Riester, et al., 2004; Xia, Wang, et al., 2004).

(d) Template Synthesis

Hectorite with PVPR, HPMC, PAN, PDDA, PANI were used as raw materials to prepare the nanocomposites. This was done by putting the inorganic material in an aqueous solution containing layered structure of the polymer by in situ formation and water soluble polymer acts as a template for the formation of layers. However, the broad range of the synthesis of LDH nanocomposites is less developed for layered silicates. The advantages of this method is large scale production and an easy procedure, while the limited applications are based on water soluble polymers which are contaminated by side products (Tomasko et al., 2003; Watkins & Mccarthy, 1994).

(e) Mixing polymerization

PVA)/Ag; PMMA/Pd Polyester/TiO_2 were used to prepare the nanocomposites. This was done by mixing of either polymer or monomer with reinforcing materials.

(f) In situ polymerization

PET/$CaCO_3$, Epoxy vinyl ester/Fe_3O_4; Epoxy vinyl ester/γ-Fe_2O_3; Poly (acrylic acid) (PAA)/Ag, PAA/Ni and PAA/Cu, $AgNO_3$, $NiSO_4$ and $CuSO_4$ were used to prepare the nanocomposites. Introduction of inorganic particles into a precursor of the polymeric matrix (monomer) takes place in the polymerization of the mixture by the addition of an appropriate catalyst and processing of this material by conventional molding technologies and using ultrasonic dispersion in epoxy systems (Aymonier et al., 2003; Mbhele et al., 2003).

(g) Sol-Gel Process

Polyimide/SiO_2; 2-hydroxyethyl acrylate (HEA)/SiO_2, polyimide/silica. PMMA/ SiO_2, polyethylacrylate/ SiO_2, polycarbonate/SiO_2 and poly (amide-imide)/TiO_2 were used to prepare the nanocomposites. Introduction of organic groups by formation of chemical bonds resulted in in situ formation of sol-gel matrix within the polymer and/ or simultaneous generation of inorganic/organic networks with inserting of organic molecules and monomers on sol-gel matrices (Jackson et al., 1996; Liu et al., 2002).

(h) Direct Mixing

Thermoset Resins were dispersion into CNTs and cure to produce the nanocomposites.

(i) Solution Mixing

Thermoplastic Resins (PS/Epoxy) were dispersion of 0.2-1% CNTs, (100 nm dia, 10 μm long) and cure to produce the nanocomposites. After polymeration, the solvent was removed, then precipitation of polymer cured for the final product (Ding et al., 2003; Qian et al., 2000). The remarkable changes of the properties are modification of polymer behavior, synergistic effect, and shape memory nanocomposites.

(j) Melt Mixing

Polymers, N6 were used to produce nanocomposites. This occurred by mechanical mixing of CNTs with prepolymer melt followed by extrusion, injection, or compression molding. Use of 0.2%–2.0% MWCNT, twin screw mixer (Andrews et al., 2002; Tang et al., 2003).

(k) In situ Polymerization

Polyaniline-CNT, MMA-CNT, Epoxy-CNT, Poly(ether-ester) were used to produce nanocomposites.

The dispersion in monomer/matrix which is used of ultrasonics and prepare the polymer with CNT which show the good chemical bonding (Maser et al., 2003; Park, Bernet, et al., 2003; Park, Cho, et al., 2003).

11.3.8 Conducting polymer nanocomposites preparation

Table 11.1 shows the classification of conducting polymer nanocomposites (inorganic-in-organic and organic-in-inorganic) (reproduced from Gangopadhyay & Amitabha, 2000 with the permission of the authors and the American Chemical Society, United States).

Table 11.1 Inorganic particle, significant characterization and it applications

Polymer of interest (Shell)	Inorganic particle (core)	Significant characterization/ applications
PPy and PAn	SiO2 (1 μm, 35 nm, 20 nm), SnO2-Sb (10 nm), Stringy SiO2 (40–300 nm long)	Stable colloidal form, "raspberry morphology" and inorganic stable rich surface
PPy and PAn	CeO2 (0.52 μm), CuO (1.6 μm), α-Fe2O3 (Sph, Polyhedral and spindle shaped), NiO (3.8 μm), SiO2 (0.46 μm)	Colloidally stable nanocomposite with low dc conductivity and formed without a polymerization initiator
PPy and PAn	BaSO4(20 nm), Colloidal gold (7–9 nm), Al2O3 membrane	In situ formation of colloidal nanocomposite within the microemulsion or inside the Al2O3 membrane
Ppy, Pan, NVC and PPV	ZrO2 (20–30 nm), Fe2O3 (25–50 nm), SiO2, n-TiO2 (~10 nm), Al2O3 (35–40 nm), MgO (2–4 μm), CB	Nanocomposites in macroscopic precipitate form or with limited colloidal stability but improved thermal and electrical properties and novel transport properties.
PPy and PAOABSA	MS (15–30 nm), *FexOy* (14 nm), Fe2O3 (~15–50 nm), γ-Fe2O3 (85 nm)	Nanocomposites with significant magnetic susceptibility
PPy and PAn	BT (~1 μm), LiMnO2, LiMnO4, V2O5, β-MnO2, PMo12, H3PMo12O40, CB, Fe2O3 (4 nm, 40 nm)	Nanocomposites with important charge storage and dielectric properties, suitable for cathode applications

Continued

Table 11.1 **Continued**

Polymer of interest (Shell)	Inorganic particle (core)	Significant characterization/ applications
Ppy, Pan, PTh, and PEDOT	Pt (~4 nm), PtO2, Pt, Cu, Pd, SiO2 (20 nm), and bimetallic couples	Nanocomposites with catalytically important metals; catalytic applications
PPy and PAn	SiO2 (20 nm)	Grafted surface nanocomposites - important for immunodiagnostic assays
PPy	SiO2, PB, MnO2, Ta2O5, TiO2	Electrochemically synthesized composite films with improved charge storage proper

Table 11.2 **The polymer-matrix nanocomposites and their properties (Evora & Shukla, 2003; Liu et al., 2002)**

Matrix/reinforcement	Properties
Polypropylene/ montmorillonite	Tensile strength, tensile stress, Young's modulus, strain at break and stiffness were improved
Nylon-6/layered-silicates	Tensile stresses, storage modulus, tensile modulus, HDT were improved and reduced the flammability
Polylactide/layered-silicates	Bending strength, bending modulus, distortion at break, storage modulus, gas barrier properties and biodegradability were improved
Epoxy/layered-silicates	Tensile strength and modulus were improved
Polyimide/ montmorillonite	Tensile strength, elongation at break and gas barrier properties were improved
Polystyrene/layered-silicates	Tensile stress and reduced flammability were improved
Polyethylene oxide/ layered-silicates	Ionic conductivity was improved
Poly(methyl methacrylate)/Pd	Thermal stability was improved
Polyester/TiO2	Tensile strength and fracture toughness were improved
Epoxy/SiC	Storage modulus, micro hardness and elastic moduli were improved
Polypropylene/ montmorillonite	Tensile strength, strain at break, stiffness, Young's modulus and tensile stress were improved

11.3.9 Polymer matrix nanocomposites

Table 11.2 shows structure and property correlations in polymer nanocomposites, which describes the mechanical properties of polymers based on nanostructure and morphology.

The production of polymer nanocomposites is classified according to their dimensions based on the reinforcing materials, whereas the three dimensions are in the nanometer scale and they are called dimensional nanoparticles, and include spherical silica, metal particles, and semiconductor nanoclusters 18, 219. Based on the reinforcement materials, nanotubes or whiskers contain two dimensions in the nanometer scale and one larger which forms an elongated structure such as carbon nanotubes and cellulose whiskers, extensively studied as reinforcing nanofillers that is the second category reinforcement. Another type of reinforcement is the third type of reinforcement which is characterized by only one dimension in the nanometer range that the filler contains sheets one to a few nanometers thick and hundreds to thousands of nanometers long and are known as polymer-layered nanocomposites (Favier et al., 1997; Ogawa & Kuroda, 1997a, 1997b). The intercalation of the polymer technique (or a monomer subsequently polymerized) used to prepare these materials which are obtained inside the galleries of the layered host, whereas many synthetic and natural crystalline hosts that are able, under specific conditions, to intercalate a polymer such as graphite, metal chalcogenides, clays, layered silicate (montmorillonite, hectorite, saponite, fluoromica, fluorohectorite, vermiculite and kaolinite), and layered double hydroxides. The well-known intercalation chemistry among the clay and polymer has been widely investigated due to the availability of clay starting materials based on clay and layered silicates (Chen et al., 1997).

11.3.10 Physical properties of nanocomposites

11.3.10.1 Fourier transform infrared spectroscopy (FTIR) analysis

FTIR analysis was conducted to analyze the chemical bonding formed between the nanofillers and the polymer matrixs. There will be no significant changes in the FTIR testing if the materials are not homogenously mixed (Qu, Gao, Wu, & Zhang, 2010). Fig. 11.1 shows the FTIR spectra of pure polypropylene. The C—H region of both pure polyethylene and polypropylene are slightly different as according to PerkinElmer (2011), by focusing on the C—H stretch region around 2950 cm^{-1}, the difference between polyethylene and polypropylene can be observed due to the differing ratios of both CH_2 and CH_3 groups. Thus, the pure polypropylene showed the peak intensity at 2846.93 cm^{-1} due to C—H stretching at 3647.39 cm^{-1} and —OH stretching at 3392.79 cm^{-1}. Nanofiller like nanoclay was introduced to improve the water uptake resistance of the polymer matrix. Hence, the analysis of —OH region for different filler contents of nanoclay/polypropylene nanocomposites was made in order to analyze the improvement of the water uptake resistance.

The different percentages of nanoclay loading to pure polypropylene are shown in Figs. 11.2–11.5, respectively. All of these four figures share the almost similar infrared spectrum, while hydroxyl functional group shows the different due to the clay content. As compared to the peak intensity of —OH stretching of pure polypropylene, 0.5 and 1.0 wt%, nanoclay/polypropylene nanocomposite showed better water uptake resistance, whereas 0.5 wt% nanoclay/polypropylene nanocomposite showed peak intensity at 3749.62 and 3566.38 cm^{-1} due to —OH stretching which is shown in Fig. 11.2. However, 1.0 wt% nanoclay/polypropylene showed the peak intensity at

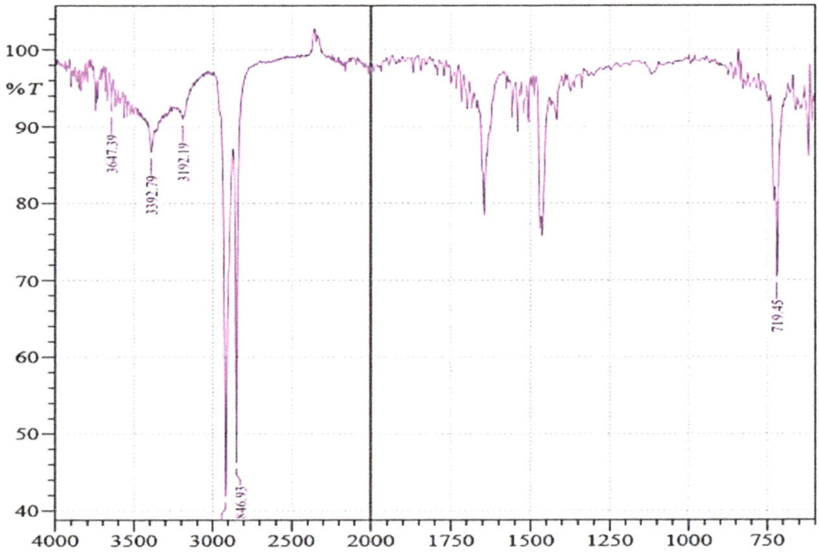

Fig. 11.1 FTIR spectra of pure polypropylene.

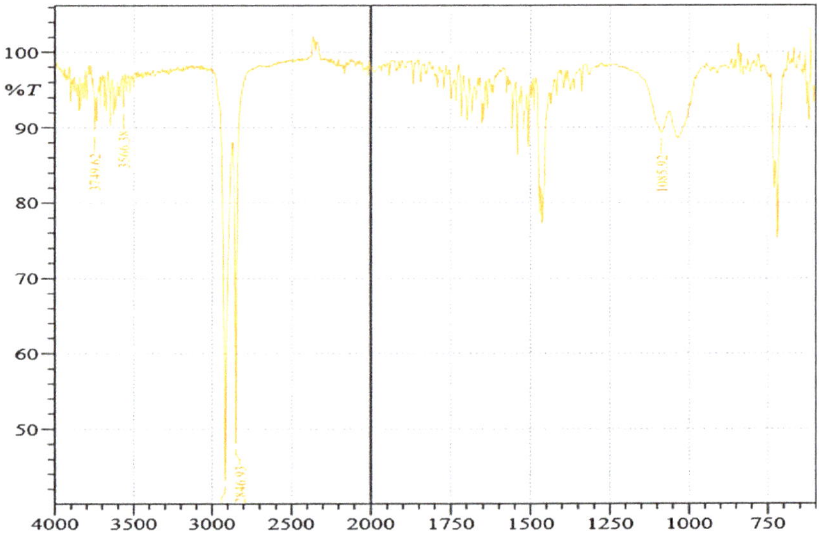

Fig. 11.2 FTIR spectra of 0.5 wt% nanoclay/polypropylene nanocomposite.

3734.19 and 3647.39 cm^{-1} due to —OH stretching which shown in Fig. 11.3. The peak intensity was further decreasing at —OH stretching which was observed from 1.5wt% nanoclay/polypropylene nanocomposites. The peak intensity of -OH stretching at 3749.62 and 3647.39 cm^{-1} as shown in Fig. 11.4. Based on Fig. 11.5, the peak intensity at 3766.98 and 3670.54 cm^{-1} was observed at a low frequency, indicating the presence of the 2.0 wt% nanocaly. The decreasing of C—H absorption peak

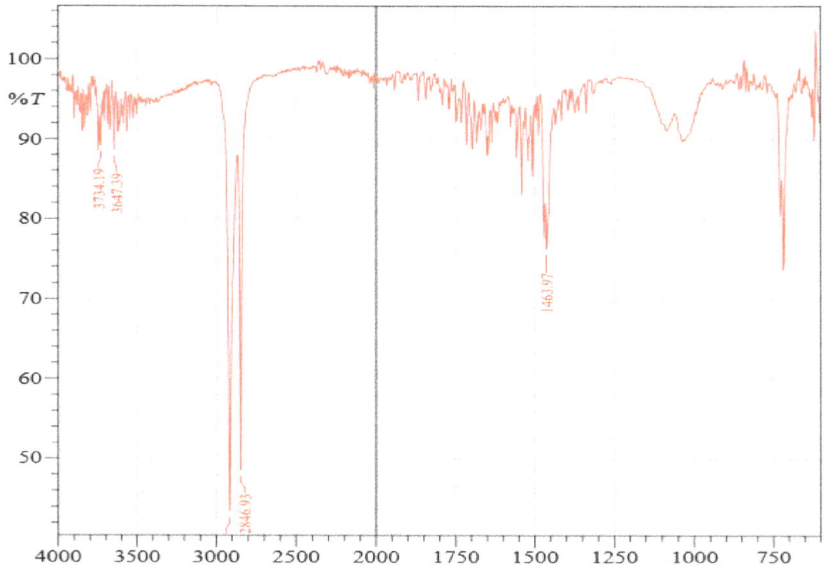

Fig. 11.3 FTIR spectra of 1.0 wt% nanoclay/polypropylene nanocomposite.

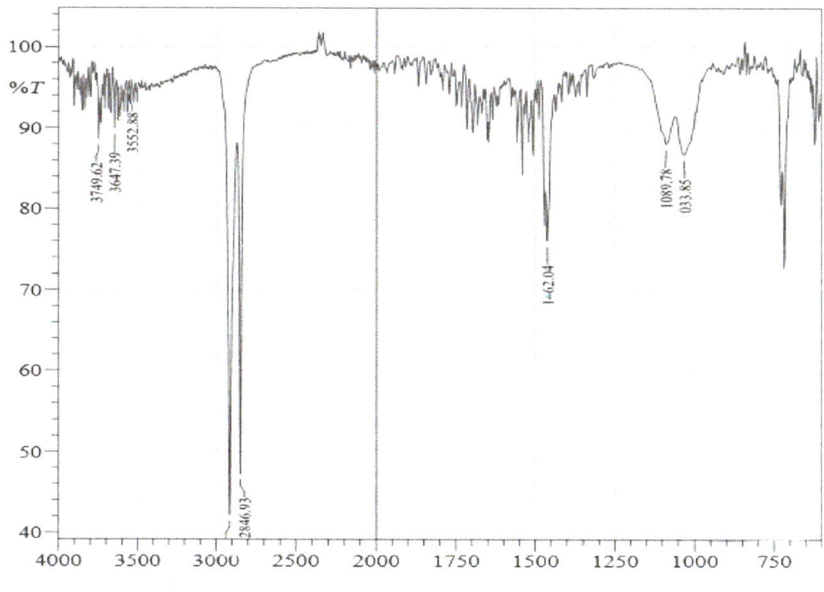

Fig. 11.4 FTIR spectra of 1.5 wt% nanoclay/polypropylene nanocomposite.

at the region of 2800–3100 cm^{-1} were detected when nanoclay was introduced to the polypropylene, which is due to nanoclay promoting the function of impeding motion of CH_2 molecular stretch (Lei & Cebe, 2009). According to that fact, the improvement of water uptake resistance exhibited by 2.0 wt% nanoclay/polypropylene nanocomposite. According to Lei et al., nanoclay based nanocomposites in which

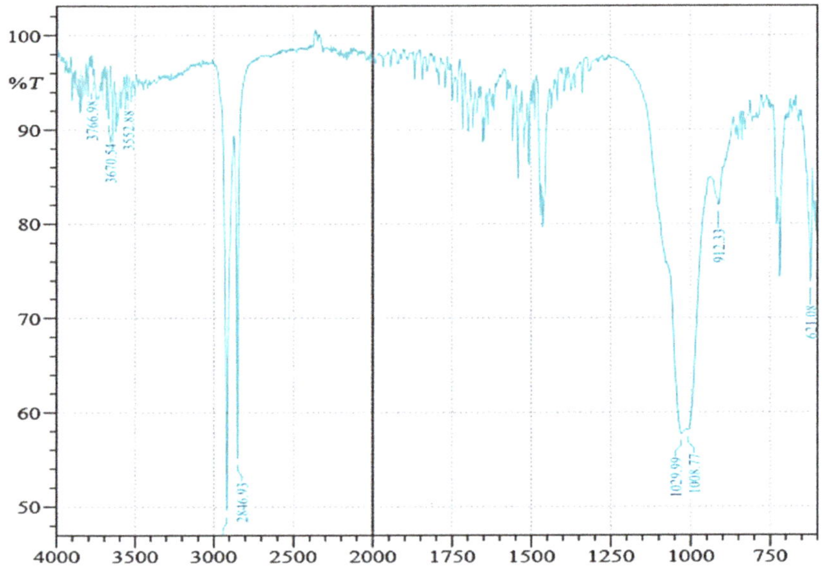

Fig. 11.5 FTIR spectra of 2.0 wt% nanoclay/polypropylene nanocomposite.

nanoclay is used as the reinforcement agent for the composites, exhibit drastic increasing in modulus, tensile, and flexural strength, barrier properties, flammability resistance, weathering resistance, water uptake resistance, and also the heat resistance, as compared with other traditional composites.

11.3.11 Mechanical properties of nanocomposites

11.3.11.1 Tensile properties of nanoclay reinforced nanocomposites

The calculated tensile strength and modulus of elasticity for both silica-based nanocomposites are tabulated in Table 11.3. Fig. 11.6 shows the tensile strength for different percentage loading of silica to low-density polyethylene, while

Table 11.3 Tensile strength and modulus of elasticity of nanoclay/polypropylene nanocomposites

Composite	Silica loading (wt%)	Tensile Strength (MPa)	Modulus of elasticity (GPa)
Nanoclay-based composite	0.5	46.2	2.11
	1.0	54.4	3.08
	1.5	65.3	3.96
	2.0	72.8	4.63

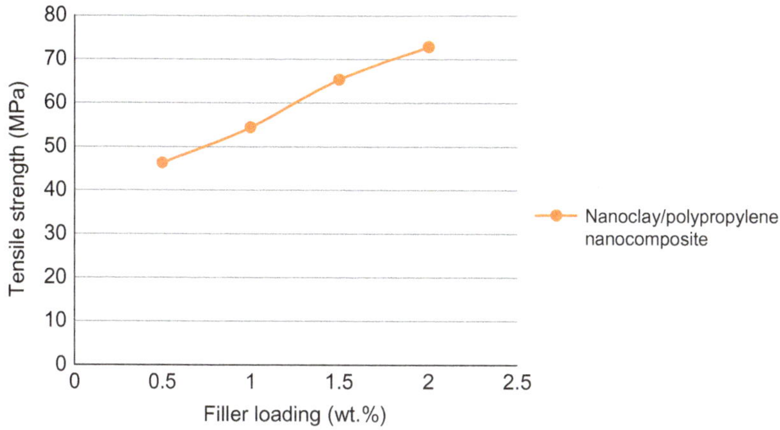

Fig. 11.6 Tensile strength of different percentage loading of nanoclay/polypropylene.

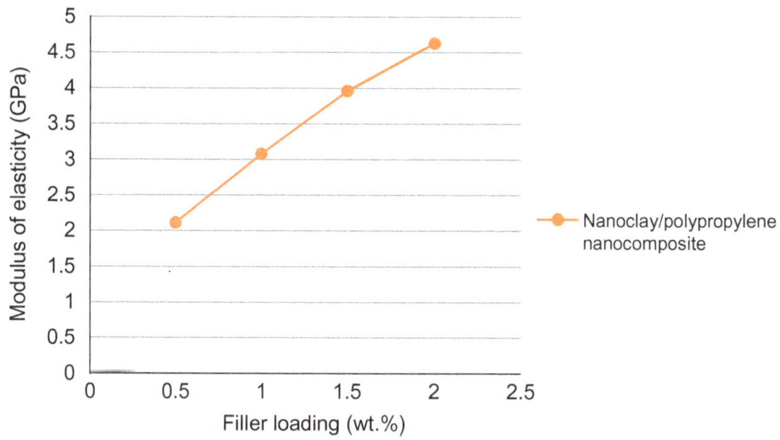

Fig. 11.7 Modulus of elasticity of different percentage loading of nanoclay/polypropylene.

Fig. 11.7 shows the modulus of elasticity for different percentage loading of silica to low-density polyethylene

The tensile strength of nanoclay-based nanocomposite is greatly improved by increasing the percentage loading of nanoclay to polypropylene. The nanoclay-based nanocomposite loaded with 2 wt% of nanoclay displayed the highest tensile strength. According to Lei et al., nanoclay based nanocomposites in which nanoclay is used as the reinforcement agent for the composites exhibit drastic increases in modulus, tensile and flexural strength, barrier properties, flammability resistance, weathering resistance, water uptake resistance, and also the heat resistance as compared with other traditional composites. The nanoclay particles used as a reinforcement agent in the polymer matrix significantly improved the mechanical properties with the addition of only a small amount of nanoclay filler (<10 wt.% and even less than 5%).

On the other hand, modulus of elasticity was showed in Fig. 11.7 which was greatly improved by the addition of nanoclay, as compared to their modulus of elasticity of polypropylene itself. High modulus of elasticity indicates the resistance of nanoclay based nanocomposites against the elastic deformation when the load is applied on the composites. Thus, the modulus of elasticity of nanoclay based nanocomposites are been improved as the percentage of filler loading increased.

11.3.11.2 Mechanical properties of layered reinforcement's nanocomposites

Polymer layered silicate (PLS improved the nanocomposites properties compared with the pure polymer and conventional micro and macrocomposites and it showed the high moduli, strength and heat resistance, decreased flammability and gas permeability and increased biodegradability (Biswas & Ray, 2001; Giannelis et al., 1999; LeBaron et al., 1999; Vaia et al., 1999). There are two types of layered silicates which are used in particular characteristics for PLS nanocomposites, whereas the ability of the silicate particles to diffuse into individual layers and the second ability to fine-tune their surface chemistry through ion exchange reactions with organic and inorganic cations.

The intercalation chemistry of polymers towards layered silicates obtained on Nylon-6 (N6)/montmorillonite (MMT) nanocomposites showed that a small concentration of layered silicate led to remarkable changes in thermal and mechanical properties (Ray & Okamoto, 2003). According to the Vaia et al., it is possible to melt-mix polymers with layered silicates without the use of organic solvents and it's called the 2:1 phyllosilicates, which is the most commonly used layered silicate in polymer nanocomposites. There are two-dimensional layers which show the tetrahedrally coordinated silicon atoms fused to an edge-shared octahedral sheet of either aluminum or magnesium hydroxide. Based on the silicate nature the layer thickness is around 1 nm and its lateral dimensions may vary from 300 Å to several micra or larger, whereas these layers generate a regular van der Waals gap which is called the "interlayer" space or the "gallery." Isomorphic substitution within the layers produces negative charges that are counterbalanced by alkaline or alkaline earth cations located in the interlayer (Ray & Okamoto, 2003). The intercalation of small molecules between the layers and Mg^{2+} or Fe^{2+} which replaces Al^{3+} or when Li^{+} replaces Mg^{2+} due to their weak van der Waals interactions (Marchal et al., 1996). The organosilicates, such as alkylammonium or alkylphosphonium cations lower the surface energy of the inorganic host and improve the wetting characteristics of the polymer matrix and results in a larger interlayer spacing can be exchanged by ion-exchange reactions with cationic surfactants including primary, secondary, tertiary, and quaternary alkylammonium or alkylphosphonium. To improve the strength of the interface between the inorganic component and the polymer matrix, the cations influence the functional groups that can react with the polymer matrix or initiate the polymerization of monomers (Blumstein, 1965; Krishnamoorti et al., 1996; Passaglia et al., 2008).

The most useful layered silicates are montmorillonite, hectorite, and saponite and their formulae which are presented in Fig. 11.8, whereas the hydrated cations are ion-

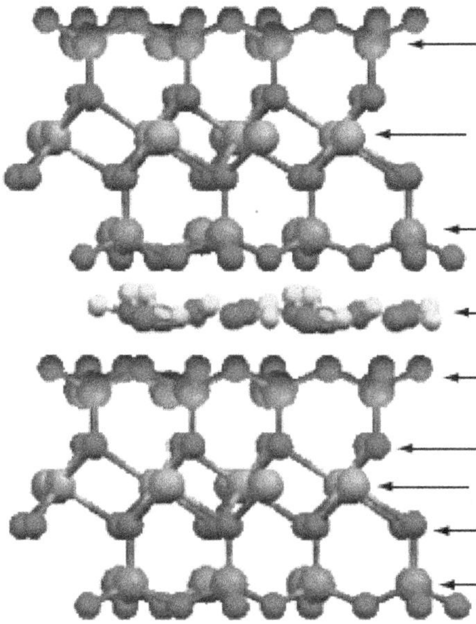

Fig. 11.8 Schematic representation of the structure of 2:1 phyllosilicates.

exchanged with bulkier organic cations and there is a larger interlayer spacing which is usually obtained from the structure.

The stronger interfacial interaction between the matrix and the silicate is due to the remarkable improvements of their properties which is observed in polymer/layered-silicate nanocomposites. The mechanical properties were a significant improvement due to the incorporation of MMT (montmorillonite) into a Nylon-6 matrix. The Young's modulus of pure Nylon-6 (1.11 GPa) was strongly improved when the nanocomposite was formed while the Nylon-6/MMT with a filler content of 4.1 wt. (%) contributed a value of 2.25 GPa, which corresponds to an increase of 102.7% (Kojima et al., 1993a, 1993b; Kojima, et al., 1993). Fig. 11.9 illustrates the dependence of tensile modulus E, at 393 K (120°C) on clay content for organo-modified montmorillonite and saponite-based nanocomposites that the increase in Young's modulus in these systems is related to the average length of the layers and the aspect ratio of the dispersed nanoparticles, as well as the extent of their exfoliation (Kojima et al., 1993a, 1993b; Kojima, et al., 1993). This behavior can also be supported by maleic anhydride modified by propylene (PP-MA)/LS nanocomposites that are a strong interaction between matrix and silicate layers occurs via formation of hydrogen bonds.

Table 11.4 represents the variation of the tensile strength for both the nano and the microcomposite that shows higher Young's modulus than the pure matrix.

The stress at break was also observed in Polymer/LS systems which are important improvements on thermoplastic-based nanocomposites which express the ultimate strength that the material can bear before breaking on the nature of the interactions

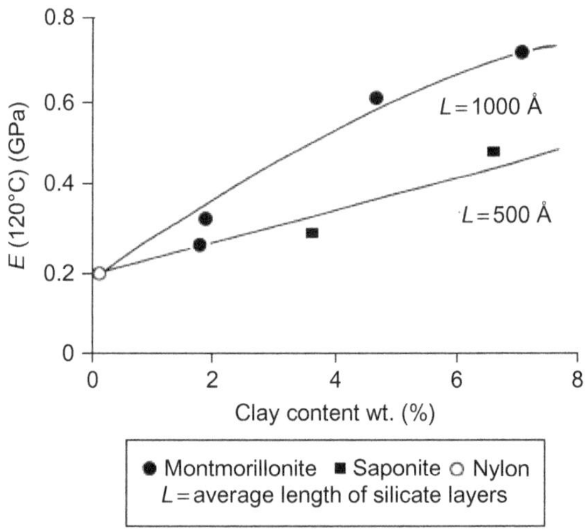

Fig. 11.9 Dependence of tensile modulus (E) at 120°C on clay content for organomodified montmorillonite-, saponite-based nanocomposites.
Reproduced from reference Kojima, Y., Usuki, A., Kawasumi, M., Okada, A., Fukushima, Y., Karauchi, T., Kamigaito, O. (1993). Mechanical properties of nylon-6-clay hybrid. *Journal of Materials Research*, *8*(5), 1185–1189. Kojima, Y., Usuki, A., Kawasumi, M., Okada, A., Kurauchi, T., Kamigaito, O. (1993). One pot synthesis of nylon-6-clay hybrid. *Journal of Polymer Science and Polymer Chemistry*, *31*(7), 1755–1758. Kojima, Y., Usuki, A., Kawasumi, M., Okada, A., Kurauchi, T., Kamigaito, O. (1993). Synthesis of nylon-6-clay hybrid by montmorillonite intercalated with ε-caprolactum. *Journal of Polymer Science and Polymer Chemistry*, *31*(7), 983–986. with the kind permission of the authors, the Materials Research Society.

between the matrix and the filler. Table 11.4 shows the tensile stress in different nanocomposite systems in which exfoliated Nylon-6 and intercalated PMMA nanocomposites exhibited an extreme increase in the stress at break (Lee & Jang, 1996). Nylo(n-6 grafted onto the layers which interact between the matrix and the silicate layers at ionic level and its increase larger in Nylon-6 nanocomposites (Hasegawa et al., 1998; Lee & Jang, 1996). Moreover, propylene-based nanocomposites showed only a slight enhancement in tensile stress which can be explained by the lack of interfacial adhesion between nonpolar PP and polar-layered silicates, while the maleic anhydride-modified polypropylene matrix has been confirmed to be effective in the intercalation of the PP chains and the maintenance of the ultimate stress at an acceptable level (Alexandre & Dubois, 2000). The authors have concluded that these weak interactions at the polystyrene-clay interface significantly decreased the ultimate tensile stress compared to that given by the PP matrix.

When the thermoplastics are reinforced by LS and produce the PMMA, PS or PP the elongation at break is reduced 150% in pure PP matrix, to 105% for a 6.9 wt. (%)

Table 11.4 Tensile stress evolution for nanocomposites based on various thermoplastic matrices

Matrix	Tensile stress (MPa)	Nanofiller content (wt.(%))	Structure	Tensile stress (MPa)
Nylon-6	68.6	4.7	Exfoliated	97.2
Nylon-6	68.6	5.3	Exfoliated	97.3
Nylon-6	68.6	4.1	Exfoliated	102
PMMA	53.9	12.6	Intercalated	62.0
PMMA	53.9	20.7	Intercalated	62.0
PP-MA 7.2 wt.(%)	31.4	5.0	Intercalated	29.5
PP-MA 21.6 wt.(%)	32.6	4.8	Intercalated	31.7
PS	28.7	11.3	Intercalated	21.7
PS	28.7	17.2	Intercalated	23.4
PS	28.7	24.6	Intercalated	16.6
PS	28.7	34.1	Intercalated	16.0

Reproduced from Alexandre, M, Dubois, P. (2000). Polymer-layered silicate nanocomposites: preparation, properties and uses of a new class of materials. *Materials Science and Engineering 28*(1–2), 1–63 with permission of the authors and Elsevier.

nonintercalated clay microcomposite 251,252. On the other hand, if the nanocomposite contained 5 wt. (%) silicate layers, then it will drop 7.5% elongation at break while ultimate elongation did not occur in elastomeric epoxy or polyol polyurethane matrices (Wang et al., 1998). The elongation at break slight improvement can be observed for the conventional composite which is prepared from magadiite modified with methyl-octadecylammonium or trimethyloctadecylammonium ions, while a drop in in this property can be observed for the intercalated nanocomposite (Wang et al., 1998).

The elongation at break largely increased with exfoliated nanoclay reinforced nanocomposites due to the plasticizing effect of the galleries and conformational effects at the clay-matrix interface. The high performance elastomeric nanocomposites materials produce the new avenue due to their improved stiffness (Young's modulus), toughness (stress at break), and elasticity (strain at break).

Polyimide polymer matrix nanocomposites material showing an increase in both stress and elongation at break which filled with montmorillonite exchanged with hexadecylammonium, these properties increase with the filler loading at least up to 5 wt. (%), whereas both properties drop toward values described for the filler-free matrix due to the formation of nonexfoliated aggregates, which make these composites more brittle. (Yano et al., 1997).

Nylon-6/protonated aminododecanoic nanocomposite system has reduced the impact resistance due to the in situ intercalative polymerization and had its Izod impact strength reduced from 20.6 to 18.1 J/m compared with the pure matrix when 4.7 wt.(%) of nanoclay was incorporated and similar reduction can observed from charpy impact testing which showed a 6.21–6.06 kJ/m^2 for the 4.7 wt. (%) (Liu et al., 1999). Table 11.4 showed the effect of MMT content and Nylon-6 molecular weight on the tensile modulus of and improvement in stiffness can be seen, which increases with increasing matrix molecular weight (at any given concentration) for all Nylon-6 nanocomposites, i.e., low (LMW), medium (MMW) and high molecular weights (HMW). A slightly larger modulus can be observed if the slow cooling of the specimen after injection molding which is resulting in a higher degree of crystallinity giving faster crystallization kinetics, whereas the HMW- and MMW-based nanocomposites show a stable growth in strength with the increasing content of clay, the LMW-based nanocomposites show a less noticeable effect. The high resistance to impact, Young's modulus, flexural modulus, and a notable improvement in the heat distortion temperature are going from 338°K (65°C) for pure Nylon-6 compared to 423°K (150°C) for the nanocomposite, which have been acceptable for this material to replace glass fiber-reinforced nylon or polypropylene in the production of timing belt covers of automotive engines, but the strain reductions for all nanocomposites when compared to the pure matrix polymer (Oakada & Usuki, 1995).

11.3.12 Thermal properties of nanocomposites

11.3.12.1 Thermogravimetric analysis (TGA) of nanocomposites

TGA was conducted to calculate the thermal stability and decomposition process for nanoclay/polypropylene nanocomposites. Fig. 11.10 shows the TGA curves nanoclay/polypropylene nanocomposite with different percentages of filler loading and Table 11.5 shows the weight loss for different nanoclay content added to polypropylene. Drastic drop of weight loss occurred at 400°C for the nanoclay/polypropylene nanocomposites. The peak degradation temperature of nanoclay/polypropylene nanocomposites started almost at the similar temperature, which is at 400°C. This is due to the presence of barrier effect resulting from the remaining silicate layers

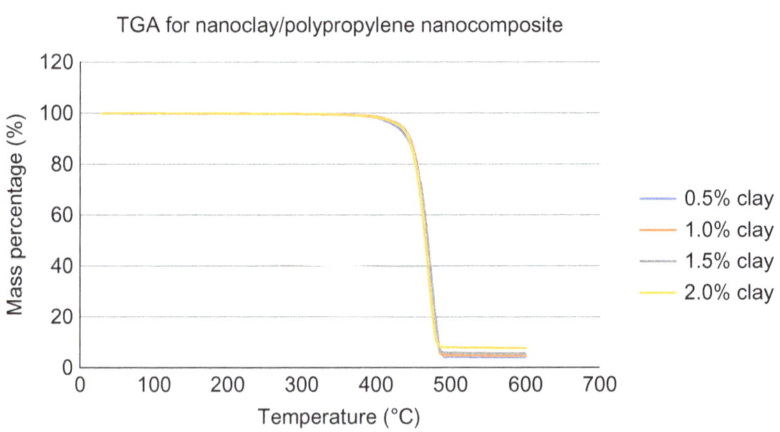

Fig. 11.10 TGA curves of nanoclay/polypropylene nanocomposite with different filler loading.

Table 11.5 Weight loss for different nanoclay loading nanocomposites

Nanocomposite	Weight loss (%)					
	100°C	200°C	300°C	400°C	500°C	600°C
0.5 wt%nanoclay/PP	1	1	1	2.1	95.84	95.98
1.0 wt%nanoclay/PP	1	1	1	2.0	95.17	95.36
1.5 wt%nanoclay/PP	0	0	1	1.6	94.19	94.47
2.0 wt%nanoclay/PP	0	0	1	1.4	92.06	92.35

which act as the diffusion barriers for the oxygen (Leszczynska, Njuguna, Pielichowski, & Banerjee, 2007). The thermal stability of nanocaly/polypropylene was increased with increasing of the nanoclay content. The mass loss of 2.0 wt% nanoclay/polypropylene nanocomposite was less than other nanocomposites which are loaded with the nanoclay contents of 0.5, 1.0, and 1.5 wt% at 600°C. According to Al-Samhan, Samuel, Al-Attar, and Abraham (2017), clay particles that are introduced into all polypropylene nanocomposites usually acted as a shelter during the formation of char which resulted in the second phase degradation of nanocomposites at high temperatures compared to pure polymer. Moreover, Al-Samhan et al. (2017) also stated that the small volatiles produced during the process of thermal decomposition may not be able to permeate, but these volatiles can be avoided by the clay layers which resulting in the addition of nanoclay helps to minimize the release rate of decomposed by-products and thus improve the thermal stability of the nanocomposites. On the other hand, nanoclay/polypropylene nanocomposite exhibits better thermal stability properties which is due to the versatile properties of nanoclay, like large surface area and layered structure provides good mechanical, fire resistant and barrier properties to the nanocomposites (Rodriguez-Llamazares et al., 2011; Zhang, Manias, & Wilkie, 2008).

However, the enhancement of thermal stability of the nanoclay based nanocomposites is achieved due to the presence of the lamination of silicate layers of clay and layered formation, as well as intercalation structures helping to delay the oxygen penetration into the polymer frame (Nafchi, Abdouss, Najafi, Gargari, & Mazhar, 2015). Hence, silica/polyethylene and nanoclay/polypropylene nanocomposites showed enhanced thermal performance and high decomposition temperature as compared to pure polymer.

11.3.12.2 Differential scanning calorimetry (DSC)

Fig. 11.11 shows the DSC thermograms of nanoclay/polypropylene nanocomposite with different filler loading. There are four heating curves included in Fig. 11.11 indicating four different percentages of nanoclay were loaded to polypropylene. The melting temperature, T_m, of different nanoclay content nanocomposites were analyzed from the heating curves. All four heating curves clearly showed that the differences of melting temperature of nanoclay/polypropylene nanocomposites loaded different

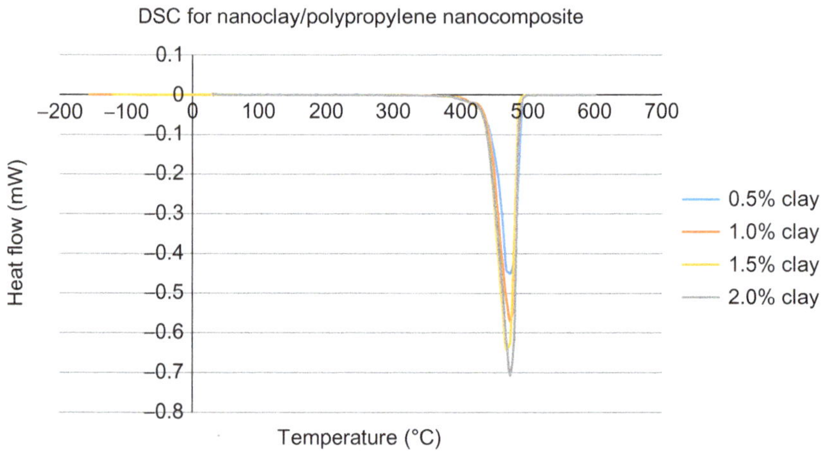

Fig. 11.11 DSC thermograms of nanoclay/polypropylene nanocomposite with different filler loading.

nanolay content. Based on Fig. 11.11, the nanoclay/polypropylene nanocomposites with 2 wt% of nanoclay loading showed the highest melting temperature, while the lowest melting temperature was observed by the 0.5 wt% nanoclay/plypropylene nanocomposites. Thus, the results obtained showed that better melting temperature can be achieved by increasing the nanoclay content to polypropylene and the high melting temperature of the nanocomposites also indicated high crystallization temperature (T_c). Much research has been conducted and proved that the effects of adding the fillers/fibers (carbon, glass, PET, etc) and different types of nanofillers (CNTs/fibers or layered silicates) included the crystallization behavior was greatly affected and the resulting crystalline morphology of polypropylene matrix, and also altered the kinetics of the crystallization process due to their nucleating action (Raka, Bogoeva-Gaceva, Lu, & Loos, 2009). Thus, the crystalline structure formed in the clay/polypropylene nanocomposites significantly affected the thermal properties of the nanocomposites due to the presence of nanoclay.

According to Al-Samhan et al. (2017), the melting temperature and the crystallization temperature (T_c) of nanoclay/polypropylene nanocomposites was enhanced and proved much higher compared to pure polypropylene due to the effect of montmorillonite clay layers which act as a nucleating for pure polypropylene, while the crystallization rate was increased as well. Although the addition of nanoclay can improve the melting temperature and increase the crystallization rate, as well as degree of pure polypropylene, but further increments of nanoclay content to polypropylene no longer affect the polymer and this also leads to negative impacts such as increased unnecessary viscosity and chain folding energy to the nanoclay/polypropylene nanocomposites (Li, Wang, Zhang, & Zhang, 2006). Thus, both silica/low-density polyethylene and nanoclay/polypropylene nanocomposites exhibited enhanced melting and crystallization temperature as compared to the pure polymers.

11.3.12.3 Dynamic mechanical analysis (DMA)

Fig. 11.12 shows the dynamic mechanical analysis (DMA) measures of nanocomposites. It was observed that three main parameters, which are the storage modulus (E'), corresponding to elastic response to deformation; the loss modulus (E''), corresponding to plastic response to deformation and tan δ, that is the (E'/E'') ratio, useful for determining the glass transition temperature. DMA analysis was used to characterize PS nanocomposites and it was observed that there is no significant difference in E' of the pure PS and a nanocomposite intercalated with 17.2 wt.(%) of Na-montmorillonite in the temperature range of 293–413 K (20–140°C). It indicated that lack of influence of intercalated nanocomposites on the elastic properties of the matrix, while the tan δ values toward higher temperatures for the nanocomposite suggests an increase in the glass transition temperature and some broadening of this transition, which is restricted segmental motions at the organic-inorganic interface of the intercalated systems.

However, symmetric styrene-butadiene-styrene [SBS] block copolymers showed the different properties on DMA which was chemically modified matrix, while melt-blended of SBS with a montmorillonite modified by dimethyldioctadecylammonium, a nanocomposite can be produce intercalate within the layered silicates that is called PS blocks (Laus et al., 1997). This material showed the large improvement of the storage modulus at 25°C, which shown in Fig. 11.12, whereas the values of the storage moduli for two sets of samples in which the reinforcement content varied from 0 to 30 wt. (%). Under the same conditions, melt-blending techniques apply to prepare the SBS

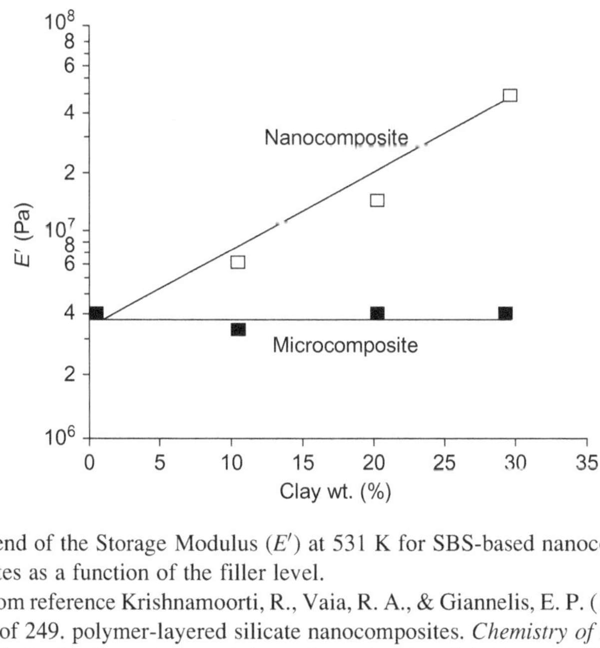

Fig. 11.12 Trend of the Storage Modulus (E') at 531 K for SBS-based nanocomposites, microcomposites as a function of the filler level.
Reproduced from reference Krishnamoorti, R., Vaia, R. A., & Giannelis, E. P. (1996). Structure and dynamics of 249. polymer-layered silicate nanocomposites. *Chemistry of Materials*, 8(8), 1728–1734 with the kind permission of the authors, the Materials Research Society, United States.

nanocomposites which filled with organo modified clay and Na-montmorillonite, while the value of elastic modulus increase for nanocomposites and microcomposites.

A two-step chemical method (173°–323°K and from 323° to 573°K) was used to prepare nanocomposites by bifunctional diglycidyl ether of bisphenol-A (DGEBA) containing different amounts (2–10 wt.(%)) of layered silicate and widely investigated the DMA properties. Results showed that the relaxation temperatures of the cured system decreased when increasing organic clay content in the composite, while both toughness and stiffness of the matrix were improved in the presence of modified layered silicate (Becker et al., 2002).

Furthermore, storage modulus, especially the T_g, was noticably improved when organoclays exfoliated within cross-linked matrices such as the epoxy/montmorillonite 4 vol.% nanocomposite showed below T_g value (Messersmith & Giannelis, 1994). Various factors influence to enhance the storage modulus, an interesting result could be observed in nitrile rubber/organoclay nanocomposites (Oakada & Usuki, 1995). The storage modulus increased three-fold in the simple dispersion/exfoliation of 10 parts of organoclay per 100 parts of rubber and that value can be obtained with the same matrix filled with 40 parts of carbon black per 100 parts of rubber. Overall, exfoliated nanocomposites which were filled with layered silicates showed the enhanced storage elastic modulus at temperatures above T_g.

11.3.13 Morphological properties of nanocomposites

11.3.13.1 Scanning electron microscopy (SEM) Analysis

The interfacial bonding of different percentages of filler loadings for nanoclay based nanocomposites was investigated through SEM analysis. Fig. 11.13 shows the SEM images for the pure polypropylene. Ductile features with some void fraction were displayed from the surface morphologies of pure polypropylene as shown in Fig. 11.13.

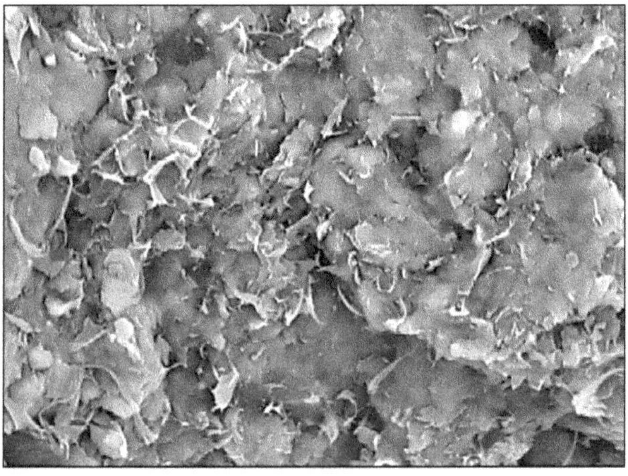

Fig. 11.13 Surface morphologies of pure polypropylene.

Figs. 11.13–11.17 shows the SEM images of different filler loading ductile to less brittle fracture of nanoclay-based polypropylene nanocomposite

Fig. 11.14 shows the surface morphologies of 0.5 wt% clay-based nanocomposite. With the 0.5 wt% loading of clay, it improved the ductile feature, as well as decreased the void fractions of the pure polypropylene. The void fraction existing at the pure polypropylene was slightly decreased as the holes were filled by nanoclay. Although there is less matrix voiding, the polypropylene with the 05 wt% nanocaly loading proved that the weakest tensile properties among all the different percentage of nanoclay loaded nanocomposites.

The surface morphologies of 1.0 wt% nanoclay/polypropylene nanocomposite showed in Fig. 11.15. There was better adhesion between the nanoclay filler and

Fig. 11.14 Surface morphologies of 0.5 wt% nanoclay/polypropylene nanocomposite.

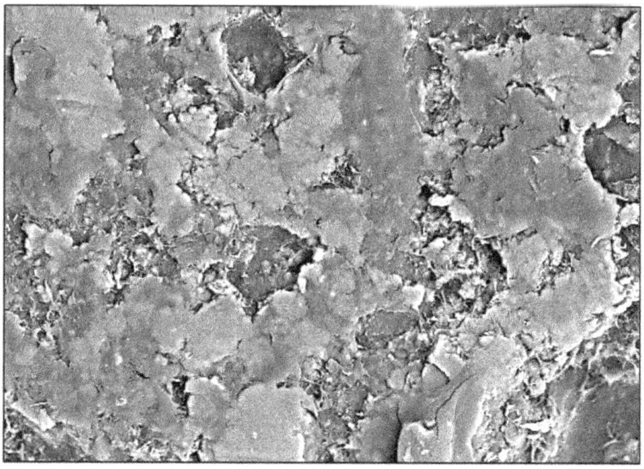

Fig. 11.15 Surface morphologies of 1.0 wt% nanoclay/polypropylene nanocomposite.

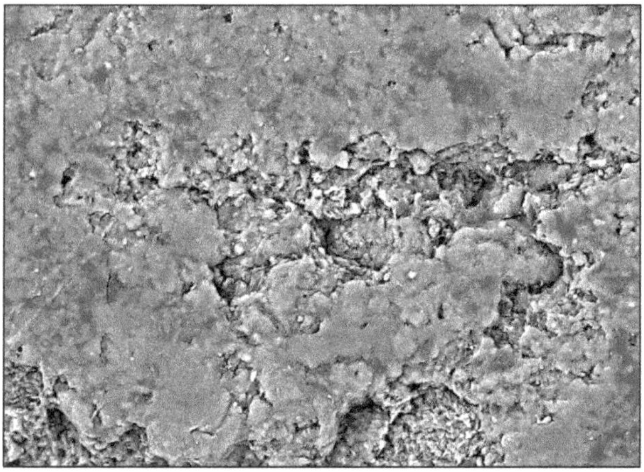

Fig. 11.16 Surface morphologies of 1.5 wt% nanoclay/polypropylene nanocomposite.

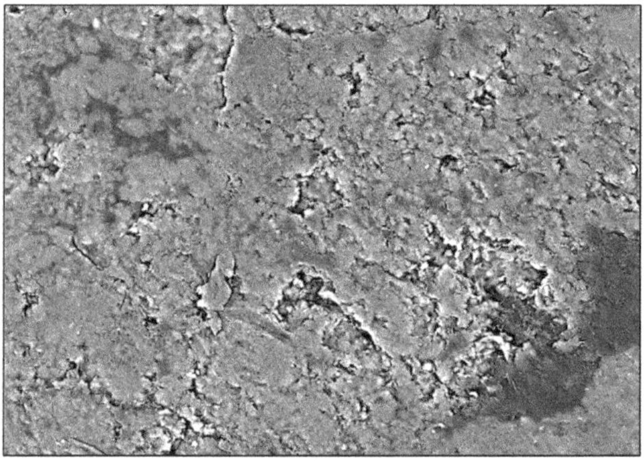

Fig. 11.17 Surface morphologies of 2.0 wt% nanoclay/polypropylene nanocomposite.

the matrix associated with even less void fraction. Besides that, the surface roughness of the 1.0 wt% nanoclay/polypropylene nanocomposite was enhanced into a smoother surface as compared to 0.5 wt% nanoclay/polypropylene nanocomposite. There are some holes existing on the surface of 1.0 wt% nanoclay/polypropylene nanocomposite which indicated the uncovered area by the nanoclay due to the insufficient loading of nanoclay to the composite. However, the polypropylene with the addition of 1.5 wt% of nanoclay showed better surface roughness comparing 1.0 wt% nanoclay/polypropylene nanocomposite. With an extra 0.5 wt% of nanoclay loading, there are more holes that were covered or filled up by the nanoclay and surface roughness were improve of the 1.5 wt% nanoclay/polypropylene nanocomposite which enhance the tensile strength that shown in Fig. 11.16.

Fig. 11.17 shows the surface morphologies of 2.0 wt% nanoclay/polypropylene nanocomposite. Flat and smooth surface was observed from this composite due to the uniform dispersion of nanoclay in matrix which contributed the strong bonding between the nanoclay and matrix. 2.0 wt% nanoclay/polypropylene nanocomposite showed the smoothest surface, as well as the highest tensile properties among all different nanoclay contents composites. Moreover, 2.0 wt% nanoclay/polypropylene nanocomposite also showed less brittle fracture as compared to the ductile of the pure polypropylene. According to Chen, Wong, et al., 2002; Chen, Lee, et al., 2002, it is an obvious alteration of fracture behavior from a slightly ductile matrix voiding and fibrous deformation to less brittle cleavage feature as the clay loading is increased. Thus, this supports that the pure polypropylene changes from ductile to less brittle fracture when the addition of nanoclay content to polypropylene is increased.

11.3.14 Optical properties of nanocomposites

11.3.14.1 Optical properties of PP/Clay nanocomposites

Knowing that polymers often suffer the problem of oxidative degradation when they are exposed to light which can lead to the discoloration and fragility of the polymer material together with both physical and mechanical properties of packaging as well as causing the degradation of foods that are in contact with these kind of materials (Coltro & Borghetti, 2007) causes concern. Although both of the pure polypropylene and polyethylene are able to absorb at the wavelengths that are longer than 290 nm, the radical-based oxidative processes may be affected or disturbed by the absorption of sunlight by:

- **(i)** The internal in-chain impurities like carbonyls formed during processing or storage.
- **(ii)** External impurities like residues of the polymerization catalysts or metal traces from the processing equipment.
- **(iii)** The charge transfer complexes former between both of the polymer and oxygen (Fernando, Christensen, Egerton, & White, 2009). However, the damage done by Ultraviolet radiation (UV) to the polymer chain can be reduced by introducing the reinforcing agent like organoclay into polypropylene (Zehetmeyer, Scheibel, & Soares, 2013). Also, this phenomenon can be improved by adding the filler which acquires the nanometer size, as well as high-aspect ratio (Horst, Tuckart, Blanco, Failla, & Quinzani, 2012). Moreover, according to Zehetmeyer et al. (2013), most of the literature has proven that the fact of the permeability of the polypropylene/organoclay nanocomposites is much lower as compared to pure polymer matrix and about 30%–50% are able to decrease by increasing the concentration of the clay reinforcing agent.

The UV light barrier properties of the film generated from both pure polypropylene and the PP/clay nanocomposites are being analyzed via the spectroscopy analysis in the ultraviolet region. The spectrum shows the characteristic band of the clay reinforcing agent where a broad absorption with the maximum wavelength of 275 nm can be able to observed due to the presence of Fe^{2+} ions (in the substitution to the interclay Ca^{2+}) as these types of metals proved the fact of the strong absorption phenomenon occurs in the UV region (Zehetmeyer et al., 2013). The UV-Vis absorption

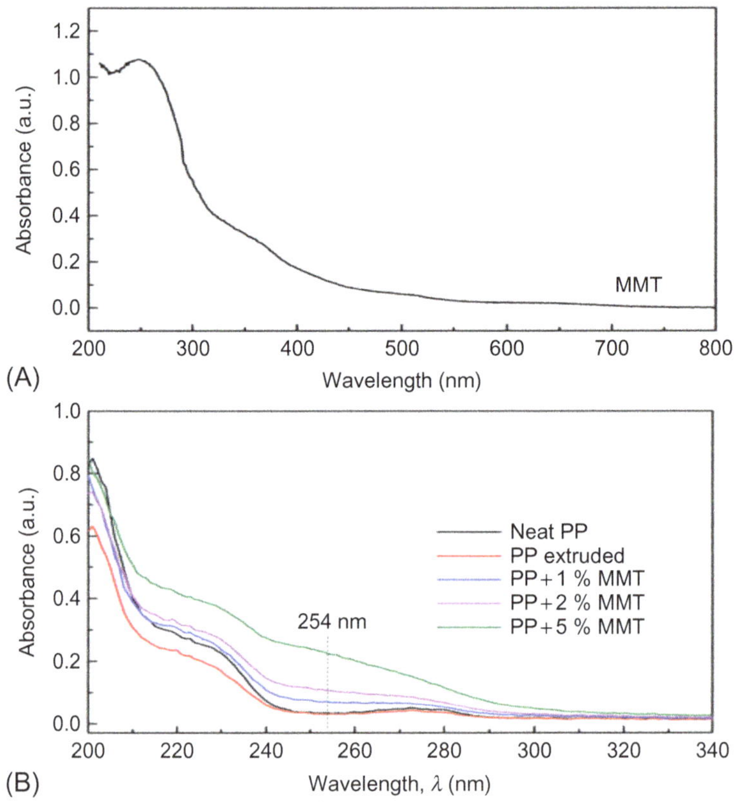

Fig. 11.18 UV-Vis spectra of (A) MMT and (B) Polypropylene/clay nanocomposites (Zehetmeyer et al., 2013).

spectrum for the presence of MMT is showed in Fig. 11.18A while Fig. 11.18B shows the absorption spectra for pure polypropylene film together with various concentrations of PP/clay nanocomposites which are produced by melt compounding method in the wavelength range of 200–340 nm. Based on Fig. 11.18, the responses of irradiated material showed almost identical behavior to the pure clay reinforcing agent at the high absorptions occurred at the wavelengths between 215 and 254 and 280 nm. An exception is observed for the nanocomposites which contain a total amount of 5 wt % organoclay. According to the Zehetmeyer et al. (2013), the transparency of film in the extruded polypropylene after processing showed great enhancement due to the UV absorbers tending to degrade during the process. However, a better absorbance at 254 nm region is observed (as shown in Fig. 11.19) with a higher concentration of MMT in the system. Thus, the influence of the MMT on the absorption of radiation becomes greater at low wavelengths which was specific in the region from UV in the wavelength range of 190–280 nm (Zehetmeyer et al., 2013).

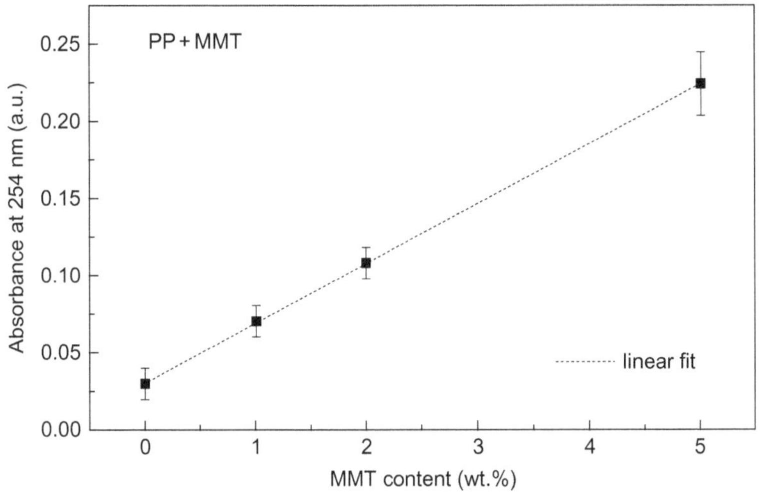

Fig. 11.19 UV-Vis absorption at 254 nm as a function of MMT content for Polypropylene/clay nanocomposites (Zehetmeyer et al., 2013).

References

Agnihotri, S., Rostam-Abadi, M., & Rood, M. J. (2004). Temporal changes in nitrogen adsorption properties of single-walled carbon nanotubes. *Carbon*, *42*(12–13), 2699–2710.

Ajayan, P. M., Schadler, L., & Braun, P. V. (2003). *Nanocomposite science and technology*. Weinheim: Wiley-VCH, Verlag Gmbh & Co. KgaA.

Akita, H., & Hattori, T. (1999). Studies on molecular composite. I. Processing of molecular composites using a precursor polymer for poly (P-Phenylene benzobisthiazole). *Journal of Polymer Science: Part B: Polymer Physics*, *37*(3), 189–197.

Akita, H., & Kobayashi, H. (1999). Studies on molecular composite. III. Nano composites consisting of poly (P-phenylene benzobisthiazole) and thermoplastic polyamide. *Journal of European Ceramic Society*, *37*(3), 209–218.

Akita, H., Kobayashi, H., Hattori, T., & Kagawa, K. (1999). Studies on molecular composite. II. Processing of molecular composites using copolymers consisting of a precursor of poly (P-phenylene benzobisthiazole) and aromatic polyamide. *Journal of European Ceramic Society*, *37*(3), 199–207.

Alexandre, M., & Dubois, P. (2000). Polymer-layered silicate nanocomposites: preparation, properties and uses of a new class of materials. *Materials Science & Engineering*, *28*(1–2), 1–63.

Al-Samhan, M., Samuel, J., Al-Attar, F., & Abraham, G. (2017). Comparative effects of MMT clay modified with two different cationic surfactants on the thermal and rheological properties of polypropylene nanocomposites. *International Journal of Polymer Science*. https://doi.org/10.1155/9484.

An, J. W., You, D. H., & Lima, D. S. (2003). Tribological properties of hot-pressed alumina–CNT composites. *Wear*, *255*(1–6), 677–681.

Ananthakumar, S., Prabhakaran, A. K., Hareesh, U. S., Manoharan, P., & Warrier, K. G. K. (2004). Gel casting process For Al2O3-SiC nanocomposites and its creep characteristics. *Materials Chemistry and Physics*, *85*(1), 151–157.

Andrews, R., Jacques, D., Minot, M., & Rantell, T. (2002). Fabrication of carbon multiwall nanotube/polymer composites by shear mixing. *Macromolecular Materials and Engineering, 287*(6), 395–403.

Andrews, R., & Weisenberger, M. C. (2004). Carbon nanotube polymer composites. *Current Opinion in Solid State and Materials Science, 8*(1), 31–37.

Anya, C. C. (1999). Microstructural nature of strengthening and toughening in Al2O3-SiC(P) nanocomposites. *Journal of Materials Science, 34*(22), 5557–5567.

Aranda, P., & Ruiz-Hitzky, E. (1992). Poly (ethylene oxide)-silicate intercalation materials. *Chemistry of Materials, 4*(6), 1395–1403.

Aruna, S. T., & Rajam, K. S. (2003). Synthesis, characterisation and properties of Ni/PSZ and Ni/YSZ nanocomposites. *Scripta Materialia, 48*(5), 507–512.

Athawale, A. A., Bhagwat, S. V., Katre, P. P., Chandwadkar, A. J., & Karandikar, P. (2003). Aniline as a stabilizer for metal nanoparticles. *Materials Letters, 57*(24-25), 3889–3894.

Avadhani, C. V., & Chujo, Y. (1997). Polyimide-silica gel hybrids containing metal salts: preparation via the sol-gel reaction. *Applied Organometalic Chemistry, 11*(2), 153–161.

Awaji, H., Choi, S. M., & Yagi, E. (2002). Mechanisms of toughening and 36. strengthening in cermaic-based nanocomposites. *Mechanics of Materials, 34*(7), 411–422.

Aymonier, C., Bortzmeyer, D., Thomann, R., & Lhaupt, R. M. (2003). Poly (methyl methacrylate)/palladium nanocomposites: synthesis and characterization of the morphological, thermomechanical, and thermal properties. *Chemistry of Materials, 15*(25), 4874–4878.

Azioune, A., Peck, K., Soudi, B., Chehimi, M. M., Mccarthy, G. P., & Armes, S. P. (1999). Adsorption of human serum albumin onto polypyrrole powder and polypyrrole-silica nanocomposites. *Synthetic Metals, 102*(1 3), 1419–1420.

Bafna, A., Beaucage, G., Mirabella, F., & Mehta, S. (2002). Shear induced orientation and associated property enhancement in polymer/clay nanocomposites. In *Proceedings Nanocomposites* (pp. 23–25). San Diego. USA: ECM Publication.

Bafna, A., Beaucage, G., Mirabella, F., & Mehta, S. (2003). 3D hierarchical orientation in polymer-clay nanocomposite films. *Polymer, 44*(4), 1103–1115.

Bai, J. B., & Allaoui, A. (2003). Effect of the length and the aggregate size of mwnts on the improvement efficiency of the mechanical and electrical properties of nanocomposites: experimental investigation. *Composites Part A—Applied Science and Manufacturing, 34*(8), 689–694.

Baiju, K., Sibu, C. P., Rajesh, K., Pillai, P. K., Mukundan, P., Warrier, K. G. K., et al. (2005). An aqueous sol-gel route to synthesize nanosized lanthana doped titania having an increased anatase phase stability for photocatalytic application. *Materials Chemistry and Physics, 90*(1), 123–127.

Bajwa, S., Rainfoth, W. M., & Lee, W. E. (2005). Sliding wear behaviour of SiC-Al2O3 nanocomposites. *International Conference on Wear of Materials, 259*(1-6), 553–561.

Baker, C., Ismat Shah, S., & Hasanain, S. K. (2004). Magnetic behavior of iron and iron-oxide nanoparticle/polymer composites. *Journal of Magnetism and Magnetic Materials, 280*(2–3), 412–418.

Balázsi, C. S., Kónya, Z., Wéber, F., Biró, L. P., & Arató, P. (2003). Preparation and characterization of carbon nanotube reinforced silicon nitride composites. *Materials Science and Engineering: C, 23*(6 8), 1133–1137.

Balázsi, C. S., Shen, Z., Kónya, Z., Kaztovszky, Z. A., Wéber, F., Vértesy, Z., et al. (2005). Processing of carbon nanotube reinforced silicon nitride composites by spark plasma sintering. *Composites Science and Technology, 65*(5), 727–733.

Baron, B., Kumar, C. S., Le Gonidec, G., & Hampshire, S. (2002). Comparison of different alumina powders for the aqueous processing and pressureless sintering of Al2O3-SiC nanocomposites. *Journal of the European Ceramic Society, 22*(9–10), 1543–1552.

Becher, P. F. (1991). Microstructural design of toughened ceramics. *Journal of the American Ceramic Society*, *74*(2), 255–269.

Becker, O., Varley, R., & Simon, G. (2002). Morphology, thermal relaxations and mechanical properties of layered silicate nanocomposites based upon high-functionality epoxy resins. *Polymer*, *43*(16), 4365–4373.

Belin, T., & Epron, F. (2005). Characterization methods of carbon nanotubes: a review. *Materials Science and Engineering*, *119*(2), 105–118.

Bharadwaj, R. K. (2001). Modeling the barrier properties of polymer layered. Silicate nanocomposites. *Macromolecules*, *34*(26), 9189–9192.

Bhattacharya, V., & Chattopadhyay, K. (2001). Microstructure and tribological behaviour of nano-embedded al-alloys. *Scripta Materialia*, *44*(8–9), 1677–1682.

Bhattacharya, V., & Chattopadhyay, K. (2004). Microstructure and wear behaviour of aluminium alloys containing embedded nanoscaled lead dispersoids. *Acta Materialia*, *52*(8), 2293–2304.

Biercuk, M. J., Llaguno, M. C., & Radosvljevicm, H. J. (2002). Carbon nanotube composites for thermal management. *Appied Physics Letters*, *80*(15), 2767–2769.

Biswas M, Ray SS. Recent progress in synthesis and evaluation of polymer–montmorillonite nanocomposites. New polymerization techniques and synthetic methodologies. Book series: Advances in polymer science; 2001. p. 167–221.

Blumstein, A. (1965). Polymerization of adsorbed monolayers. II. Thermal degradation of the inserted polymers. *Journal of Polymer Science Part A—General Papers*, *3*(7PA), 2665–2673.

Borsa, C. E., & Brook, R. J. (1995). Fabrication of Al2O3/SiC nanocomposites using a polymeric precursor for SiC. In: H. Hausner, G. L. Messing, & S. Horano (Eds.), *Ceramic Transactions: Vol 51 Proceedings of the International Conference of Ceramic Processing Science and Technology; September 11–14, 1994, Friedrichshafen,* (pp. 653–658) Westerville, Germany: The American Ceramic Society.

Bourbigot, S., LeBras, M., Dabrowski, F., Gilman, J. W., & Kashiwagi, T. (2000). PA-6 clay nanocomposite hybrid as char forming agent in intumescent formulations. *Fire and Materials*, *24*(4), 201–208.

Branagan, D. J. (2000). In D. E. Alman & J. W. Newkirk (Eds.), *Powder metallurgy, particulate materials for industrial applications*. St. Louis: TMS Publication.

Branagan, D. J., & Tang, Y. (2002). Developing extreme hardness (>15 GPa) in iron based nanocomposites. *Composites Part A—Applied Science and Manufacturing*, *33*(6), 855–859.

Braun, T., Schubert, A., & Sindelys, Z. (1997). Nanoscience and nanotechnology on the balance. *Scientometrics*, *38*(2), 321–325.

Camargo, P. H. C., Nunes, G. G., Friedermann, G. R., Evans, D. J., Leigh, G. J., GSEL, T. -F., et al. (2004). Single-source precursor and homometal approaches to the sol–gel synthesis of iron and titanium oxides. In: *Progress in Colloid and Polymer Science SeriesVol. 128. Surface and Colloid Science* (pp. 221–226). ISBN 978-3-540-21247-8, pp. 221–226. https://doi.org/10.1007/b97089.

Camargo, P. H. C., Nunes, G. G., Friedermann, G. R., Evans, D. J., Leigh, G. J., Tremiliosi-Filho, G. S. E. L., et al. (2003). Titanium and iron oxides produced by sol-gel processing of [FeCl{Ti-2(Opri)Q}]: structural, spectroscopic and morphological features. *Materials Research Bulletin*, *38*(15), 1915–1928.

Carpenter, E. E., Kumbhar, A., Wiemann, J. A., Srikanth Wiggins, H. J., Zhou, W., & O'Connor, C. J. (2000). Synthesis and magnetic properties of gold–iron–gold nanocomposites. *Materials Science and Engineering*, *286*(1), 81–86.

Carrado, K. A., & Xu, L. Q. (1998). In situ synthesis of polymer-clay nanocomposites from silicate gels. *Chemistry of Materials*, *10*(5), 1440–1445.

Cha, S. I., Kim, K. T., Lee, K., Mo, C. B., & Hong, S. H. (2005). Strengthening and toughening of carbon nanotube reinforced alumina nanocomposite fabricated by molecular level mixing process. *Scripta Materialia*, *53*(7), 793–797.

Chaisan, W., Yimnirun, R., & Ananta, S. (2009). Preparation and characterization of ceramic: nanocomposites in the PZT–BT system. *Ceramics International*, *35*(1), 121–124.

Chang, J. H., & An, Y. U. (2002). Nanocomposites of polyurethane with various organoclays: thermomechanical properties, morphology, and gas permeability. *Journal of European Ceramic Society*, *40*(7), 670–677.

Chazeau, L., Cavaille, J. Y., Canova, G., Dendievel, R., & Boutherin, B. (1999). Viscoelastic properties of plasticized PVC reinforced with cellulose whiskers. *Journal of Applied Polymer Science*, *71*(11), 1797–1808.

Chen, W. X., Lee, J. Y., & Liu, Z. (2002). Electrochemical lithiation and de-lithiation of carbon nanotube-Sn2Sb nanocomposites. *Electrochemical Communications*, *4*(3), 260–265.

Chen, W. X., Lee, J. Y., & Liu, Z. (2003). The nanocomposites of carbon nanotube with Sb and SnSb0.5 as Li-ion battery anodes. *Carbon*, *41*(5), 959–966.

Chen, L., Wong, S. C., & Pisharath, S. (2002). Fracture properties of nanoclay-filled polypropylene. *Journal of Applied Polymer Science*, *88*, 3298–3305.

Chen, X., Xia, J., Peng, J., Li, W., & Xie, S. (2000). Carbon-nanotube metal-matrix composites prepared by electroless plating. *Composites Science and Technology*, *60*(2), 301–306.

Chen, L., Yang, W. J., & Yang, C. Z. (1997). Preparation of nanoscale iron and Fe3O4 powders in a polymer matrix. *Journal of Materials Science*, *32*(13), 3571–3575.

Chen, W., Zhang, J., & Cai, W. (2003). Sonochemical preparation of Au, Ag, Pd/SiO2 mesoporous nanocomposites. *Scripta Materialia*, *48*(2), 1061–1066.

Choa, Y. H., Yang, J. K., Kim, B. H., Jeong, Y. K., Lee, J. S., Nakayama, T., et al. (2003). Preparation and characterization of metal: ceramic nanoporous nanocomposite powders. *Journal of Magnetism and Magnetic Materials*, *266*(1-2), 12–19.

Choi, S. M., & Awaji, H. (2005). Nanocomposites: a new material design concept. *Science and Technology of Advanced Materials*, *6*(1), 2–10.

Chow, G. M., Holtz, R. L., Pattnaik, A., Edelstein, A. S., Schlesinger, T. E., & Cammarata, R. C. (1990). Alternative approach to nanocomposite synthesis by sputtering. *Applied Physical Letters*, *56*(19), 1853–1855.

Choy, K. L. (2003). Chemical vapour deposition of coatings. *Progress in Materials Science*, *48* (2), 57–170.

Chung, D. D. L. (2004). Cement-matrix structural nanocomposites. *Metals & Materials International*, *10*(1), 55–67.

Coltro, L., & Borghetti, J. (2007). Plastic packages for personal care products—evaluation of light barrier properties. *Polimeros*, *17*, 56–61.

Contreras, A., López, V. H., & Bedolla, E. (2004). Mg/TiC composites manufactured by pressureless melt infiltration. *Scripta Materialia*, *51*(3), 249–253.

Curtin, W. A., & Sheldon, B. W. (2004). CNT-reinforced ceramics and metals. *Materials Today*, *7*(11), 44–49.

Cushing, B. L., Kolesnichenko, V. L., & O'Connor, C. J. (2004). Recent advances in the liquid-phase syntheses of inorganic nanoparticles. *Chemical Reviews*, *104*(9), 3893–3946.

Dabrowskii, F., Bourbigot, S., Delobel, R., & Lebras, M. L. (2000). Kinetic molding of the thermal degradation of polyamide-6 nanocomposite. *European Polymer Journal*, *36*(2), 273–284.

Dalton, A. B., Coolins, S., Muñoz, E., Razal, J. M., Ebron, V. H., Ferraris, J. P., et al. (2003). Super-tough carbon-nanotube fibres: these extraordinary composite fibres can be woven into electronic textiles. *Nature*, *423*(6941), 703.

Darkrim, F. L., Malbrunot, P., & Tartaglia, G. P. (2002). Review of hydrogen storage by adsorption in carbon nanotubes. *Internation Journal of Hydrogen energy*, *27*(2), 193–202.

Di Lorenzo ML, Errico ME, Avella M. Thermal and morphological characterization of poly (ethylene terephthalate)/calcium carbonate nanocomposites. Journal of Materials Science. 2002; 37(11):2351 2358.

Ding, W., Eitan, A., Fisher, F. T., Chen, X., Dikin, D. A., Andrews, R., et al. (2003). Direct observation of polymer sheathing in carbon nanotube-polycarbonate composites. *Nano Letters*, *3*(11), 1593–1597.

Do Nascimento, G. M., Constantino, V. R. L., & Temperini, M. L. A. (2002). Spectroscopic characterization of a new type of conducting polymer-clay nanocomposite. *Macromolecules*, *35*(20), 7535–7537.

Dresselhaus, M. S., Dresselhaus, G., Saito, R., & Jorio, A. (2005). Raman spectroscopy of carbon nanotubes. *Physics Report—Review Section of Physics Letters*, *409*(2), 47–99.

Dresselhaus, M. S., Dresslhaus, G., & Avouris, P. (2001). *Carbon nanotubes: Synthesis, structure, properties and applications*. Berlin: Springer Verlag.

Dresselhaus, M. S., Dresslhaus, G., & Eklund, P. C. (1996). *Science of fullerenes: Carbon nanotubes*. San Diego: Academic Press.

Dutta A, Das D, Grilli ML, Di Bartolomeo E, Traversa E, Chakravorty D. Preparation of sol-gel nano-composites containing copper oxide and their gas sensing properties. Journal of Sol-Gel Science and Technology 2003; 26(1-3):1085-1094.

El-Eskandarany, M. S. (1998). Mechanical solid state mixing for synthesizing of SiCp/Al nanocomposites. *Journal of Alloys and Compounds*, *279*(2), 263–271.

Endo, M., Kim, Y. A., Ezaka, M., Oshida, K., Yanagisawa, T., & Dresselhaus, M. S. (2003). Selective and efficient impregnation of metal nanoparticles on cup-stacked-type carbon nanofibers. *Nano Letters*, *3*(6), 723–726.

Endo M, Kim YA, Hayashi T, Fukai Y, Oshida K, Terronnes M et al. Structural characterization of cup-stacked-type carbon nanofibers with an entirely hollow core. Applied Physical Letters 2002; 80(7):1267 1269.

Ennas, G., Mei, A., Musinu, A., Piccaluga, G., Pinna, G., & Solinas, S. (1998). Sol-gel preparation and characterization of Ni-SiO2 nanocomposites. *Journal of Non-Crystalline Solids*, *232–234*, 587–593.

Evora, V. M. F., & Shukla, A. (2003). Fabrication, characterization, and dynamic behavior of polyester/TiO2 nanocomposites. *Materials Science and Engineering*, *361*(1-2), 358–366.

Favier, V., Canova, G. R., Shrivastava, S. C., & Cavaille, J. V. (1997). Mechanical percolation in cellulose whisker nanocomposites. *Polymer Engineering Science*, *37*(10), 1732–1739.

Fernando, S. S., Christensen, P. A., Egerton, T. A., & White, J. R. (2009). Humidity dependence of carbon dioxide generation during photodegradation of bixally oriented polypropylene in oxygen. *Polymer Degradation and Stability*, *94*(1), 83–89. https://doi.org/10.1016/j.polymdegradstab.2008.10.007.

Fernando, W., & Satyanarayana, K. G. (2005). Functionalization of single layers and nanofibers: a new strategy to produce polymer nanaocomposites with optimized properties. *Journal of Colloid and Interface Science*, *285*(1), 532–543.

Ferroni, L. P., Pezzotti, G., Isshiki, T., & Kleebe, H. J. (2001). Determination of amorphous interfacial phases in Al2O3/SiC nanocomposites by computer-aided high-resolution electron microscopy. *Acta Materialia*, *49*(11), 2109–2113.

Finegan, I. C., Tibbetts, G. G., & Gibson, R. F. (2003). Modeling and characterization of damping in carbon nanofiber/polypropylene composites. *Composites Science and Technology*, *63*(11), 1629–1635.

Finegan, I. C., Tibbetts, G. G., & Glasgow, D. G. (2003). Surface treatment for improving the mechanical properties of carbon nanofiber/thermoplastic composites. *Journal of Materials Science*, *38*(16), 3485–3490.

Fischer, H. (2003). Polymer nanocomposites: from fundamental research to specific applications. *Materials Science and Engineering: C*, *23*(6-8), 763–772.

Fornes, T. D., Yoon, P. J., Keskkula, H., & Paul, D. R. (2001). Nylon nanocomposites: the effect of matrix molecular weight. *Polymer*, *42*(25), 9929–9940.

Francis, C. W. (1973). Adsorption of polyvinylpyrrolidone on reference clay minerals. *Soil Science*, *115*(1), 40–54.

Frankl, S. J. V., Harik, V. M., Odegard, G. M., Brenner, D. W., & Gates, T. S. (2003). The stress–strain behavior of polymer–nanotube composites from molecular dynamics simulation. *Composites Science and Technology*, *63*(11), 1655–1661.

Fredrickson, G. H., & Bicerano, J. (1999). Barrier properties of oriented disk composites. *Journal of Chemical Physics*, *110*(4), 2181–2188.

Gall, K., Dunn, M. L., Liu, Y., Finch, D., Lake, M., & Munshi, N. A. (2002). Shape memory polymer nanocomposites. *Acta Materialia*, *50*(20), 5115–5126.

Gangopadhyay, R., & Amitabha, D. (2000). Conducting polymer nanocomposites: a brief overview. *Chemistry of Materials*, *12*(7), 608–622.

Gary, A. (2003). Nanotubes of advanced polymer products. *Plastics, Additives and Compounding*, *5*(4), 12.

Ge, Q. L., Lei, T. C., & Zhou, Y. (1991). Microstructure and mechanical-properties of hot-pressed Al2O3-ZrO2 ceramics prepared from ultrafine powders. *Materials Science and Technology*, *7*(6), 490–494.

Ghose, S., Watson, K. A., Delozier, D. M., Working, D. C., Siochi, E. J., & Connell, J. W. (2006). Incorporation of multi-walled carbon nanotubes into high temperature resin using dry mixing techniques. *Composites Part A—Applied Science and Manufacturing*, *37*(3), 465–475.

Ghosh, N. N., & Pramanik, P. (2001). Aqueous sol–gel synthesis of nanosized ceramic composite powders with metal-formate precursors. *Materials Science and Engineering: C*, *16*(1–2), 113–117.

Giannelis, E. P. (1996). Polymer layered silicate nanocomposites. *Advanced Materials*, *8*(1), 29–35.

Giannelis, E. P. (1998). Polymer-layered silicate nanocomposites: synthesis, properties and applications. *Applied Organometalic Chemistry*, *12*(10–11), 675–680.

Giannelis EP, Krishnamoorti R, Manias E. Polymer-silica nanocomposites: model systems for confined polymers and polymer blends. In: Polymers in Confined Environments. Book Series: Advances in Polymer Science. 1999. p. 107-147.

Gilman, J. W., Jackson, C. L., Morgan, A. B., Harris, R., Jr., Manias, E., Giannelis, E. P., et al. (2000). Flammability properties of polymer-layered silicate nanocomposites: propylene and polystyrene nanocomposites. *Chemistry of Materials*, *12*(7), 1866–1873.

Gilmann, J. W. (1999). Flammability and thermal stability studies of polymer-layered-silicate (clay) nanocomposites. *Appied Clay Science*, *15*(1-2), 31–49.

Gleiter, H. (1992). Materials with ultrafine microstructures: retrospectives and perspectives. *Nanostructured Materials*, *1*(1), 1–19.

Gojny, F. H., Wichmann, M. H. G., Kopke, U., Fiedler, B., & Schulte, K. (2004). Carbon nanotube-reinforced epoxy-composites: enhanced stiffness and fracture toughness at low nanotube content. *Composites Science and Technology*, *64*(15), 2363–2371.

Goujon, C., & Goeuriot, P. (2001). Solid state sintering and high temperature compression properties of Al-alloy5000/AlN nanocomposites. *Materials Science and Engineering: A*, *315* (1–2), 180–188.

Greenland, D. J. (1963). Adsorption of polyvinylalcohols by montmorillonite. *Journal of Colloid Science*, *18*(7), 647–664.

Haque, A., & Ramasetty, A. (2005). Theoretical study of stress transfer in carbon nanotube reinforced polymer matrix composites. *Composites Structure*, *71*(1), 68–77.

Harmer, M., Chan, H. M., & Miller, G. A. (1992). Unique opportunities for microstructural engineering with duplex and laminar ceramic composites. *Journal of the American Ceramic Society*, *75*(2), 1715–1728.

Hasegawa, N., Kawasumi, M., Kato, M., Usuki, A., & Okada, A. (1998). Preparation and mechanical properties of polypropylene-clay hybrids using a maleic anhydride-modified polypropylene oligomer. *Journal of Applied Polymer Science*, *67*(1), 87–92.

Haubold, T., & Gertsman, V. (1992). On the structure and properties of nanostructured copper-tungsten alloys. *Nanostructured Materials*, *1*(4), 303–312.

Hench, L. L., & West, J. K. (1990). The sol-gel process. *Chemical Review*, *90*(1), 33–72.

Herron, N., & Thorn, D. L. (1998). Nanoparticles: uses and relationships to molecular cluster compounds. *Advanced Materials*, *10*(15), 1173–1184.

Holtzt, R. L., & Provenzano, V. (1994). Enhanced microhardness of copper-niobium nanocomposites. *Nanostructured Materials*, *4*(2), 241–256.

Hori, T., Kuramoto, N., Tayagaya, H., Karasu, M., Kadokawa, J. I., & Chiba, K. (1999). Preparation of conducting film composed of polyaniline and metal oxide by sol-gel method. *Journal of Materials Research*, *14*(1), 5–7.

Horst, M. F., Tuckart, W., Del Blanco, L., Failla, M. D., & Quinzani, L. M. (2012). Effect of clay concentration on the wear behavior and permeability of polypropylene/clay nanocomposites. Journal of Applied Polymer Science 125 (SI): E495-E502. https://doi.org/10.1002/app.36414.

Hu, Y., Jang, I., & Sinnott, S. B. (2003). Modification of carbon nanotube-polystyrene matrix composites through polyatomic-ion beam deposition: predictions from molecular dynamics simulations. *Composites Science and Technology*, *63*(11), 1663–1669.

Hussain, M., & Simon, G. P. (2003). Fabrication of phosphorus-clay polymer nanocomposites for fire performance. *Journal of Materials Science Letters*, *22*(21), 1471–1475.

Iijima, S. (1991). Helical microtubes of graphitic carbon. *Nature*, *354*(6348), 56–58.

Jackson, C. L., Bauer, B. J., Nakatani, A. I., & Barnes, J. D. (1996). Synthesis of hybrid organic-inorganic materials from interpenetrating polymer network chemistry. *Chemistry of Materials*, *8*(3), 727–733.

Jang, Y., Kim, S., Lee, S., Kim, D., & Um, M. (2005). Fabrication of carbon nano-sized fiber reinforced copper composite using liquid infiltration process. *Composites Science and Technology*, *65*(5), 781–802.

Jeon, H. G., Jung, H. T., Lee, S. W., & Hudson, S. D. (1998). Morphology of polymer silicate nanocomposites. *Polymer Bulletin*, *41*(1), 107–113.

Jiang, L., & Gao, L. (2003). Carbon nanotubes magnetite nanocomposites from solvothermal processes: formation, characterization, and enhanced electrical properties. *Chemistry of Materials*, *15*(14), 2848–2853.

Jimenez, G., Ogata, N., Kawai, H., & Ogihara, T. (1997). Structure and thermal/mechanical properties of poly (ε-caprolactone) - clay blend. *Journal of Applied Polymer Science*, *64*, 2211–2220.

Jordan, J., Jacob, K. I., Tannenbaum, R., Sharaf, M. A., & Jasiuk, I. (2005). Experimental trends in polymer nanocomposites: a review. *Materials Science and Engineering: A*, *393*(1–2), 1–11.

Joseph, M. C., Tsotsos, C., Baker, M. A., Kench, P. J., Rebholz, C., Matthews, A., et al. (2005). Characterisation and tribological evaluation of nitrogen-containing molybdenum-copper pvd metallic nanocomposite films. *Surface and Coating Technology*, *190*(2–3), 345–356.

Kamat, P. V., Flumiani, M., & Dawson, A. (2002). Metal-metal and metal-semiconductor composite nanoclusters. *Colloid Surface A—Physicochemical and Engineering Aspects, 202* (2–3), 269–279.

Kamigaito, O. (1991). What can be improved by nanometer composites? *Journal of Japan Society of Powder Metalurgy, 38*, 315–321.

Karger-Kocsis, J., & Zhang, Z. (2004). In J. F. Palta Calleja & G. Michler (Eds.), *Mechanical properties of polymers based nano structure morphology*. New York: Marcel Dekker.

Kawasumi, M., Hasegawa, N., Kato, M., Usuki, A., & Okada, A. (1997). Preparation and mechanical properties of polypropylene-clay hybrids. *Macromolecules, 30*(20), 6333–6338.

Kickelbick, G. (2003). Concepts for the incorporation of inorganic building 24. blocks into organic polymers on a nanoscale. *Progress in Polymer Science, 28*(1), 83–114.

Kim, J. Y., Kim, M., & Choi, J. H. (2003). Characterization of light emitting devices based on a single-walled carbon nanotube–polymer composite. *Synthetic Metals, 139*(3), 565–568.

Kim, J. W., Kim, S. G., Choi, J. H., & Jhon, M. S. (1999). Synthesis and electrorheological properties of polyaniline-na+-montmorillonite suspensions. *Macromolecular Rapid Communications, 20*(8), 450–452.

Koerner, H., Price, G., Pearce, N. A., Alexer, M., & Vaia, R. A. (2004). Remotely actuated polymer nanocomposites-stress-recovery of carbon-nanotube filled thermoplastic elastomers. *Nature Materials, 3*(2), 115–119.

Kojima, Y., Usuki, A., Kawasumi, M., Okada, A., Fukushima, Y., Karauchi, T., et al. (1993). Mechanical properties of nylon-6-clay hybrid. *Journal of Materials Research, 8*(5), 1185–1189.

Kojima, Y., Usuki, A., Kawasumi, M., Okada, A., Kurauchi, T., & Kamigaito, O. (1993a). One pot synthesis of nylon-6-clay hybrid. *Journal of Polymer Science and Polymer Chemistry, 31*(7), 1755–1758.

Kojima, Y., Usuki, A., Kawasumi, M., Okada, A., Kurauchi, T., & Kamigaito, O. (1993b). Synthesis of nylon-6-clay hybrid by montmorillonite intercalated with ε-caprolactum. *Journal of Polymer Science and Polymer Chemistry, 31*(7), 983–986.

Krishnamoorti, R., Vaia, R. A., & Giannelis, E. P. (1996). Structure and dynamics of polymer-layered silicate nanocomposites. *Chemistry of Materials, 8*(8), 1728–1734.

Kubayashi K, Hayashi S. Woven fabric made of shape memory polymers. United States. Patent 5128. 1992.

Kundu, T. K., Mukherjee, M., Chakravorty, D., & Sinha, T. P. (1998). Growth of nano-alpha-Fe2O3 in a titania matrix by the sol gel route. *Journal of Matererials Science, 33*(7), 1759–1763.

Kupfer, M. (2001). Electronic properties of carbon nanostructures. *Surface Science Reports, 42* (1-2), 1–74.

Kuznetsov, D., & Balazs, A. C. (2000). Scaling theory for end-functionalized polymers confined between two surfaces: predictions for fabricating polymer nanocomposites. *Journal of Chemical Physics, 112*(9), 4365–4375.

Kuzumaki, T., Miyazawa, K., Ichinose, H., & Ito, K. (1998). Processing of carbon nanotube reinforced aluminum composite. *Journal of Materials Research, 13*(9), 2445–2449.

Lan, T., Kavitatna, P. D., & Pinnavia, T. J. (1994). On the nature of polyimide clay hybrid composites. *Chemistry of Materials, 6*(5), 573–575.

Lange, F. F. (1973). Effect of microstructure on strength of si3n4-sic composite system. *Journal of the American Ceramic Society, 56*(9), 445–450.

Laurent, C., Peigney, A., Dumortier, O., & Rousset, A. (1998). Carbon nanotubes Fe- alumina nanocomposites. Part II: microstructure and mechanical properties of the hot-pressed composites. *Journal of the European Ceramic Society, 18*(14), 2005–2013.

Laus, M., Francesangeli, O., & Sandrolini, F. (1997). New hybrid nanocomposites based on an organophilic clay and poly (styrene-b-butadiene) copolymers. *Journal of Materials Research*, *12*(11), 3134–3139.

LeBaron, P. C., Wang, Z., & Pinnavaia, T. J. (1999). Polymer-layered silicate nanocomposites: an overview. *Applied Clay Science*, *15*(1–2), 11–29.

Lee, J. Y., Baljon, A. R. C., Loring, R. F., & Panagiopoulos, A. Z. (1998). Simulation of polymer melt intercalation in layered nanocomposites. *Journal of Chemical Physics*, *109*(23), 10321–10330.

Lee, D. C., & Jang, L. W. (1996). Preparation and characterization of PMMA-clay composite by emulsion polymerization. *Journal of Applied Polymer Science*, *61*(7), 1117–1122.

Lee, D. Y., Lee, M. H., Kim, K. J., Heo, S., Kim, B. Y., & Lee, S. J. (2005). Effect of multiwalled carbon nanotube (M-CNT) loading on M-CNT distribution behavior and the related electromechanical properties of the M-CNT dispersed ionomeric nanocomposites. *Surface and Coating Technology*, *200*(5–6), 1920–1925.

Lee, J., Takekoshi, T., & Giannelis, E. P. (1997). Fire retardant polyetherimide nanocomposites. *Materials Research Society Symposium Proccedings*, *457*, 513–518.

Lei, Y., & Cebe, P. (2009). Crystal polymorphism in electrospun composite nanofibers of poly(vinylindene fluoride) with nanoclay. *Polymer*, *50*, 2133–2141.

Leszczynska, A., Njuguna, J., Pielichowski, K., & Banerjee, J. R. (2007). Polymer/montmorillonite nanocomposites with improved thermal properties: Part I. Factors influencing thermal stability and mechanisms of thermal stability improvement. *Thermochimica Acta*, *453*(2), 75–96.

Li, G. J., Huang, X. X., & Guo, J. K. (2003). Fabrication, microstructure and mechanical properties of Al2O3/Ni nanocomposites by a chemical method. *Material Research Bulletin*, *38* (11–12), 1591–1600.

Li, Y., Wang, S., Zhang, Y., & Zhang, Y. (2006). Crystalization behavior of carbon black-filled polypropylene and polypropylene/epoxy composites. *Journal of Aplied Polymer Science*, *102*, 104–118.

Lijie, C. I., & Jinbo, B. (2006). The reinforcement role of carbon nanotubes in epoxy composites with different matrix stiffness. *Composites Science and Technology*, *66*(3–4), 599–603.

Lim, D. S., An, J. W., & Lee, H. J. (2002). Effect of carbon nanotube addition on the tribological behavior of carbon/carbon composites. *Wear*, *252*(5 6), 512–517.

Lim, D. S., You, D. H., Choi, H. J., Lim, S. H., & Jang, H. (2005). Effect of CNT distribution on tribological behavior of alumina-CNT composites. International Conference on Wear of Materials Special Issue *259*(1–6), 539–544.

Lin, Y., Zhou, B., Fernando, K. A. S., Liu, P., Allard, L. F., & Sun, Y. P. (2003). Polymeric carbon nanocomposites from carbon nanotubes functionalized with matrix polymer. *Macromolecules*, *36*(19), 7199–7204.

Liu, J., Gao, Y., Wang, F., Li, D., & Xu, J. (2002). Preparation and characteristic of a new class of silica/polyimide nanocomposites. *Journal of Materials Science*, *37*(14), 3085–3088.

Liu, T. X., Phang, I. Y., Shen, L., Chow, S. Y., & Zhange, W. D. (2004). Morphology and mechanical properties of multiwalled carbon nanotubes reinforced nylon-6 composites. *Macromolecules*, *37*(19), 7214–7222.

Liu, L. M., Qi, Z. N., & Zhu, X. G. (1999). Studies on nylon 6/clay nanocomposites by melt-intercalation process. *Journal of Applied Polymer Science*, *71*(7), 1133–1138.

Liu, S. H., Qian, X. F., Yuan, J. Y., Yin, J., He, R., & Zhu, Z. K. (2003). Synthesis of monodispersed CdSe nanocrystals in poly(styrene-alt-maleic anhydride) at room temperature. *Materials Research Bulletin*, *38*(8), 1359–1366.

Liu, H. H., Wang, L., Wang, A., Lou, T., Ding, B., & Hu, Z. (1997). Study of SiC/Al nanocomposites under high pressure. *Nanostructured Materials*, *9*(1-8), 225–228.

Livage, J. (1997). Sol-gel processes. *Current Opinon in Solid State and Materials Science*, *2*(2), 132–136.

Lu, G. T., & Huang, Y. (2002). Synthesis of polymaleimide/silica nanocomposites. *Journal of Materials Science*, *37*(11), 2305–2308.

Lucas, M., & Young, R. J. (2004). Raman spectroscopic study of the effect of strain on the radial breathing modes of carbon nanotubes in epoxy/SWNT composites. *Composites Science and Technology*, *64*(15), 2297–2302.

Lwan, B. R. (2002). Ceramic-based layer structures for biomechanical applications. *Current Opinion in Solid State and Materials Science*, *8*(3), 229–235.

Ma, R. Z., Wu, J., Wei, B. Q., Liang, J., & Wu, D. H. (1998). Processing and properties of carbon nanotubes-nano-SiC ceramic. *Journal of Materials Science*, *33*(21), 5243–5246.

Ma, H. A., Zeng, J. J., Realff, M. L., Kumar, S., & Schiraldi, D. A. (2003). Processing, structure, and properties of fibers from polyester/carbon nanofiber composites. *Composites Science and Technology*, *63*(11), 1617–1628.

Mabuchi, H., Tsuda, H., Ohtsuka, T., Matsui, T., & Morii, K. (1999). In-situ synthesis of Si3N4-SiC composites by reactive hot-pressing high temperatures-high pressures. *High Temperatures-high Pressures*, *31*(5), 499–506.

Manias, E., Chen, E., Krisnhamoorti, R., Genzer, J., Kramer, E. J., & Giannelis, E. P. (2000). Intercalation kinetics of long polymers in 2 nm confinements. *Macromolecules*, *33*(21), 7955–7966.

Marchal, Y., Delannay, F., & Froyen, L. (1996). The essential work of fracture as a means for characterizing the influence of particle size and volume fraction on the fracture toughness of plates of Al/SiC composites. *Scripta Materialia*, *35*(2), 193–198.

Maser, W. K., Benito, A. M., Callejas, M. A., Seeger, T., Martínez, M. T., Schreiber, J., et al. (2003). Synthesis and characterization of new polyaniline/nanotube composites. *Materials Science and Engineering*, *23*(1-2), 87–91.

Mathur, S., Veith, M., Shen, H., Hufner, S., & Jilavi, M. (2002). Structural and optical properties of NdAlO3 nanocrystals embedded in an Al2O3 matrix. *Chemistry of Materials*, *14*(2), 568–582.

Mbhele, Z. H., Salemane, M. G., Van Sittert, C. G. C. E., Nedeljkovic, J. M., Djokovic, V., & Luyt, A. S. (2003). Fabrication and characterization of silver-polyvinyl alcohol nanocomposites. *Chemistry of Materials*, *15*(26), 5019–5024.

Meda, L., Marra, G., Galfetti, L., Inchingalo, S., Severini, F., & De Luca, L. (2005). Nanocomposites for rocket solid propellants. *Composites Science and Technology*, *65*(5), 769–773.

Merkulov, V. I., Lowndes, D. H., Wei, Y. Y., Eres, G., & Voelkl, E. (2000). Patterned growth of individual and multiple vertically aligned carbon nanofibers. *Applied Physical Letters*, *76* (24), 3555–3557.

Messersmith, P. B., & Giannelis, E. P. (1994). Synthesis and characterization of layered silicate-epoxy nanocomposites. *Chemistry of Materials*, *6*(10), 1719–1725.

Mohanty, A. K., Drzal, L. T., & Misra, M. (2003). Nano reinforcement of Bio-based polymers—The Hope and The Reality. In: *88. Polymeric Materials Science & Engineering Presented at 225th ACS National Meeting New Orleans, March 2003* (pp. 60–61). U665-665, Abstract: 33-PMSE.

Morgan, A. B., Harris, R. H., Kashiwagi, T., Chyall, L. J., & Gilman, J. W. (2002). Flammability of polystyrene layered silicate (clay) nanocomposites: carbonaceous char formation. *Fire and Materials*, *26*(6), 247–253.

Nafchi, H. R., Abdouss, M., Najafi, S. K., Gargari, R. M., & Mazhar, M. (2015). Effects of nanoclay particles and oxidied polypropylene polymers on improvements of the thermal properties of wood plastic composite. *Maderas Ciencia y technologia, 17(1)*, 45–54 https://doi.org/4067/S0718-221X2015005000005.

Nakahira, A., & Niihara, K. (1992). Strctural ceramics-ceramic nanocomposites by sintering method: roles of nano-size particles. *Journal of the Ceramic Society of Japan, 100*(4), 448–453.

Nalwa, H. S. (2000). *Handbook of nanostructured materials and technology*. New York: Academic Press.

Natile, M. M., & Glisenti, A. (2003). New NiO/Co3O4 and Fe2O3/ Co3O4 nanocomposite catalysts: synthesis and characterization. *Chemistry of Materials, 15*(13), 2502–2510.

Nawa, M., Sekino, T., & Niihara, K. (1994). Fabrication and mechanical-behavior of Al2O3/Mo nanocomposites. *Journal of Materials Science, 29*(12), 3185–3192.

Nielsen, L. E. (1967). Models for the permeability of filled polymer systems. *Journal of Macromolecular Science Part A: Pure and Applied Chemistry, 1*(5), 929–942.

Niihara, K. (1991). New design concept of strctural ceramics-ceramic nanocomposite. *Journal of the Ceramic Society of Japan (Nippon Seramikkusu Kyokai Gakujutsu Ronbunshi), 99* (6), 974–982.

Ning, J., Zhang, J., Pan, Y., & Guo, J. (2003). Fabrication and mechanical properties of SiO2 matrix composites reinforced by carbon nanotube. *Materials Science and Engineering: A—Structural Materials Properties Microstructure and Processing, 357*(1-2), 392–396.

Noguchi, T., Magario, A., Fuzukawa, S., & Shimizu, S. (2004). Carbon nanotube/aluminium composites with uniform dispersion. *Materials Transactions, 45*(2), 602–604.

Noh, M. W., & Lee, D. C. (1999). Synthesis and characterization of ps-clay nanocomposite by emulsion polymerization. *Polymer Bulletin, 42*(5), 619–626.

Oakada, A., & Usuki, A. (1995). The chemistry of polymer-clay hybrids. *Materials Science and Engineering, 3*(2), 109–115.

Ogasawara, T., Ishida, Y., Ishikawa, T., & Yokota, R. (2004). Characterization of multi-walled carbon nanotube/phenylethynyl terminated polyimide composites. *Composites Part A—Applied Science, 35*(1), 67–74.

Ogata, N., Jimenez, G., Kawai, H., Ogihara, T., & Ogata, N. (1997). Structure and thermal/ mechanical properties of poly (l-lactide)— clay blend. *Journal of Polymer Science Part B: Polymer Physics, 35*(2), 389–396.

Ogawa, M., & Kuroda, K. (1997a). Preparation of inorganic composites through intercalation of organoammoniumions into layered silicates. *Bulletin of the Chemical Society of Japan, 70* (11), 2593–2618.

Ogawa, M., & Kuroda, K. (1997b). Preparation of inorganic–organic nanocomposites through intercalation of organoammonium ions into layered silicates. *Bulletin of the Chemical Society of Japan, 70*(11), 2593–2618.

Okada, A., Kawasumi, M., Kurauchi, T., & Kamigaito, O. (1987). Synthesis and characterization of a nylon 6-clay hybrid. *Polymer Preprints, 28*(2), 447–448.

Okada, A., Kawasumi, M., Usuki, A., Kojima, Y., Kurauchi, T., & Kamigaito, O. (1990). In: D. W. Schaefer & J. E. Mark (Eds.), *Polymer-based molecular composites. Proceedings of the MRS Symposium; 1990; Pittsburgh. USA* (pp. 18–45).

Okamoto, M., Morita, S., & Kotaka, T. (2001). Dispersed structure and ionic conductivity of smectic clay/polymer nanocomposites. *Polymer, 42*(6), 2685–2688.

Okamoto, M., Morita, S., Taguchi, H., Kim, Y. H., Kotaka, T., & Tateyama, H. (2000). Synthesis and structure of smectic clay/poly(methyl methacrylate) and clay/polystyrene nanocomposites via in situ intercalative polymerization. *Polymer, 41*(10), 3887–3890.

Oriakhi, C. O., Farr, I. V., & Lerner, M. M. (1997). Thermal characterization of poly (styrene sulfonate)/layered double hydroxide nanocomposites. *Clays and Clay Minerals*, *45*(2), 194–202.

Ounaies, Z., Park, C., Wise, K. E., Siochi, E. J., & Harrison, J. S. (2003). Electrical properties of single wall carbon nanotube reinforced polyimide composites. *Composites Science and Technology*, *63*(11), 1637–1646.

Pandey, J. K., Kumar, A. P., Misra, M., Mohanty, A. K., Drzal, L. T., & Singh, R. P. (2005). Recent advances in biodegradable nanocomposites. *Journal of Nanoscience and Nanotechnology*, *5*(4), 497–526.

Pandey, J. K., Reddy, K. R., Kumar, A. P., & Singh, R. P. (2005). An overview on the degradability of polymer nanocomposites. *Polymer Degradation and Stability*, *88*(2), 234–250.

Park, S. S., Bernet, N., De La Roche, S., & Hanh, H. T. (2003). Processing of iron oxide-epoxy vinyl ester nanocomposites. *Journal of Composite Materials*, *37*(5), 465.

Park, S. J., Cho, M. S., Lim, L. T., Choi, H. J., & Jhon, M. S. (2003). Synthesis and dispersion characteristics of multi-walled carbon nanotube composites with poly(methyl methacrylate) prepared by in-situ bulk polymerization. *Macromolecular Rapid Communications*, *24*(18), 1070–1073.

Park, A. Y., Kwon, H., Woo, A. J., & Kim, S. J. (2005). Layered double hydroxide surface modified with (3-aminopropyl) triethoxysilane by covalent bonding. *Advanced Materials*, *17* (1), 106–109.

Park, C., Ounaies, Z., Watson, K. A., Crooks, R. E., Smith, J. J., Lowther, S. E., et al. (2002). Dispersion of single wall carbon nanotubes by in situ polymerization under sonication. *Chemical Physics Letters*, *364*(3-4), 303–308.

Passaglia, E., Bertoldo, M., Ciardelli, F., Prevosto, D., & Lucchesi, M. (2008). Evidences of macromolecular chains confinement of ethylene-propylene copolymer in organophilic montmorillonite nanocomposites. *European Polymer Journal*, *44*(5), 1296–1308.

Pathak, J. P., Tiwari, S. N., & Malhotra, S. L. (1986). On the wear characteristics of leaded aluminum bearing alloys. *Wear*, *112*(3–4), 341–353.

Peigney, A., Flahaut, E., Lautent, C. H., Chastel, F., & Rousset, A. (2002). Aligned carbon nanotubes in ceramic-matrix nanocomposites prepared by high-temperature extrusion. *Chemical Physics Letters*, *352*(1–2), 20–25.

Peigney, A., et al. (2000). Carbon nanotubes in novel ceramic matrix nanocomposites. *Ceramic International*, *26*(6), 677–683.

Penn, S. G., He, L., & Natan, M. J. (2003). Nanoparticles for bioanalysis. *Current Opinion in Chemical Biology*, *7*(5), 609–615.

PerkinElmer (2011). *Polymer idenfication using mid infrared spectroscopy. Available at: https://www.perkinelmer.com/labsolutions/resources/docs/APP_PolymerIdentification MidInfaredSpectroscopy.pdf.*

Philip, B., Xie, J., Abraham, J. K., & Varadan, V. K. (2005). Polyaniline - carbon nanotube composites: starting with phenylamino functionalized carbon nanotubes. *Polymer Bulletin*, *53*(2), 127–138.

Pokropivnyi, V. V. (2002a). Two-dimensional nanocomposites: photonic crystals and nanomembranes (review): types and preparation. *Powder Metallurgy and Metal Ceramics*, *41*(5–6), 264–272.

Pokropivnyi, V. V. (2002b). Two-dimensional nanocomposites: photonic crystals and nanomembranes (review). II: properties and applications. *Powder Metallurgy and Metal Ceramics*, *41*(7–8), 369–381.

Provenzano, V., Louat, N. P., Imam, M. A., & Sadananda, K. (1992). Ultrafine superstrength materials. *Nanostructured Materials*, *1*(1), 89–94.

Qian, D., Dickey, E. C., Andrews, R., & Rantell, T. (2000). Load transfer and deformation mechanisms in carbon nanotube-polystyrene composites. *Applied Physical Letters*, *76*(20), 2868–2870.

Qian, D., Wagner, G. J., Liu, W. K., Yu, M. F., & Ruoff, R. S. (2002). Mechanics of carbon nanotubes. *Applied Mechanics Review*, *55*(6), 495–533.

Qu, P., Gao, Y., Wu, G., & Zhang, L. (2010). PLA/cellulose nanocomposites. *BioResources*, *5*(3), 1811–1823.

Quin, D., Dickey, E. C., Andrews, R., & Rantell, T. (2000). Load transfer and deformation mechanisms in carbon nanotube-polystyrene composites. *Applied Physical Letters*, *76*(20), 2868–2870.

Raka, L., Bogoeva-Gaceva, G., Lu, K., & Loos, J. (2009). Characterization of latex-based isotactic polypropylene/clay nanocomposites. *Journal of Polymer Science Part B Polymer Physics*, *50*, 3739–3746.

Ray, S. S., & Bousmina, M. (2005). Biodegradable polymers and their layered silicate nanocomposites: in greening the 21st century materials world. *Progress in Matererials Science*, *50*(8), 962–1079.

Ray, S. S., & Okamoto, M. (2003). Polymer—layered silicate nanocomposites: a review from preparation to processing. *Progress in Polymer Science*, *28*(11), 1539–1641.

Riedel, R., Seher, M., & Becker, G. (1989). Sintering of amorphous polymer-derived Si, N and C containing composite powders. *Journal of European Ceramic Society*, *5*(2), 113–122.

Riedel, R., Seher, M., Mayer, J., & Szabo, D. (1995). Polymer-derived Si-based bulk ceramics. Part I: preparation, processing and properties. *Journal of European Ceramic Society*, *15*(8), 703–715.

Riedel, R., Strecker, K., & Petzow, G. (1989). In situ polysilane-derived silicon-carbide particulates dispersed in silicon nitride composite. *Journal of the American Ceramic Society*, *72*(11), 2071–2077.

Rodriguez-Llamazares, S., Rivas, B. L., Perez, M., Perrin-Sarazin, F., Maldonado, A., & Venegas, C. (2011). The effect of clay type and of clay-masterbatch product in the preparation of polypropylene/clay nanocomposites. *Journal of Applied Polymer Science*, *122*(3), 2013–2025.

Roy, R., Roy, R. A., & Roy, D. M. (1986). Alternative perspectives on "quasi-crystallinity": non-uniformity and nanocomposites. *Materials Letters*, *4*(8 9), 323–328.

Ruhle, M. (2003). In-situ formation of carbon nanotubes in an alumina-nanotube composite by spray pyrolysis. *Carbon*, *41*(14), 2737–2741.

Sakka, Y., Bidinger, D. D., & Aksay, A. (1995). Processing of silicon carbide-mullite-alumina nanocomposites. *Journal of the American Ceramic Society*, *78*(2), 479–486.

Sandler, J., Shaffer, M. S. P., Prasse, T., Bauhofer, W., Schulte, K., & Windle, A. H. (1999). Development of a dispersion process for carbon nanotubes in an epoxy matrix and the resulting electrical properties. *Polymer*, *40*(21), 5967–5971.

Sandler, J., Windle, A. H., Werner, P., Altstädt, V., Es, M. V., & Shaffer, M. S. P. (2003). Carbon-nanofibre-reinforced poly (ether ether ketone) fibres. *Journal of Materials Science*, *38* (10), 2135–2141.

Sasaki, G., Suga, T., Yanai, T., Suganuma, K., & Niihara, K. (1994). Microstructure of B4C/TiB2 composite fabricated by reaction sintering of B4C and TiC. *Journal of the Ceramic Society of Japan*, *102*(1184), 321–325.

Sawaguchi, A., Toda, K., & Niihara, K. (1991). Mechanical and electrical properties of silicon-silicon nitride-silicon carbide nanocomposite material. *Journal of the American Ceramic Society*, *74*(2), 1142–1144.

Schadler, L. S., Giannaris, S. C., & Ajayan, P. M. (1998). Load transfer in carbon nanotube epoxy composites. *Applied Physical Letters*, *73*(26), 3842–3844.

Scherer, C. (1999). *PA film grade with improved barrier properties for flexible food packaging applications. Proceedings of the 99th New Plastics. February 2–4, London.*

Schmidt, D., Shah, D., & Giannelis, E. P. (2002). New advances in polymer/layered silicate nanocomposites. *Current Opinion in Solid State & Materials Science*, *6*(3), 205–212.

Sealy, C. (2004). Stronger by a hair—composites: glassy route to ultrahard ceramics: composites: uncommon behavior in ceramic composites: composites. *Materials Today*, *7*(10), 15.

Sen, S., Choudharya, R. N. P., & Pramanik, P. (2004). Synthesis and characterization of nanostructured ferroelectric compounds. *Materials Letters*, *58*(27-28), 3486–3490.

Serp, P., Corrias, M., & Kalck, P. (2003). Carbon nanotubes and nanofibers in catalysis. *Appied Catalysis A: General*, *253*(2), 337–358.

Shofner, M. L., Rodríguez, –. M. F. J., Vaidyanathan, R., & Barrera, E. V. (2003). Single wall nanotube and vapor grown carbon fiber reinforced polymers processed by extrusion freeform fabrication. *Composites Part A: Applied Science and Manufacturing*, *34*(12), 1207–1217.

Siegel, R. W., Chang, S. K., Ash, B. J., Stone, J., Ajayan, P. M., Doremus, R. W., et al. (2001). Mechanical behaviour of polymer and ceramic matrix nanocomposites. *Scripta Materialia*, *44*(8–9), 2061–2064.

Sinnott, S. B., Andrews, R., Qian, D., Rao, A. M., Mao, Z., Dickey, E. C., et al. (1999). Model of carbon nanotube growth through chemical vapor deposition. *Chemical Physics Letters*, *315*(1–2), 25–30.

Sivakumar, S., Sibu, C. P., Mukundan, P., Pillai, P. K., & Warrier, K. G. K. (2004). Nanoporous titania-alumina mixed oxides: an alkoxide free sol-gel synthesis. *Materials Letters*, *58*(21), 2664–2669.

Srinivasan, D., & Chattopadhyay, K. (2004). Hardness of high strength nanocomposite Al–X–Zr (X = Si,Cu,Ni) alloys. *Materials Science and Engineering A: Structural Materials Properties Microstructure and Processing*, *375–377*(Special Issue), 1228–1234.

Srivastava, D., Wei, C., & Cho, K. (2003). Nanomechanics of carbon nanotubes and composites. *Applied Mechanics Review*, *56*(2), 215–230.

Stearns, L. C., Zhao, J., & Harmer, M. P. (1992). Processing and microstructure development in Al2O3-SiC 'nanocomposites'. *Journal of European Ceramic Society*, *10*(3), 473–477.

Sternitzke, M. (1997). Review: Structural ceramic nanocomposites. *Journal of European Ceramic Society*, *17*(9), 1061–1082.

Suzuki, Y., Sekino, T., & Niihara, K. (1995). Effects of ZrO2 addition on microstructure and mechanical properties of MoSi2. *Scripta Metallurgica and Materialia*, *33*(1), 69–74.

Tai, W. P., Kim, Y. S., & Kim, J. G. (2003). Fabrication and magnetic properties of Al2O3/Co nanocomposites. *Materials Chemistry and Physics*, *82*(2), 396–400.

Tai, N. H., Yeh, M. K., & Liu, H. H. (2004). Enhancement of the mechanical properties of carbon nano tube/phenolic composites using a carbon nanotube network as the reinforcement. *Carbon*, *42*(12–13), 2774–2777.

Tang, W., Santare, M. H., & Advani, S. G. (2003). Melt processing and mechanical property characterization of multi-walled carbon nanotube/high density polyethylene (MWNT/HDPE) composite films. *Carbon*, *41*(14), 2779–2785.

Theng, B. K. G. (1974). *The chemistry of clay-organic reactions.* New York: Wiley.

Theng, B. K. G. (1979). *Formation, properties of clay–polymer complexes.* Amsterdam: Elsevier.

Thompson, C. M., Herring, H. M., Gates, T. S., & Connel, J. W. (2003). Preparation and characterization of metal oxide/polyimide nanocomposites. *Composites Science and Technology*, *63*(11), 1591–1598.

Thostenson, E. T., & Chou, T. W. (2002). Aligned multi-walled carbon nanotube-reinforced composites: processing and mechanical characterization. *Journal of Physics D Applied Physics*, *35*(16), L77–L80.

Thostenson, E. T., & Chou, T. W. (2003). On the elastic properties of carbon nanotube-based composites: modelling and characterization. *Journal of Physics D Applied Physics*, *36*(5), 573–582.

Thostenson, E. T., Li, C., & Chou, T. W. (2005). Nanocomposites in context. *Composites Science and Technology*, *65*(3–4), 491–516.

Thostenson, E. T., Ren, Z., & Chou, T. W. (2001). Advances in the science and technology of carbon nanotubes and their composites: a review. *Composites Science and Technology*, *61* (13), 1899–1912.

Timms, L. A., Ponton, C. B., & Strangwood, M. (2002). Processing of Al2O3/SiC nanocomposites—Part 2: green body formation and sintering. *Journal of the European Ceramic Society*, *22*(9-10), 1569–1586.

Tjong, S. C., & Wang, G. S. (2004). High-cycle fatigue properties of Al-based 41. composites reinforced with in situ TiB2 and Al2O3 particulates. *Materials Science and Engineering: A*, *386*(1-2), 48–53.

Tomasko, D. L., Han, X., Liu, D. H., & Gao, W. (2003). Supercritical fluid applications in polymer nanocomposites. *Current Opinion in Solid State and Materials Science*, *7*(4–5), 407–412.

Trindade T, Neves MC, Barros AMV. Preparation and optical properties of CdSe/polymer nanocomposites. Scripta Materialia 2000; 43(6): 567 571.

Trindade, T., & O'Brien, P. (1996). A single source approach to the synthesis of CdSe. nanocrystallites. *Advanced Materials*, *8*(2), 161–163.

Usuki, A., Kawasumi, M., Kojima, Y., Okada, A., Kurauchi, T., & Kamigaito, O. (1993). Swelling behaviour of montmorrinollitecation exchanged for Ω-amino acid by ε–Caprolactum. *Journal of Materials Research*, *8*(5), 1174–1178.

Usuki, A., Kojima, Y., Kawasumi, M., Okada, A., Fukushima, Y., Kurauchi, T., et al. (1993). Synthesis of Nylon-6-clay hybrid. *Journal of Materials Research*, *8*(5), 1179–1183.

Vaia, R. A., & Giannelis, E. P. (1997a). Lattice of polymer melt intercalation in organically modified layered silicates. *Macromolecules*, *30*(25), 7990–7999.

Vaia, R. A., & Giannelis, E. P. (1997b). Polymer melt intercalation in organically modified layered silicates: model predictions and experiment. *Macromolecules*, *30*(25), 8000–8009.

Vaia, R. A., Ishii, H., & Giannelis, E. P. (1993). Synthesis and properties two-dimensional nanostructures by direct intercalation of polymer melts in layered silicates. *Chemistry of Materials*, *5*(12), 1694–1696.

Vaia, R. A., Price, G., Ruth, P. N., Nguyen, H. T., & Lichtenhan, J. (1999). Polymer/layered silicate nanocompsosites as high performance ablative materials. *Applied Clay Science*, *15*(1-2), 67–92.

Vaia, R. A., Vasudevan, S., Krawiec, W., Scanlon, L. G., & Giannelis, E. P. (1995). New polymer electrolyte nanocomposites: melt intercalation of poly (ethylene oxide) in mica-type silicates. *Advanced Materials*, *7*(2), 154–156.

Vaia, R. A., & Wagner, H. D. (2004). Framework for nanocomposites. *Materials Today*, *7*(11), 32–37.

Valentini, L., Biagiotti, J., Kenny, J. J., & Santucci, S. (2003). Morphological characterization of single-walled carbon nanotubes-PP composites. *Composites Science and Technology*, *63*(8), 1149–1153.

Valentini, L., Puglia, D., Frulloni, E., Armentano, I., Kenny, J. M., & Santucci, S. (2004). Dielectric behavior of epoxy matrix/single-walled carbon nanotube composites. *Composites Science and Technology*, *64*(1), 23–33.

Veith, M., Mathur, S., Lecerf, N., Bartz, K., Heintz, M., & Huch, V. (2000). Synthesis of a NdAlO3/Al2O3 ceramic-ceramic composite by single-source precursor CVD. *Chemistry of Materials*, *12*(2), 271–274.

Veith, M., Mathur, S., Shen, H., Lecerf, N., Huffner, S., & Jilavi, M. (2001). Single-step preparation of oxide-oxide nanocomposites: chemical vapor synthesis of LnAlO3/Al2O3 (Ln = Pr, Nd) thin films. *Chemistry of Materials*, *13*(11), 4041–4052.

Venkatraman, B., & Sundararajan, G. (1996). The sliding wear behaviour of Al-SiC particulate composites—I. Macrobehaviour. *Acta Materialia*, *44*(2), 451–460.

Viart, N., Richard-Plouet, M., Muller, D., & Pourroy, G. (2003). Synthesis and characterization of Co/ZnO nanocomposites: towards new perspectives offered by metal/piezoelectric composite materials. *Thin Solid Films*, *437*(1–2), 1–9.

Voevodin, A. A., & Zabinski, J. S. (2005). Nanocomposite and nanostructured tribological materials for space applications. *Composites Science and Technology*, *65*(5), 741–748.

Vorotilov, K. A., Yanovskaya, M. I., Turevskaya, E. P., & Sigov, A. S. (1999). Sol-gel derived ferroelectric thin films: avenues for control of microstructural and electric properties. *Journal of Sol-Gel Science and Technology*, *16*(2), 109–118.

Wang, C., Guo, Z. X., Fu, S., Wu, W., & Zhu, D. (2004). Polymers containing fullerene or carbon nanotube structures. *Progress in Polymer Science*, *29*(11), 1079–1141.

Wang, S., Long, C., Wang, X., Li, Q., & Qi, Z. (1998). Synthesis and properties of silicone rubber/organomontmorillonite hybrid nanocomposites. *Journal of Applied Polymer Science*, *63*(8), 1557–1561.

Wang, X. T., Padture, N. P., & Tanaka, H. (2004). Contact-damage-resistant ceramic/single-wall carbon nanotubes and ceramic/graphite composites. *Nature Materials*, *3*(8), 539–544.

Wang, Z., & Pinnavaia, T. J. (1998a). Nanolayer reinforcement of elastomeric polyurethane. *Chemistry of Materials*, *10*(12), 3769–3771.

Wang, Z., & Pinnavaia, T. J. (1998b). Hybrid organic-inorganic nanocomposites: exfoliation of magadiite nanolayers in an elastomeric epoxy polymer. *Chemistry of Materials*, *10*(7), 1820–1826.

Warrier, K. G. K., & Anilkumar, G. M. (2001). Densification of mullite-SiC nanocomposite sol-gel precursors by pressureless sintering. *Materials Chemistry and Physics*, *67*(1–3), 263–266.

Watkins, J. J., & Mccarthy, T. J. (1994). Polymerization in supercritical fluid-swollen polymers: a new route to polymer blends. *Macromolecules*, *27*(17), 4845–4847.

Watkins, J. J., & Mccarthy, T. J. (1995a). Polymer/metal nanocomposite synthesis in supercritical CO2. *Chemistry of Materials*, *7*(11), 1991–1994.

Watkins, J. J., & Mccarthy, T. J. (1995b). Polymerization of styrene in supercritical CO2-swollen poly (chlorotrifluoroethylene). *Macromolecules*, *28*(12), 4067–4074.

Watkins JJ, Mccarthy TJ. Chemistry in supercritical fluid-swollen polymers: direct synthesis of metal-polymer nanocomposites: Part 2. Abstract of Papers of the American Chemical Society 1995c; 210:84 84.

Weimer, A. W., & Bordia, R. K. (1999). Processing and properties of nanophase In-situ synthesis of SiC/ Si3N4 composites. *Composites Part B: Engineering*, *30*(7), 647–655.

West, R., Wang, Y., & Goodson, T. (2003). Nonlinear absorption properties in novel gold nanostructured topologies. *Journal of Physical Chemistry B*, *107*(15), 3419–3426.

Wiesenberger, M. C., Grulke, E. A., Jacques, D., Ramtell, T., & Andrews, R. (2003). Enhanced mechanical properties of polyacrylonitrile: multiwall carbon nanotube composite fibers. *Journal of Nanoscience and Nanotechnology*, *3*(6), 535–539.

Wilson, O. C., Olorunyolemi, T., Jaworski, A., Borum, L., Young, D., Siriwat, A., et al. (1999). Surface and interfacial properties of polymer-intercalated layered double hydroxide nanocomposites. *Appied Clay Science*, *15*(1–2), 265–279.

Wong M, Paramsothy M, Xu XJ, Ren Y, Liao K. Physical interactions at carbon nanotube-polymer interface. Polymer 2003; 44(25):7757 7764.

Wunderlich, W., Padmaja, P., & Warrier, K. G. K. (2004). TEM characterization of sol-gel-processed alumina–silica and alumina–titania nano-hybrid oxide catalysts. *Journal of European Ceramic Society*, *24*(2), 313–317.

Wypych, F., Adad, L. B., & Grothe, M. C. (1998). Synthesis and characterisation of the nanocomposites K-0, K0,1(PEO)xMoS2 (X = 0,5; 1,2). *Quimica Nova*, *21*(6), 687–692.

Wypych, F., Seefeld, N., & Denicolo, I. (1997). Preparation of nanocomposites based on the encapsulation of conducting polymers into 2H-MoS2 and 1T-TiS2. *Quimica Nova*, *20* (4), 356–360.

Xia, Z., Riester, L., Curtin, W. A., Li, H., Sheldon, B. W., Liang, J., et al. (2004). Direct observation of toughening mechanisms in carbon nanotube ceramic matrix composites. *Acta Materialia*, *52*(4), 931–944.

Xia, H. S., Wang, Q., Li, K. S., & Hu, G. H. (2004). Preparation of polypropylene/carbon nanotube composite powder with a solid-state mechanochemical pulverization process. *Journal of Applied Polymer Science*, *93*(1), 378–386.

Xiaochun, L., Yang, Y., & Cheng, X. (2004). Ultrasonic-assisted fabrication of metal matrix nanocomposites. *Journal of Materials Science*, *39*(9), 3211–3212.

Xie, X. L., Mai, Y. W., & Zhou, X. P. (2005). Dispersion and alignment of carbon nanotubes in polymer matrix: a review. *Materials Science and Engineering: R*, *49*(4), 89–112.

Xu, R., Manias, E., Snyder, A. J., & Runt, J. (2001). New biomedical poly (urethane uera)-layered silicate nanocomposites. *Macromolecules*, *34*(2), 337–339.

Xu, Y. S., Ray, G., & Abdel-Magid, B. (2006). Thermal behavior of single-walled carbon nanotube polymer–matrix composites. *Composites part A: Applied Science and Manufacturing*, *37*(1), 114–121.

Xu, C. L., Wei, B. Q., Ma, R. Z., Liang, J., Ma, X. K., & Wu, D. H. (1999). Fabrication of aluminium-carbon nanotube composites and their electrical properties. *Carbon*, *37*(1), 855–858.

Xu, X., Yin, Y., Ge, X., Wu, H., & Zhang, Z. (1998). Γ-radiation synthesis of poly (acrylic acid)–metal nanocomposites. *Materials Letters*, *37*(6), 354–358.

Yang, J., & Schaller, R. (2004). Mechanical spectroscopy of mg reinforced with Al2O3 short fibers and carbon nanotubes. *Materials Science and Engineering*, *370*(1–2), 512–515.

Yang, Y., Zhu, Z. K., Yin, J., Wang, X. Y., & Qi, Z. E. (1999). Preparation and properties of hybrids of organo-soluble polyimide and montmorillonite with various chemical surface modification methods. *Polymer*, *40*(15), 4407–4414.

Yano, K., Usuki, A., & Okada, A. (1997). Synthesis and properties of polyimide-clay hybrid films. *Journal of Polymer Science and Polymer Chemistry*, *35*(11), 2289–2294.

Yano, K., Usuki, A., Okada, A., Kurauchi, T., & Kamigaito, O. (1993). Synthesis and properties of polimidie-clay hybrid. *Journal of Polymer Science Part A: Polymer Chemistry*, *31*(10), 2493–2498.

Yao, K. J., Song, M., Hourston, D. J., & Luo, D. Z. (2002). Polymer/layered clay nanocomposites: 2- polyurethane nanocomposites. *Polymer*, *43*(3), 1017–1020.

Ying, D. Y., & Zhang, D. L. (2000). Processing of Cu–Al2O3 metal matrix nanocomposite materials by using high energy ball milling. *Materials Science and Engineering*, *286* (1), 152–156.

Yoon, E. S., Lee, J. S., Oh, S. T., & Kim, B. K. (2002). Microstructure and sintering behavior of W–Cu nanocomposite powder produced by thermo-chemical process. *International Journal of Refractory Metals and Hard Materials*, *20*(3), 201–206.

Zavyalov, S. A., Pivkina, A. N., & Schoonman, J. (2002). Formation and characterization of metal-polymer nanostructured composites. *Solid State Ionics*, *147*(3–4), 415–419.

Zehetmeyer, G., Scheibel, J. M., & Soares, R. M. D. (2013). Morphological, Optical, and Barrier Properties of PP/MMT Nanocomposites. *Polymer Bulletin*, *70*, 2181–2191.
Zhan, G. D., Kuntz, J. D., Wan, J., Garay, J. E., & Mukherjee, A. K. (2003). Electrical properties of nanoceramics reinforced with ropes of single-walled carbon nanotubes. *Applied Physical Letters*, *83*(6), 1228–1230.
Zhan, G. D., Kuntz, J. D., Wan, J., & Mukherjee, A. K. (2003). Single-wall carbon nanotubes as attractive toughening agents in alumina-based nanocomposites. *Nature Materials*, *2*(1), 38–42.
Zhang, J., Manias, E., & Wilkie, C. A. (2008). Polymerically Modified Layered Silicates: an Effective Route to Nanocomposites. *Journal of Nanoscience and Nanotechnology*, *8*(4), 1597–1615.
Zhao, X., Urano, K., & Ogasawara, S. (1989). Adsorption of polyethylene glycol from aqueous solutions on monmorillonite clays. *Colloid and Polymer Science*, *267*(10), 899–906.
Zhu, J., Morgan, A. B., Lamelas, F. J., & Wilkie, C. A. (2001). Fire properties of polystyrene-clay nanocomposites. *Chemistry of Materials*, *13*(10), 3774–3780.

Further reading

Anon. (2004). Special feature: Tiny tubes boost for metal matrix composites. *Metal Powder Report*, *59*(7), 40–43.
Azonano, n.d. Nanotechnology information by application. Available from: <http://www.azonano.com/Applications.asp>. Accessed: 20 November 2008.
Balazs, A. C., Singh, C., & Zhulina, E. (1998). Modeling the interactions between polymers and clay surfaces through self consistent field theory. *Macromolecules*, *31*(23), 8370–8381.
Chaudhry, A. U., & Mittal, V. (2015). High-density polyethylene nanocomposites using masterbatches of chlorinated polyethylene/graphene oxide. *Polymer Engineering and Science*, 78–88. https://doi.org/10.1002/pen.23241.
El-Tonsy, M. M., Fouda, I. M., Oraby, A. H., Felfel, R. M., & El-Henawey, M. I. (2014). Dependence of physical properties of linear low-density polyethylene on the silicon dioxide filler size. *Journal of Thermoplastic Composite Materials*, 1–14. https://doi.org/10.1177/0892705714533376.
Garcia-Lopez, D., Picazo, O., Merino, J. C., & Pastor, J. M. (2003). Polypropylene-clay nanocomposites: Effect of compatibilizing agents on clay dispersion. *European Polymer Journal*, *39*(5), 945–950. https://doi.org/10.1016/s0014-3057(02)00333-6.
Hussain, F., Hojjati, M., Okamoto, M., & Gorga, R. E. (2006). Review article: polymer-matrix nanocomposites, processing, manufacturing, and application: an overview. *Journal of Composite Materials*, *40*(17), 1511–1575. https://doi.org/10.1177/0021998306067321.
Kaleel, S. H., Bahuleyan, B. K., Masihullah, J., & Al-Harthi, M. (2011). Thermal and mechanical properties of polyethylene/doped-TiO_2 nanocomposites synthesized using in situ polymerization. *Journal of Nanomaterials*, 1–6. https://doi.org/10.1155/2011/964353.
Khalid, F. A., et al. (2003). Study of microstructure and interfaces in an aluminium–C60 composite material. *Acta Materialia*, *51*(15), 4575–4582.
Klitkou, R., Jensen, E. A., Christiansen, J., & C. (2012). Effect of Multiple Extrusions on the Impact Properties of Polypropylene/Clay Nanocomposites. *Journal of Applied Polymer Science*, *12*(6), 620–630. https://doi.org/10.1002/app.36639.
Kumar, V., & Singh, A. (2015). Polypropylene Clay Nanocomposites. *Reviews in Chemical Engineering*, *29*(6), 439–448. https://doi.org/10.1515/revce-2013-0014.
Lee, S. R., Park, H. M., Lim, H. L., Kang, T., Li, X., Cho, W. J., et al. (2002). Microstructure, tensile properties, and biodegradability of aliphatic polyester/clay nanocomposites. *Polymer*, *43*(8), 2495–2500.

Messersmith, P. B., & Giannelis, E. P. (1993). Polymer-layered silicate nanocomposites: in situ intercalative polymerization of ε–caprolactone in layered silicates. *Chemistry of Materials*, *5*(8), 1064–1066.

Parvinzadeh, M., Moradian, S., Rashidi, A., & Yazdanshenas, M. E. (2009). Surface Characterization of Polyethylene Terephthalate/Silica Nanocomposites. *Applied Surface Science*, *256*, 2792–2802. https://doi.org/10.1016/j.apsusc.2009.11.030.

Patwary, F., & Mittal, V. (2015). Degradable polyethylene nanocomposites with silica, silicate and thermally reduced graphene using oxo-degradabale pro-oxidant. *Heliyon*. https://doi.org/10.1016/j.heliyon.2015.e00050.

Prachum, Y., Adam Strauss, R. H., & Kiatkamjornwong, S. (2011). The physical and mechanical properties of beta-nucleated polypropylene/montmorillonite nanocomposites. *Journal of Applied Polymer Science*, *122*(2), 1066–1076.

Preschilla, N., Rasheed, A. S. A., Sahadevan, S., Biswas, A., Bellare, J. R., & Shyamroy, S. (2010). Study of layered silicate clays as synergistic nucleating agent for polypropylene. *Journal of Polymer Science Part B: Polymer Physics*, *48*(16), 1786–1794.

She, J., et al. (2000). Mechanical properties and fracture behavior of fibrous Al2O3/Sic ceramics. *Journal of European Ceramic Society*, *20*(12), 1877–1881.

Velasco-Santos, C., et al. (2003). Improvement of thermal and mechanical properties of carbon nanotube composites through chemical functionalization. *Chemistry of Materials*, *15*(23), 4470–4475.

Improvement of epoxy nanocomposites on physical, morphology, and mechanical properties as well as fracture behavior with the addition of mesoporous silica/nano-silica

12

Muhammad Khusairy Bin Bakri, Elammaran Jayamani*, Md. Rezaur Rahman†, Akshay Kakar**

*Faculty of Engineering, Computing and Science, Swinburne University of Technology Sarawak Campus, Kuching, Sarawak, Malaysia, †Department of Chemical Engineering and Energy Sustainability, Faculty of Engineering, University Malaysia Sarawak (UNIMAS), Kota Samarahan, Sarawak, Malaysia

12.1 Introduction

Since the middle 1990s, both the engineering and academic communities were attracted to the growing interest in the future of the broad field of nanotechnology, especially toward the incorporation of nanofillers into polymers. Due to that, both theoretically and experimentally, research was conducted to find the best physical, morphology, and mechanical properties which indirectly increase the performance of the nanocomposites. Most of the extensive investigations were focused on the research to determine the strength (Koval'chuk et al., 2008; Liu & Wu, 2001; Ma, Mai, Rong, Ruan, & Zhang, 2007; Reynaud, Jouen, Gauthier, Vigier, & Varlet, 2001; Szazdi, Pozsgay, & Pukanszky, 2007; Uddin & Sun, 2010), Young's modulus (Chan, Wu, Li, & Cheung, 2002; Chavarria & Paul, 2004; Chen, Justice, Schaefer, & Baur, 2008; Jiang, Zhang, & Wolcott, 2007; Koval'chuk et al., 2008; Liu & Wu, 2001; Ma et al., 2007; Reichert et al., 2000; Reynaud et al., 2001; Uddin & Sun, 2010), deformation behavior (Akbari & Bagheri, 2007; Akbari & Bagheri, 2013; Dasari et al., 2006; Gloaguen & Lefebvre, 2001; Kim, Qin, Fang, Sun, & Mather, 2003; Marouf, 2009), and fracture toughness (Boo et al., 2007; Choi, Yee, & Laine, 2004; He, Liu, Tjiu, Sue, & Yee, 2008; Khosh, Bagheri, & Zokaei, 2008; Pearson, Dittanet, Marouf, & Siotong, 2010; Reichert et al., 2000) of the nanocomposites. At present, more work is being focused on such materials by using more advanced nanoindentation devices (Li et al., 2004; Seltzer, Kim, & Mai, 2011; Wang, Liu, Yu, & Mai, 2009; Shen, Wang, Liu, & He, 2006). Furthermore, some have adopted

Silica and Clay Dispersed Polymer Nanocomposites. https://doi.org/10.1016/B978-0-08-102129-3.00012-9

available studies based on the theoretical models, especially on the strength (Frankland, Harik, Odegard, Brenner, & Gates, 2003; Szazdi et al., 2007) and stiffness (Anthoulis & Kontou, 2008; Brune & Bicerano, 2002; Cannillo et al., 2006; Fornes & Paul, 2003; Luo & Daniel, 2003; Odegard, Clancy, & Gates, 2005; Sheng et al., 2004) of the composites due to the influence of nanofillers in polymers, and their interface region mechanical properties (Ciprari, Jacob, & Tannenbaum, 2006; Desai & Haque, 2005; Miltner, Assche, Pozsgay, Pukanszky, & Mele, 2006; Odegard et al., 2005; Pukanszky, 2005).

The shape of the nanofiller is very important in creating better interface adhesion between matrix and filler. Nanofillers are classified according to their shapes, as shown in Fig. 12.1. Mostly, nanofillers are categorized into three parts; one-dimensional nanoscale, two-dimensional platelets, and three-dimensional nanoscale opposed to fibers/tubes, respectively (Schadler, Brinson, & Sawyer, 2007). By definition, it shows that clay and graphene nanoplatelets are examples of 1D nanofillers; carbon nanotubes and halloysite nanotubes are considered as 2D nanofillers; while silica and rubber nanoparticles are identified as 3D nanofillers. Even though the interface size between two-phase interface region of the materials decrees with the volume fraction, the polymer matrix properties eventually differ than those bulk polymer matrix properties.

It is well known that there is a thin layer of interphase around the particles in the polymer matrix chain which have different conformation to their counterparts in the bulk polymer matrix composites (Cech, Palesch, & Lukes, 2013; Piggott & Chua, 1987). The small size of interphase region in microcomposites does not change the mechanical properties very much. However in polymer nanocomposites, the region of the volume fraction affected the mechanical properties significantly. Thus, it was concluded that the interfacial volume fraction is directly influenced by the size, shape, and loading of the nanofiller. With consideration of the same loading, smaller 3D nanofiller have a larger interfacial volume fraction than the 2D nanofiller, which indicates that the 3D nanofiller is more effective than 2D nanofiller (Schadler et al.,

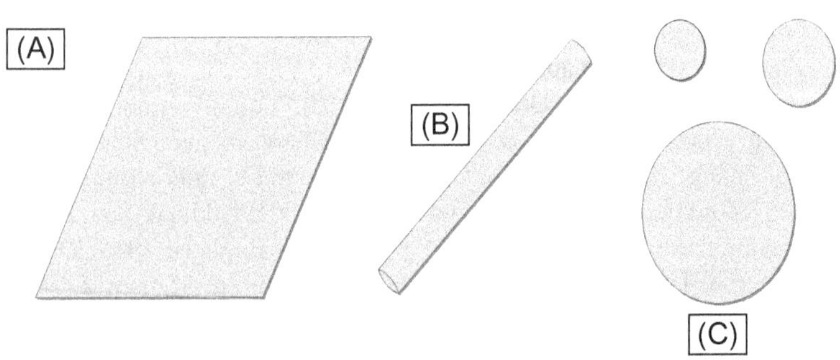

Fig. 12.1 The categorization of nanofiller: (A) 1D nano-platelets, (B) 2D nano-tubes, and (C) nanoparticles.
Source: Taken and modified from Marouf, B. T., Mai, Y.-W., Bagheri, R., Pearson, R. A., (2016). Toughening of epoxy nanocomposites: nano and hybrid effects. *Polymer Reviews*, *56*(1), 70–112.

2007). Schadler et al. (2007) also showed the positive interfacial area on the tribological, thermomechanical, and mechanical of the polymer nanocomposites properties.

Thermosets and thermoplastics polymer nanocomposites had broadly been addressed in the past few years; covering the aspects of processing, preparation, properties (functional, mechanical, thermomechanical, etc.), and applications (Alexandre & Dubois, 2000; Azeez, Rhee, Park, & Hui, 2013; Paul & Robeson, 2008; Tjong, 2006; Vaia & Wagner, 2004; Young, Kinloch, Gong, & Novoselov, 2012). Fracture toughness was attributed to the polymer nanocomposites energy absorption capability, which is very important in engineering applications (Sun, Gibson, Gordaninejad, & Suhr, 2009). As updated by Sun et al. (2009), the current progress focused on the 1D, 2D, and 3D fracture behaviors of nanofiller, owing to the high demand of epoxies in a variety of industries (i.e., transportation, construction, adhesives, energy, etc.).

In the past 10 years, nanofiller such as rubber, silica, and others have been used to toughen epoxies (Chen & Taylor, 2012; Deng, Ye, & Friedrich, 2007; Hameed, Guo, Xu, Hanley, & Mai, 2010; Hydro & Pearson, 2007; Kishi et al., 2011; Liu et al., 2008; Liu et al., 2009; Luo et al., 2013; Rebizant, Abetz, Tournilhac, Court, & Leibler, 2003; Rebizant et al., 2004; Wu et al., 2012; Wu, Guo, Kraska, Stuhn, & Mai, 2013; Wu, Guo, Zhang, & Mai, 2013; Wu, Thio, & Bates, 2005). The addition of nanofillers such as mesoporous silica and nano-silica particles (or modifiers) in epoxies is an established industry practice to overcome the inherent ductility of these materials. This book chapter critically examines the improvements in natural and synthetic epoxy nanocomposites, especially on those physical, morphological, and mechanical, as well as the fracture behavior with the addition of mesoporous silica/nano-silica.

12.2 Nanocomposites and nanofillers

In general, composites are a combination of reinforced phase (i.e., in form sheets, fibers, particles, etc.) and matrix phase (i.e., polymer, plastics, etc.) of the materials. Usually the reinforced materials are stronger and have lower densities, while the matrix is usually ductile or tough materials. By combining these two or more phase materials according to the correct fabrication and design, desirable properties can be achieved. Solid materials that have at least one of these domains in nanoscale structure and multiple phases are called nanocomposites. The novel chemical and physical properties are dependent on the component of the materials and dependent on its interfacial characteristics and morphology. Nanocomposites are also considered as one of the extended broad class of the materials, whereas its microstructures can be modulated in zero to three dimensions length scales less than 100 nm. The arrangement of the atoms in the nanosize clusters creates better grain and block in the materials. This causes an increase in the area of dispersion within a host matrix polymer, particularly.

The main desirability of nanosized particles has been found to improve as filler particle size decreases significantly. Two prospects were created, better absolute properties, or the same properties at much lower filler levels, and so a lighter product was introduced (as fillers generally have a higher specific gravity than polymers).

Most applications proposed are in the additive range of 5%–10% (either wt% or vol %), whereas compared with conventional ones, the market size for such fillers is going to be much smaller in weight terms, which undergo fair judgement in term of basis success (Rothon, 2016).

Recently introduced nanofillers have faced a number of issues, limiting their successful commercialization. Firstly, the reduction in the size always costs money, when involving nanosizes. Secondly, as the size reduced, the mixture process becomes more difficult, especially in dispersing the particles, which adds significantly more cost. Thirdly, the surface properties of the filler become more and more significant as the size reduces and causes several effects and issues, such as stabilizer deactivation and adversely affects the polymer structure. Thus, expensive coupling agents were needed due to the high amount of surface increasing significantly due to the increase in the coupling agent amount used (Rothon, 2016). Furthermore, addition of maximum level of particles size attainable decreases and prevent properties above conventional composites to be achieved.

12.3 Epoxy

Either the cured end products or the basic component of the epoxy resins had a unique colloquial name, which falls under epoxide functional group (May, 1986). Polyepoxides is another known name for epoxy resins. It contains epoxide groups, which are classed as polymers and reactive prepolymers. With a wide range of coreactants (including polyfunctional amines, phenols, alcohols, thiols, and acids (and acid anhydrides)) and through catalytic homopolymerization, enable the epoxy resins to be crosslinked or reacted with themselves. For curing, it is commonly referred to as the crosslinking reaction, while the curatives and hardeners are referred to as the coreactants. To form a thermosetting polymer, reaction of polyfunctional hardeners and polyepoxides with themselves is required, often with high temperature, chemical resistance, and high mechanical properties. Epoxy covers a wide range of applications which include structural adhesives, fiber-reinforced plastics materials, metal coatings, LED electrical/electronic components, paint brush manufacturing, and high tension electrical insulators.

Higher molecular weight polymers and low molecular weight prepolymers of epoxy resins contained at least two epoxide groups. Sometimes the epoxide groups are referred to as an oxirane or glycidyl group. Most of these wide ranges of epoxy resins are industrially produced. Today, the epoxy resin materials' raw productions are still largely derived from petroleum, even though with advances in technology, some plants are also used to produce commercialized epoxy resin (e.g., glycerol to make epichlorohydrin).

That such rarely pure substances of polymeric or semipolymeric epoxy resins exist due to the results of variable chain length resulting from the reaction of polymerization used to produce them. Take the example of certain application processes, such as a distillation purification process, pure high grades can be reproduced. However, the downfall of it is that the high liquid grade tends to form high crystalline solids structure that required melting, to enable processing due to its highly regular structure. Furthermore, the epoxide content is also one of the important criterions for epoxy resins. It is also common to express it as the epoxy equivalent weight, in which the weight in grams of resin containing 1 mol equivalent to epoxide (g/mol) or as the equivalent to weight, in which the number of epoxide equivalents in 1 kg of resin (Eq./Kg). The measurement of the epoxy can be related as stated below:

$$\text{Equivalent weight}\,(\text{g/mol}) = 1000/\text{epoxide number}\,(\text{Eq./Kg})$$

The equation above is usually used to calculate the amount of coreactant (hardener) for epoxy resins curing purposes which is usually the epoxide number or equivalent weight. Thus, to achieve its maximum physical properties, epoxies are typically cured with quantities of curative, which is stoichiometric or near-stoichiometry.

To reduce cost, one of the solutions is to blend different grades of epoxy resins, either with its own classes or other classes of thermoset polymer materials, using additives, fillers or plasticizers. Most researchers preferred blending of additives and fillers as formulating techniques.

12.4 Mesoporous silica/nano-silica

Silica is a chemical compound which is usually called silicone dioxide. It is made from oxygen and silicon, SiO_2, which are the two most abundant elements in Earth's crust. About 56% of the mass of earth is covered by silica, whereas 95% of them are commonly referred to as rocks. Quartz, tridymite, and cristobalite are three main varieties of crystalline silica. Lechatelierite, keatite, and coesite are other varieties of silica. Most of this silica sand was used in the form of sandstone, Portland cement, mortar and concrete to build houses and roads. It is also used as polishing and grinding stone and glass (i.e., as refractory materials: in manufacture of silicones, ceramics, silicon carbide and glass; in foundry molds; and as gemstones). In other applications, to remove moisture, silica gel is often used as a desiccant.

Most of the research on silica particles was concentrated in applications such as drug delivery, imaging, and catalysis (Trewyn, Nieweg, Zhao, & Lin, 2007). Whereas, the most current development of this nanotechnology has enabled the development of these mesoporous silica and nano-silica types of silica. Mesoporous silica is a silica containing pores with diameters between 2 and 50 nm, while nano-silica is silica with a nano size (<100 nm). SBA-15 and MCM-41 is the most common types of mesoporous silica (Katiyar, Yadav, Smirniotis, & Pinto, 2006). The first compounded mesoporous silica was patented around 1970 (Patent US 3383172A (1964); Patent US 3493341D A (1967); Chiola, Ritsko, & Vanderpool, 1971). However, none of this was noticed and known until it was reproduced and reintroduced back in 1997 (Di Renzo et al. 1997; Xu, Pang, Yu, Huo, & Chen, 2007). In 1990, the mesoporous silica nanoparticles were independently synthesized, later being produced by laboratories under Mobil Corporation and named under Mobil Composition of Matter or Mobil Crystalline Materials (Beck et al., 1992; Trewyn, Nieweg, et al., 2007). However, University of California,

Santa Barbara, produced silica particles with much larger scale, around 4.6–30 nm, 6 years after that (Zhao et al., 1998). The silica materials produced were named after the university location itself, Santa Barbara Amorphous (SBA-15). These particles showed a hexagonal array of pores.

These types of particles allowed most researchers to use them as molecular sieves. While today, the wide applications of mesoporous silica nanoparticles can be seen in imaging, medicine, thermal energy storage, and biosensors (Mitran, Berger, Munteanu, & Matei, 2015). Most the mesoporous silica nanoparticles are synthesized with a template made by micellar rods, which is synthesized by reacting tetraethyl orthosilicate. This caused a collection of nanosized rods and spheres, which filled with a regular pores arrangement. By washing it properly with adjusted solvent, the template can be removed (Trewyn, Slowing, Giri, Chen, & Lin, 2007). A simple sol–gel method can also be used to synthesize the mesoporous particles, Stöber process, or a drying spray method (Nandiyanto, Iskandar, & Okuyama, 2008; Nandiyanto, Kim, Iskandar, & Okuyama, 2009). With additional polymer monomer, tetraethyl orthosilicate can also be used, especially as a template. Unfortunately, by comparing TEOS and MPTMS, the effectiveness of precursor for synthesizing such particles is often advantageous to MPTMS. After this precursor, more uniform spheres are produced and the chance of aggregation is reduced.

12.5 Physical and mechanical properties

In general, physical properties are something that can be used to describe or observe matter without changing it physical state. Mechanical properties are usually used to identify and classify the materials performance. It covers the aspects of fracture, toughness, ductility, hardness, strength, impact, etc. Depending on the orientation and structure, the properties may vary for each material. According to Jumahat, Soutis, Abdullah, and Kasolang (2012), epoxy was classified as brittle, due to its low strain values during tension. Their result confirms that epoxy nanocomposites fail after the proportions, as the limit is exceeded at small elongation (about 3% of the average failure strain) and the ultimate stress values are similar to those fracture stress values (Jumahat et al., 2012). Due to the addition of nano-silica in the epoxy, the tensile stress–strain were enhanced, in which nano-silica/epoxy nanocomposites exhibited higher tensile modulus and strength (Jumahat et al., 2012), whereas the addition of 13 wt% nano-silica into the epoxy enhanced the tensile modulus by 21%, failure strain by 10%, and tensile strength by 20%. Remarkably, the highest content of nano-silica increase the strength and modulus about 24% and 38%, which through addition of 25 wt% nano-silica in the epoxy nanocomposites resulted in higher values compared with pure polymer without disturbing the strain to failure (Jumahat et al., 2012). The increase of modulus in the epoxy nanocomposites was due to the modulus of the silica, which is about 70 GPa (Johnsen, Kinloch, Mohammed, Taylor, & Sprenger, 2007). It can be seen that as the tensile load increased, the matrix is elongated, but was resisted by the deformation due to nanofiller. Compared to neat polymer, nanofiller created smaller deformation. Therefore, more loads were sustained on the nanocomposites compared with the pure one, which contribute to the higher tensile strength and modulus.

Battistella et al. (2008) did research on the nontreated fumed silica epoxy nanocomposites, 3-aminopropyltrimethoxysilane AMEO-modified silica epoxy nanocomposites, and 3-glycidyloxypropyltrimethoxysilane GLYMO-modified fumes silica epoxy nanocomposites. Battistella et al. (2008) reported that the addition of filler in the epoxy nanocomposites does not affect the strength of the materials significantly. It is noticed that the unmodified fumes silica particles decrease the strength slightly, while modified fumes silica lead to benefit restricted tensile parameters. The decreasing strength was due to the clustering which leads to matrix weakened filler. At volume content of 0.1 vol%, unmodified fumes silica were having an optimum strength at 63 MPa, while for 0.1 vol% fumes AMEO-modified silica optimum strength at 67 MPa, and 0.3 vol% fumes silica GLYMO optimum strength is at 66 MPa. Similarly, Wichmann, Cascione, Fiedler, Quaresimin, and Schulte (2006) obtained dynamic mechanical thermal analysis (DMTA) results for unmodified fumes silica epoxy nanocomposites, 3-aminopropyltrimethoxysilane-modified fumes silica epoxy nanocomposites, and modified 3-glycidyloxypropyltrimethoxysilane fumes silica epoxy nanocomposites. It showed that there is no significant influence on the modulus storage for the nanocomposites. Based on Halpin and Tsai (Halpin, 1969) equations, an increase in the tensile modulus by 1.35% is expected due to a corresponding increase in total of amount of silica content at 0.5 vol% in the nanocomposites at 35 MPa. The value obtained through the results of the tensile test is much lower than the standard deviation obtained by the neat resin modulus of elasticity ($\pm$0.081 GPa). Although it is very small, reduction in modulus was observed for all surface modification and volume contents which has an opposite effect (Williams, Donnellan, James, & Morris, 1990). It might be due to the formation of interphase layer, which have lower modulus and higher mobility than the bulk matrix. This well-known formation of interphases with reduced mechanical properties is common in conventional composites (Ishida, 1984; Ishida & Koenig, 1978; Williams et al., 1990; Wright, 1990). Reduced crosslinking density interphase lead was due to the selective adsorption of hardener component on the surface of the reinforcing phase. It is known that in the composite, the total volume is proportional to the interface area, which explained why this effect becomes much more pronounced for nanocomposites. It may due to the involvement of nanoconfinement effects near the molecules interfaces, which cause the changes in the interphase mechanical properties.

Constantinescu, Apostol, Picu, Krawczyk, and Sieberer (2017) ran a test on the unfunctionalized powdered silica epoxy nanocomposites using uniaxial traction testing, with maximum force of 10 kN. Weight percentage (wt%) of silica powdered was 0.1, 0.3, 0.5, 1.0, and 3.0 in different batches. The methods were divided into two (a) R + NP mixture were kept under a vacuum of 30 mbar for 2 h at room temperature for degassing (M1 method) and (b) Degassing was omitted to generated additional gas bubbles in the specimen (M2 methods). M1 method tests result showed that the 1 wt% of unfunctionalized nanoparticles were having 3.633 GPa of longitudinal modulus of elasticity, 85.74 MPa of ultimate strength, and 3.64% elongation at failure. While M2 method test result showed that 0.1 wt% of unfunctionalized nanoparticles were having 3.910 GPa of longitudinal modulus of elasticity, 82.86 MPa of ultimate strength, and 4.35% elongation at failure.

Cosmoiu, Apostol, Picu, Constantinescu, and Sandu (2015) also showed similar results. Constantinescu et al. (2017) summarized approximately 3%–4% elongation at failure, 80 MPa of ultimate strength (with some increase for pure epoxy M2) and Young's modulus range between 3.2 to 3.9 GPa, with exceptional exemption in which variable changes observed between given batch. The presence of silica agglomerates and gas bubbles influenced the most changes in the physical and mechanical properties of the nanocomposites. Furthermore, no significant systematic effects were observed on the silica nanopowder weight fraction. Additionally, the epoxy systems used in the nanocomposites were still believed to be unknown in chemical composition and the unfunctionalized silica nanopowder may also lead to the large scatter in the experimental data.

To solve this problem, two epoxy systems (S5 and S2) from BTO Epoxy were used as matrix Constantinescu et al. (2017). However, to prepare the nanocomposites, similar silica nanopowder properties with much smaller particles around 5–15 nm were used. The pure specimen in the epoxy system S5 was much lower in mechanical properties than S2, which include smaller ultimate strength. However, the approximately identical elongation at failure can be seen for both pure systems, S2 and S5. In traction testing at room temperature, the addition of 0.1 and 0.3 wt% does not significantly change the mechanical properties. 0.3 wt% of S2 showed larger Young's modulus, however, the ductility and strength are slightly smaller than the pure epoxy or 0.1 wt%.

Preghenella, Pegoretti, and Migliaresi (2005) performed mechanical tests on the fumed silica-epoxy nanocomposites to determine the effect of filler on the composites. Table 12.1 showed the result obtained by Preghenella et al. (2005). Elastic modulus, tensile strength, and yield stress were seen reduced at the filler loadings of 10 phr and 20 phr, which increased the strain at the breakpoint. However, at 30 phr, tensile modulus was observed to have increased, even though the ultimate tensile strength and yield strength did not improve with respect to the unfilled matrix. This was due to the postulated reduction in the crosslinking degree of the matrix and better physical interactions between polymer phase and filler occurrence at highest silica loading. Table 12.1 showed the tensile properties of the fumed silica-epoxy nanocomposites.

In order to assess the fracture toughness of the nanocomposites, Preghenella et al. (2005) also performed three point bending tests. Both of the apparent fracture toughness K_{IQ} and apparent critical energy release rate G_{IQ} were measured and assessed due to the notch tip plastic zone radius were found to exceed the limit established by

Table 12.1 Tensile properties of fumed silica-epoxy nanocomposites

Silica content (phr)	Elastic modulus E (GPa)	Stress (MPa)		Strain (%)	
		At yield	At break	At yield	At break
0	2.26 ± 0.071	35.3 ± 0.8	30.2 ± 0.9	3.8 ± 0.1	12.0 ± 1.7
10	1.5 ± 0.120	17.8 ± 0.8	19.7 ± 0.3	4.1 ± 0.1	27.2 ± 0.5
20	1.9 ± 0.150	21.3 ± 0.6	23.8 ± 1.0	3.8 ± 0.4	26.0 ± 2.1
30	2.8 ± 0.240	25.0 ± 0.6	24.9 ± 0.7	4.0 ± 1.2	8.0 ± 2.9

Source: Taken from Preghenella, M., Pegoretti, A., Migliaresi, C. (2005). Thermo-mechanical characterization of fumed silica-epoxy nanocomposites. *Polymer 46*(26), 12065–12072.

ASTM D5045 standard for linear elastic fracture mechanics (LEFM) applications. Preghenella et al. (2005) when compared with the unfilled polymer matrix, it was found that at low silica content (i.e., 10 phr), the fracture toughness of the composite exceeded and as the silica content increased, and then it decreased again.

Ahmad, Jaafar, Palaniandy, and Azizli (2008) make silica epoxy nanocomposites using mineral silica particles instead of using conventional fumed silica. It was concluded by them that at higher volume fraction, the silica nanoparticles tend to produce higher Young's modulus and stress, as change in different volume fractions of silica nanoparticles in the composites. Besides, it is also reported that improvements were seen with elongated silica nanoparticles compared with cubical silica and angular silica (Ahmad et al., 2008). More improvement was seen in Zamanian, Mortezaei, Salehnia, and Jam (2013) nanocomposites research, especially on higher weight percentage of Young's modulus. Whereas, smaller nanoparticle use showed higher improvement. Ahmad et al. (2008) also tested the three different shaped nanofillers of silica epoxy nanocomposites flexural behavior at different volume fraction. Improvement was seen drastically on the flexural results with respect to angular and cubical nanoparticles. Furthermore, it also showed that the all three different nanoparticle shape increased in flexural modulus as increase in volume fraction.

Dittanet and Pearson (2012) researched the effect of the addition of nanoparticles and microparticles in the epoxy polymer on the mechanical properties. It was found that there is no inclusion on the nanoparticles or effect on the Young's modulus, as due to existence of microparticles. But, at higher volume percentage, maximum improvement was seen on the yield stress and Young's modulus (Nicolais & Narkis, 1971). Tsai, Hsiao, and Cheng (2010) used two different kinds of polymer matrix, which created brittle and ductile silica epoxy nanocomposites. The results showed that the ductile silica nanocomposites showed increasing stress and Young's modulus as nanoparticle weight fractions increased. But, while stress increased up to an optimum level of 20 wt%, the brittle one only showed similar Young's modulus trend. Both ductile and brittle silica epoxy nanocomposites also showed linearly improvement in the flexural behavior as increase in particle loading. In addition, similar results were also seen in both types of nanocomposites in their fracture toughness. However, compared to the pure epoxy, the improvement in tensile stress for brittle matrix nanocomposites was seen only at 5%, 10%, and 15% weight fractions (Tsai et al., 2010). In the case of fracture toughness, the brittle one showed lowered fracture toughness than ductile silica epoxy nanocomposite at any nanoparticle loading (Tsai et al., 2010). In the fracture toughness of nanocomposites, small increases at higher weight fractures were also reported and significant improvement were seen at smaller weight fractions (Tsai et al., 2010).

The effects of rubber and silica nanofillers on the fracture toughness of epoxy were studied synergistically and individually by Liu, Wang, Mai, and Zeng (2011). Liu et al. (2011) measured the yield stress and Young's modulus at different weight fractions as in their research on the effect of weight fraction on silica epoxy nanocomposites. Their results revealed that gradually the yield stress and Young's modulus were increased when the weight fraction of the silica particles increased.

It was also revealed that in rubber silica epoxy and silica epoxy, rubber epoxy nanocomposites fracture toughness increased monotonically with the increase in weight fraction of nanofillers. In case of rubber epoxy nanocomposite with the silica-epoxy nanocomposite, the improvement was higher comparably. Both silica and rubber nanoparticles Young's modulus synergistic effect was present in rubber silica epoxy nanocomposite. However, no significant effect on fracture toughness showed on these two types of particles.

Mahrholz, Stangle, and Sinapius (2009) and Chen et al. (2008) also studied the effect of different weight fractions of silica on the pure resin and silica epoxy nanocomposites. It showed that the values of yield stress and Young's modulus of silica epoxy nanocomposites at 20 wt% are 1.28 and 1.22 times much stronger than the corresponding values of epoxy. Furthermore, Chen et al. (2008) showed that at 20 wt% silica nanoparticles, 1.40 times of improvement were seen compared with the pure epoxy. However, majorly, the tensile stress remains constant as the weight fraction increases from zero to 10%.

Yield stress and young modulus of rubber epoxy nanocomposites is known to decrease linearly as increase in weight fractions of its nanofillers (Liu et al., 2011), which is in different incorporation with the stiffening effect of silica epoxy nanocomposites. The nanoparticles in rubber epoxy composites cause stress concentrators in the matrix, which may lead to plastic deformation (Akbari & Bagheri, 2007). In the case of silica epoxy nanocomposites, the effect caused increase in stiffness of the matrix system due to the rigidity of the silica nanocomposites. Mahrholz et al. (2009) stated that increase of weight fraction of silica nanofillers increased the Young's modulus and the tensile stress, as the critical values of 15 wt% were observed at maximum elongation. Mahrholz et al. (2009) also stated that flexural stress linearly increases with an increase in weight fraction of nanofiller up to 25%. Respectively, when compared with the pure epoxy, the tensile stress and flexural modulus of silica epoxy nanocomposite with 25% weight fraction of nanoparticle improved by 8.9% and 29.8%.

In comparison with Young's modulus of epoxy resin (3.09 GPa), Chasiotis (2009) found that the Young's modulus of silica epoxy nanocomposites (with 12 nm particles size) monotonically increased as the as the weight percentage increased up to 15%. Chasiotis (2009) studied different weight fractions of epoxy-fumed silica nanocomposites. In fracture toughness of nanocomposites, significant improvements were seen at small weight fractions (up to 1%) of nanoparticles. However, at higher weight fractions of nanoparticles, the improvement was not significant.

Uddin and Sun (2008) used standard DGBEA epoxy and sol–gel technique to prepare at 15% weight fraction silica epoxy nanocomposite. Uddin and Sun (2008) on the other hand showed that the transverse and longitudinal stress mean values of silica epoxy nanocomposites increased to 32% and 11%, respectively, as compared to the pure epoxy. On the other hand, the Young's modulus of the nanocomposites showed little effect, especially due to the silica nanoparticles (Uddin & Sun, 2008). On comparing silica/epoxy nanocomposite with the pure epoxy, it was found that silica/epoxy nanocomposite had higher elastic modulus and compressive stress (Uddin & Sun, 2008). The nanocomposite showed an improvement of 40% in elastic modulus (Uddin & Sun, 2008). It was also seen that the compression stress for the epoxy and the silica epoxy nanocomposite at 10% elongation were about 92 and

108 MPa, respectively. After adding the nanoparticles, significant improvement was achieved (Uddin & Sun, 2008). It was observed that higher off-axis compressive stress of silica nanoparticle reinforced glass epoxy nanocomposite than the pure glass/epoxy composites, regardless of the fact that compressive stress of composites with higher volume fraction of glass fiber is higher (Sun, Li, Zhang, Du, & Burnell-Gray, 2006; Uddin & Sun, 2008).

Islam (2013) research the tensile behavior of silica epoxy nanocomposites at different variable particles sized, while having a fixed of 5 wt% samples. It was observed that the Young's modulus, 0.2% offset yield stress, and tensile stress improved, when silica nanoparticles were added into the pure epoxy. Furthermore, the yield and the ultimate stress increased by 50% and 20%, respectively over the epoxy when 20 nm nanoparticles size of silica were used in 5 wt% samples.

Sun et al. (2006) created four samples, untreated fumed silica/PVC nanocomposite, pure PVC, γ-methacryloxypropyl trimethoxy silane treated silica (KHS)/PVC nanocomposite and dimethyl dichlorosilane treated silica (DDS)/PVC nanocomposite, and launched an investigation on the effect of various weight fractions and particle modification of silica fume on the yield stress silica PVC nanocomposites. It was concluded that particle modification and particle weight both influenced the tensile strength of the polymer nanocomposites. In another experiment, Jiao, Sun, and Pinnavaia (2009) used four types of mesoporous silica to get the strain-at-break, Young's modulus, and tensile stress of silica rubbery epoxy nanocomposites. It was found that the Young's modulus and tensile stress of mesoporous epoxy nanocomposites linearly increased, as increase in weight fraction of nanoparticles. Furthermore, bigger pore size of mesoporous silica showed higher improvement compared to the smaller one.

Liang and Pearson (2010) used two different sizes (20 and 80 nm) with carboxyl terminated butadiene acrylonitrile (CTBN) to make rubber epoxy composite with different volume fractions of nano-silica. The compressive yield stress and compressive modulus of this hybrid nanocomposite reported linearly increased up to a volume fraction of 7% and beyond that it started to decrease. With increasing volume fraction of nano-silica, Kinloch et al. (2005) reported the compressive yield and compressive modulus stress increased monotonically. Liang and Pearson (2010) suggested that aggregation of the nano-silica caused unexpected results. Moreover, the nanocomposite toughness remains constant at high aspect ratio and higher weight fractions of nanoparticles (Kinloch et al., 2005; Kinloch & Taylor, 2006; Liang & Pearson, 2010).

12.6 Fracture behavior and morphological properties

One of the major challenges in fabricating nanocomposites is due to its homogeneous dispersion of nanofillers in a polymer. The homogenous dispersion of the nano-silica nanocomposites was due to the addition of high stiffness nanofiller, such as nano-silica and mesopous silica. Under stress–strain curve of the epoxy nanocomposites, the high stiffness of these nanofillers enhanced the fracture toughness of the epoxy nanocomposites by creating a larger area under the area of the graph (Jumahat et al., 2012). Uniform distributions were shown in epoxy nanocomposites, as shown in TEM micrographs in Fig. 12.2 (Jumahat et al., 2012). Fig. 12.2 shows that there were no nanoparticles of SiO_2 agglomeration at high volume fraction. Furthermore, nanoparticles of silica are spherical and have mean size 20 to 40 nm when observed under high magnification (Jumahat et al., 2012).

Comparable from the TEM results as shown in Fig. 12.3, the unmodified and the GLYMO-modified fumes silica epoxy nanocomposites showed fine degree of dispersion, which caused high attractive forces between the particles, due to its fine network structure (Battistella et al., 2008). Fig. 12.2A showed that fumed silica epoxy nanocomposites had small clusters within the matrix, whereas these aggregates tend to attract each other to form greater structure called agglomerates. The structures tend to behave like chain-branched structures (Battistella et al., 2008). Thus, this showed that the aggregates consist of tens of primary particles and dimension of tens to one hundred nanometers.

Based on Battistella et al. (2008), the fracture toughness tests were increased due to change in modification of silica in the nanocomposites. At 0.1 vol%, the unmodified fumes silica epoxy nanocomposite have 0.8 MPa × $m^{0.5}$, modified 3-aminopropyltrimethoxysilane AMEO-modified silica epoxy nanocomposites have 0.9 MPa × $m^{0.5}$, and 3-glycidyloxypropyltrimethoxysilane GLYMO-modified fumes

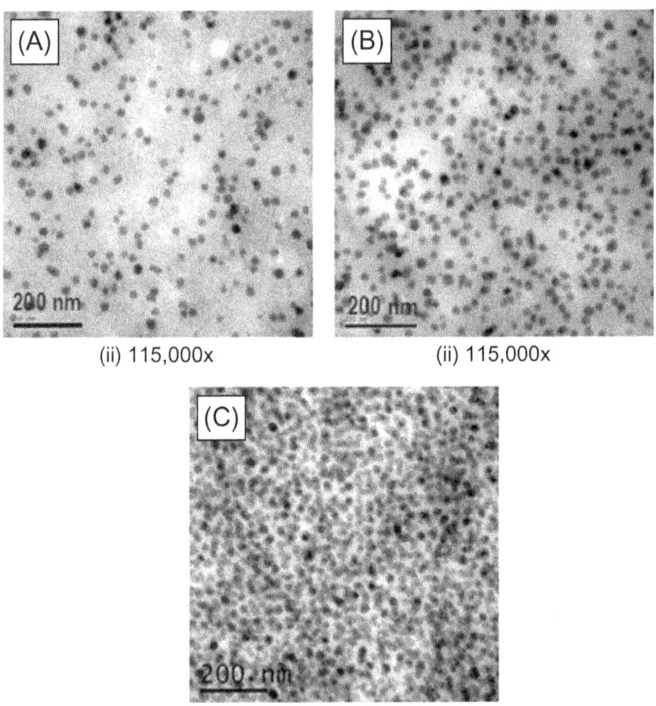

Fig. 12.2 Addition of (A) 5 wt% nano-silica, (B) 13 wt% nano-silica, and (C) 25 wt% nano-silica in epoxy nanocomposites.
Source: Taken from Jumahat, A., Soutis, C, Abdullah, S. A., Kasolang, S. (2012). Tensile properties of nano-silica/epoxy nanocomposites. *Procedia Engineering*, *41*(1), 1634–1640.

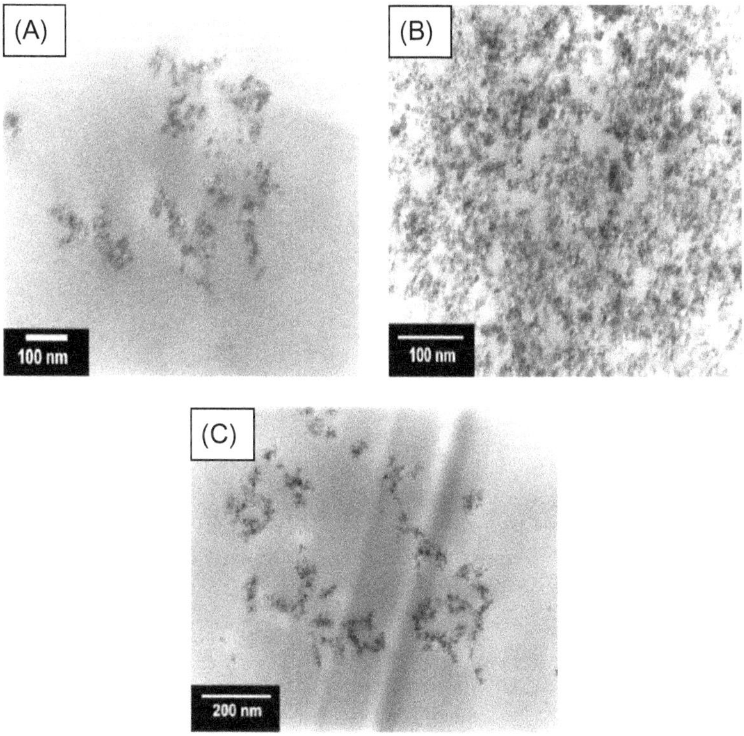

Fig. 12.3 TEM analysis of (A) 0.3 vol% unmodified fumed silica epoxy nanocomposites, (B) 0.5 vol% GLYMO-modified fumed silica epoxy nanocomposites, and (C) 0.5 vol% AMEO-modified fumed silica epoxy nanocomposites.
Source: Taken from Battistella, M., Cascione, M., Fiedler, B., Wichmann, M. H. G., Quaresimin, M., Schulte, K. (2008). Fracture behaviour of fumed silica/epoxy nanocomposites. *Composites Part A: Applied Science and Manufacturing*, *39*(12), 1851–1858.

silica epoxy nanocomposites have 1.1 MPa $\times$ m$^{0.5}$ (Battistella et al., 2008). The addition of nanofiller in the nanocomposites improved the fracture toughness of all the nanocomposites produced. Furthermore, particularly, the GLYMO-modified silica particles show a fracture toughness of 1.34 MPa $\times$ m$^{0.5}$ comparably higher by 54% than the neat epoxy (0.87 MPa $\times$ m$^{0.5}$). Whereas, through morphological analyses, it is worth noting that the indicated GLYMO-modified silica modified particles had the finest and most uniform dispersion which created strong adhesion between the matrices. Fig. 12.4 shows the tensile fracture surface of 0.5 unmodified fumed silica specimen.

Constantinescu et al. (2017) used functionalized nanopowder by coupling the azidopenylsilanes to nano-silica under specific surface area of 175–225 m^2 [determined using Brunauer-Emmett-Teller (BET) theory]. Approximately, 20–80 nm of particles were agglomerated and intergrown from bigger aggregates, which is due to the fractal structure of fumes silica. Due to its high temperature, especially in the industrial preparation, hydroxyl groups (silanols) are lost, especially on its large specific surface area. Thus, nonpolar solvents are used to minimize the condensation

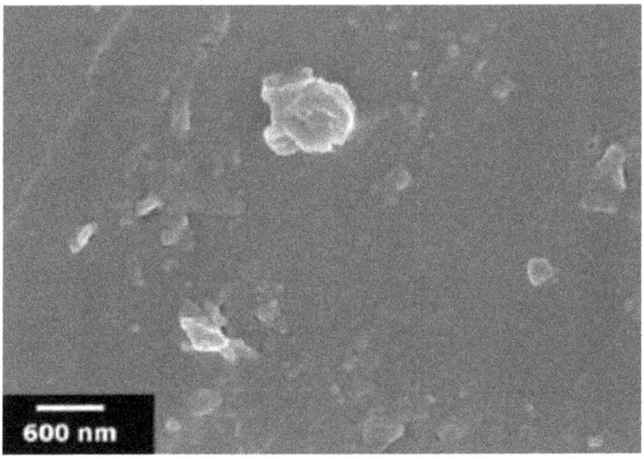

Fig. 12.4 Tensile fracture surface SEM of 0.5% unmodified fumes silica specimen.

and caused attachment of silanols to be present on the silica surface covalently with azidopheyl silanes. In this process, an acidic catalyst was used.

Similar particles were also subjected to wet impregnation, in which brief, sonication methods were used to prepare the wet ethanolic solutions, which later the solvents were evaporated to dryness, in which the procedures were repeated twice. This is to ensure the silanes hydrolyzed and the surface of the silica degree of condensation increased (Constantinescu et al., 2017). The functionalized silica had dry particle of 0.28 mol of phenylazide per 1 g. Dynamic light scatter measurements and electron microscopy had shown that the aggregates are commonly in the size of 200–500 nm, which are smaller particles with broad size distribution of 20–80 nm. It also showed that the functionalization process does not change the nanopowder structure under microscope. As the effect of sonication, this caused larger aggregates to break into small clusters. Fig. 12.5 showed the silica nanopowder epoxy nanocomposites.

Preghenella et al. (2005) showed that SEM images of the fracture surfaces of fumed silica epoxy nanocomposites and unfilled epoxy matrix. For 35 × magnifications, in 10 and 20 phr filled samples, a quite large plastic deformation were noted, especially on the crack propagation over the whole ligament section. Comparatively, much more brittle behavior can be seen in 30 phr filled samples, which is, on the contrary, similar to those observed for limited plastic deformation unfilled matrix in the region of the notch-tip region. At higher magnification of 20000 ×, the unfilled matrix fracture surface was found to be very smooth, whereas from their surface, the composites samples showed a much rougher pattern with submicrometric clusters emerging particles on the surface. Surprisingly, at 20 phr, the relative amount of aggregates is higher compared it at 30 phr, which is remarkably reduced. The trend can be explained at 30 phr filled samples, in which due to the improvement of the filler–matrix interaction and also due to the path of the crack propagation inside the epoxy matrix (Preghenella et al., 2005). Fig. 12.6 shows the 35 × magnification of the nanocomposites. Fig. 12.7 shows the 20,000 × magnification of the nanocomposites.

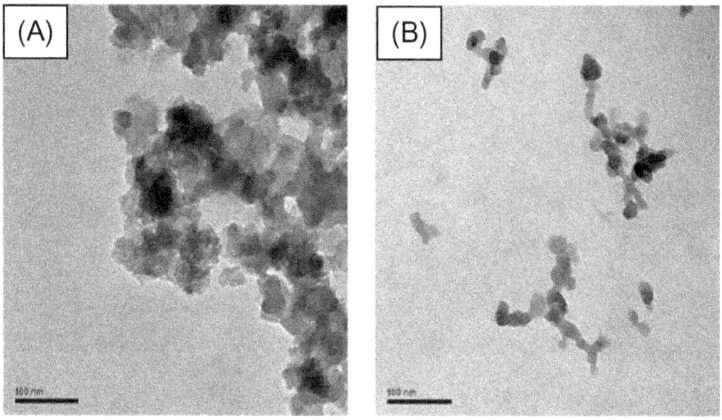

Fig. 12.5 SEM images of silica nanopowder nanocomposites, (A) unfunctionalized silica clustered with nominal particle size 5–15 nm; (B) functionalized fumed silica of 20–80 nm. *Sources:* Taken from Constantinescu, D. M., Apostol, D. A., Picu, C. R., Krawczyk, K., Sieberer, M. (2017). Mechanical properties of epoxy nanocomposites reinforced with functionalized silica nanoparticles. *Procedia Structural Integrity*, *5*(1), 647–652.

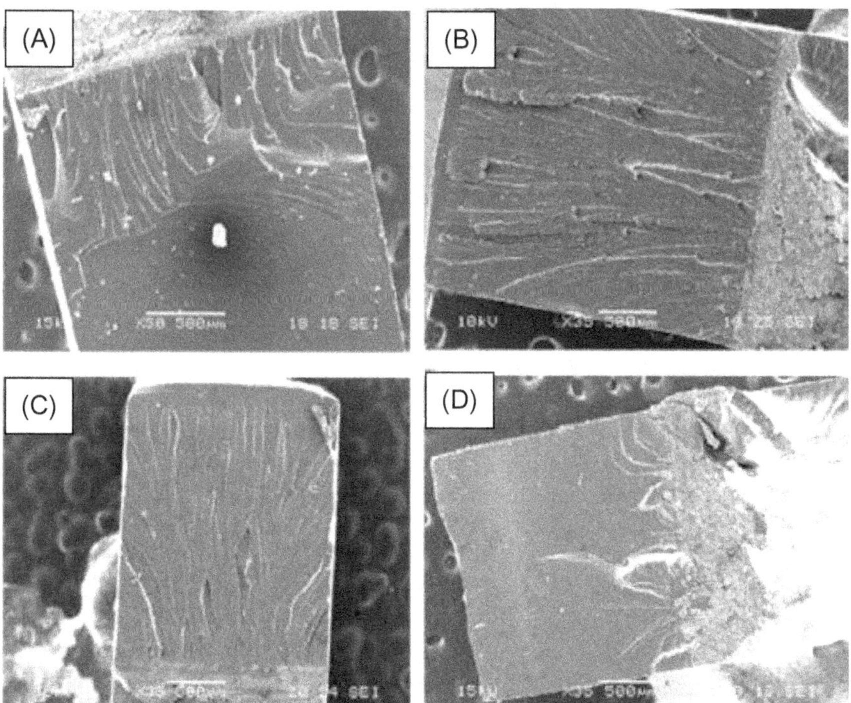

Fig. 12.6 35 × magnification SEM images of fracture surface of nanocomposites for three-point bending (A) unfilled matrix, (B) 10 phr silica, (C) 20 phr silica, and (d) 30 phr silica. *Sources:* Taken from Preghenella, M., Pegoretti, A., Migliaresi, C. (2005). Thermo-mechanical characterization of fumed silica-epoxy nanocomposites. *Polymer 46*(26), 12065–12072.

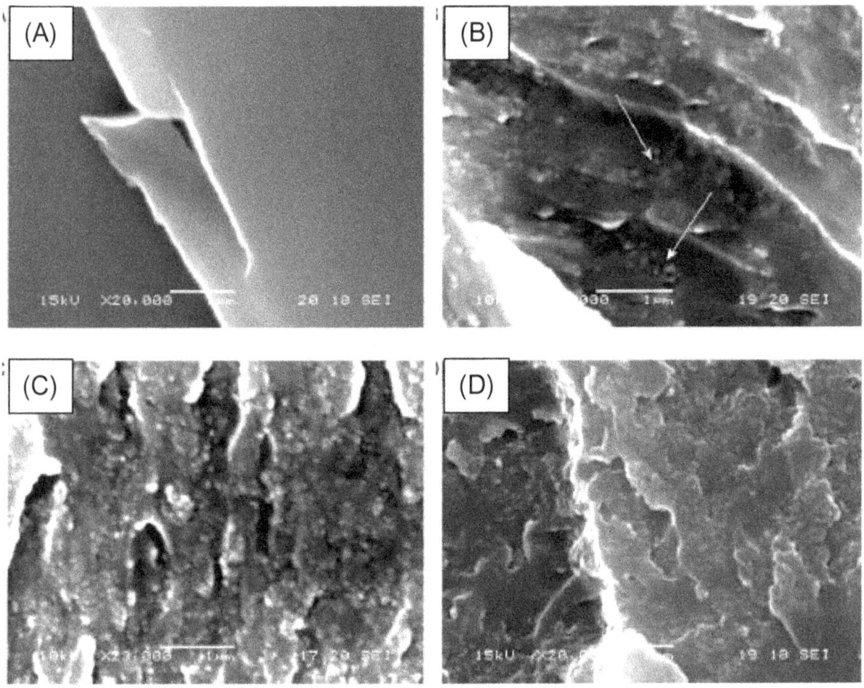

Fig. 12.7 20,000 × magnification SEM images of fracture surface of nanocomposites for three-point bending (A) unfilled matrix, (B) 10 phr silica, (C) 20 phr silica, and (D) 30 phr silica. *Sources:* Taken from Preghenella, M., Pegoretti, A., Migliaresi, C. (2005). Thermo-mechanical characterization of fumed silica-epoxy nanocomposites. *Polymer 46*(26), 12065–12072.

12.7 Summary

In this book chapter, the recent research on mechanical, physical, fracture, and morphology behavior of mesoporous silica and nano-silica epoxy composites were reviewed. In the mechanical and physical properties, the tensile stress, Young's modulus and yield stress were incorporated with the rigid size, structure, and properties of the mesoporous silica and nano-silica. Increases in the particle loading showed linearly increase in the physical and mechanical properties. Compared with other types of nanoparticles or mesoporous particles, such as rubber, it showed opposite results. In terms of fracture and morphology behavior, more agglomerations were seen at higher particle loading and more formation of micro voids. Thus, more research is needed on silica, especially on mesoporous silica and nano-silica, as there is much development needed to be done in order to improve the existing nanocomposites properties.

References

Ahmad, F. N., Jaafar, M., Palaniandy, S., & Azizli, K. A. M. (2008). Effect of particle shape of silica mineral on the properties of epoxy composites. *Composites Science and Technology*, *68*(2), 346–353.

Akbari, B., & Bagheri, R. (2007). Deformation mechanism of epoxy/clay nanocomposite. *European Polymer Journal*, *43*(3), 782–788.

Akbari, B., & Bagheri, R. (2013). Influence of nanoclay on morphology, mechanical properties and deformation mechanism of polystyrene. *Polymer-Plastics Technology and Engineering*, *53*(2), 156–161.

Alexandre, M., & Dubois, P. (2000). Polymer-layered silicate nanocomposites: preparation, properties and uses of a new class of materials. *Materials Science and Engineering: R: Reports*, *28*(1–2), 1–63.

Anthoulis, G. I., & Kontou, E. (2008). Micromechanical behaviour of particulate polymer nanocomposites. *Polymer*, *49*(7), 1934–1942.

Azeez, A. A., Rhee, K. Y., Park, S. J., & Hui, D. (2013). Epoxy clay nanocomposites—processing, properties and applications: a review. *Composites Part B: Engineering*, *45*(1), 308–320.

Battistella, M., Cascione, M., Fiedler, B., Wichmann, M. H. G., Quaresimin, M., & Schulte, K. (2008). Fracture behaviour of fumed silica/epoxy nanocomposites. *Composites Part A: Applied Science and Manufacturing*, *39*(12), 1851–1858.

Beck, J. S., Vartuli, J. C., Roth, W. J., Leonowicz, M. E., Kresge, C. T., Schmitt, K. D., et al. (1992). A new family of mesoporous molecular sieves prepared with liquid crystal templates. *Journal of the American Chemical Society*, *114*(27), 10834–10843.

Boo, W. J., Sun, L. Y., Liu, J., Clearfield, A., Sue, H. J., Mullins, M. J., et al. (2007). Morphology and mechanical behavior of exfoliated epoxy/a-zirconium phosphate nanocomposites. *Composites Science and Technology*, *67*(2), 262–269.

Brune, D. A., & Bicerano, J. (2002). Micromechanics of nanocomposites: comparison of tensile and compressive elastic moduli, and prediction of effects of incomplete exfoliation and imperfect alignment on modulus. *Polymer*, *43*(2), 369–387.

Cannillo, V., Bondioli, F., Lusvarghi, L., Montorsi, M., Avella, M., Errico, M. E., et al. (2006). Modeling of ceramic particles filled polymer–matrix nanocomposites. *Composites Science and Technology*, *66*(7–8), 1030–1037.

Cech, V., Palesch, E., & Lukes, J. (2013). The glass fiber–polymer matrix interface/interphase characterized by nanoscale imaging techniques. *Composites Science and Technology*, *83*(1), 22–26.

Chan, C. -M., Wu, J., Li, J. -X., & Cheung, Y. -K. (2002). Polypropylene/calcium carbonate nanocomposites. *Polymer*, *43*(10), 2981–2992.

Chasiotis, I. (2009) Effect of nanoscale fillers on the local mechanical behavior of polymer nanocomposites, *Final Performance Report*, University of Illinois at Urbana–Champaign.

Chavarria, F., & Paul, D. R. (2004). Comparison of nanocomposites based on nylon 6 and nylon 66. *Polymer*, *45*(25), 8501–8515.

Chen, C., Justice, R. S., Schaefer, D. W., & Baur, J. W. (2008). Highly dispersed nanosilica-epoxy resins with enhanced mechanical properties. *Polymer*, *49*(17), 3805–3815.

Chen, J., & Taylor, A. (2012). Epoxy modified with triblock copolymers: morphology, mechanical properties and fracture mechanisms. *Journal of Materials Science*, *47*(11), 4546–4560.

Chiola, V., Ritsko, J. E., Vanderpool, C.D. 1971. Process for producing low-bulk density silica, Application No. US 3556725D A filed on February 26, 1969; Publication No. US 3556725 A published on January 19, 1971.

Choi, J., Yee, A. F., & Laine, R. M. (2004). Toughening of cubic silsesquioxane epoxy nanocomposites using core-shell rubber particles: a three-component hybrid system. *Macromolecules*, *37*(9), 3267–3276.

Ciprari, D., Jacob, K., & Tannenbaum, R. (2006). Characterization of polymer nanocomposite interphase and its impact on mechanical properties. *Macromolecules*, *39*(19), 6565–6573.

Constantinescu, D. M., Apostol, D. A., Picu, C. R., Krawczyk, K., & Sieberer, M. (2017). Mechanical properties of epoxy nanocomposites reinforced with functionalized silica nanoparticles. *Procedia Structural Integrity*, *5*(1), 647–652.

Cosmoiu, I., Apostol, D. A., Picu, C. R., Constantinescu, D. M., & Sandu, M. (2015). Manufacturing and testing of nanocomposites with carbon nanotubes and nanoparticles. *UPB Scientific Bulletin*, *77*, 107–119.

Dasari, A., Yu, Z. -Z., Yang, M., Zhang, Q. -X., Xie, X. -L., & Mai, Y. -W. (2006). Micro- and nano-scale deformation behavior of nylon 66-based binary and ternary nanocomposites. *Composites Science and Technology*, *66*(16), 3097–3114.

Deng, S., Ye, L., & Friedrich, K. (2007). Fracture behaviours of epoxy nanocomposites with nano-silica at low and elevated temperatures. *Journal of Materials Science*, *42*(8), 2766–2774.

Desai, A. V., & Haque, M. A. (2005). Mechanics of the interface for carbon nanotube–polymer composites. *Thin-Walled Structures*, *43*(11), 1787–1803.

Dittanet, P., & Pearson, R. A. (2012). Effect of silica nanoparticle size on tougheningmechanisms of filled epoxy. *Polymer*, *53*(9), 890–1905.

Fornes, T. D., & Paul, D. R. (2003). Modeling properties of nylon 6/clay nanocomposites using composite theories. *Polymer*, *44*, 4993–5013.

Frankland, S. J. V., Harik, V. M., Odegard, G. M., Brenner, D. W., & Gates, T. S. (2003). The stress–strain behavior of polymer–nanotube composites from molecular dynamics simulation. *Composites Science and Technology*, *63*(11), 1655–1661.

Gloaguen, J. M., & Lefebvre, J. M. (2001). Plastic deformation behaviour of thermoplastic/clay nanocomposites. *Polymer*, *42*(13), 5841–5847.

Halpin, J.C., (1969) Environmetnal factors in composite material design, *US Air Force Materials Laboratory Report AFML-TR-67-423*.

Hameed, N., Guo, Q., Xu, Z., Hanley, T. L., & Mai, Y. -W. (2010). Reactive block copolymer modified thermosets: highly ordered nanostructures and improved properties. *Soft Matter*, *6*(24), 6119–6129.

Hanns, B., Gottfried, K (1964), Process for producing silica in the form of hollow spheres, Application No. US 342525 A filed on February 04, 1964; Publication No. US 3383172 A published on May 14, 1968.

Hanns, B., Gottfried, K, (1967), Porous silica particles containing a crystallized phase and method, Application No. US 3493341D A filed on January 23, 1967; Publication No. US 3493341 A published on February 03, 1970

He, C., Liu, T., Tjiu, W. C., Sue, H. -J., & Yee, A. F. (2008). Microdeformation and fracture mechanisms in polyamide-6/organoclay nanocomposites. *Macromolecules*, *41*(1), 193–202.

Hydro, R. M., & Pearson, R. A. (2007). Epoxies toughened with triblock copolymers. *Journal of Polymer Science Part B: Polymer Physics*, *45*(12), 1470–1481.

Ishida, H. (1984). A review on the recent progress in the studies of molecular and microstructure of silane coupling agents in composite materials, adhesive joints and coatings. *Polymer Composites*, *5*(2), 101–123.

Ishida, H., & Koenig, J. L. (1978). The reinforcement mechanism of fibreglass reinforced plastics under wet conditions. *Polymer Engineering and Science*, *18*(2), 128–144.

Islam, M.S. (2013) The effect of nanoparticle size and percentage on mechanical properties of silica-epoxy nanocomposites, Master Thesis. Philadelphia University, Philadelphia, PA.

Jiang, L., Zhang, J., & Wolcott, M. P. (2007). Comparison of polylactide/nano-sized calcium carbonate and polylactide/montmorillonite composites: reinforcing effects and toughening mechanisms. *Polymer*, *48*(26), 7632–7644.

Jiao, J., Sun, X., & Pinnavaia, T. J. (2009). Mesostructured silica for the reinforcement and toughening of rubbery and glassy epoxy polymers. *Polymer*, *50*(4), 983–989.

Johnsen, B. B., Kinloch, A. J., Mohammed, R. D., Taylor, A. C., & Sprenger, S. (2007). Toughening mechanicms of nanoparticle-modified epoxy polymers. *Polymer*, *48*(2), 530–541.

Jumahat, A., Soutis, C., Abdullah, S. A., & Kasolang, S. (2012). Tensile properties of nanosilica/epoxy nanocomposites. *Procedia Engineering*, *41*(1), 1634–1640.

Katiyar, A., Yadav, S., Smirniotis, P. G., & Pinto, G. N. (2006). Synthesis of ordered large pore SBA-15 spherical particles for adsorption of biomolecules. *Journal of Chromatography*, *1122*(1–2), 13–20.

Khosh, R. L., Bagheri, R., & Zokaei, S. (2008). Sequences of fracture toughness micromechanisms in PP/$CaCO_3$ nanocomposites. *Journal of Applied Polymer Science*, *110*(6), 4040–4048.

Kim, G. M., Qin, H., Fang, X., Sun, F. C., & Mather, P. T. (2003). Hybrid epoxy-based thermosets based on polyhedral oligosilsesquioxane: cure behavior and toughening mechanisms. *Journal of Polymer Science Part B: Polymer Physics*, *41*(15), 3299–3313.

Kinloch, A. J., Mohammed, R. D., Taylor, A. C., Eger, C., Sprenger, S., & Egan, D. (2005). The effect of silica nano particles and rubber particles on the toughness of multiphase thermosetting epoxy polymers. *Journal of Materials Science*, *40*(18), 5083–5086.

Kinloch, A. J., & Taylor, A. C. (2006). The mechanical properties and fracture behaviour of epoxy-inorganic micro- and nanocomposites. *Journal of Materials Science*, *41*(11), 3271–3297.

Kishi, H., Kunimitsu, Y., Imade, J., Oshita, S., Morishita, Y., & Asada, M. (2011). Nano-phase structures and mechanical properties of epoxy/acryl triblock copolymer alloys. *Polymer*, *52*(3), 760–768.

Koval'chuk, A. A., Shevchenko, V. G., Schegolikhin, A. N., Nedorezova, P. M., Klyamkina, A. N., & Aladyshev, A. M. (2008). Effect of carbon nanotube functionalization on the structure and mechanical properties of polypropylene/MWCNT composites. *Macromolecules*, *41*(20), 7536–7542.

Li, X., Gao, H., Scrivens, W. A., Fei, D., Xu, X., Sutton, M. A., et al. (2004). Nanomechanical characterization of single-walled carbon nanotube reinforced epoxy composites. *Nanotechnology*, *15*(11), 1416–1423.

Liang, Y. L., & Pearson, R. A. (2010). The toughening mechanism in hybrid epoxy-silica-rubber nanocomposites (HESRNs). *Polymer*, *51*(21), 4880–4890.

Liu, H. Y., Wang, G. T., Mai, Y. W., & Zeng, Y. (2011). On fracture toughness of nano-particle modified epoxy. *Composite Part B: Engineering*, *42*(8), 2170–2175.

Liu, J., Sue, H. -J., Thompson, Z. J., Bates, F. S., Dettloff, M., Jacob, G., et al. (2008). Nanocavitation in self-assembled amphiphilic block copolymer-modified epoxy. *Macromolecules*, *41*(20), 7616–7624.

Liu, J., Sue, H. -J., Thompson, Z. J., Bates, F. S., Dettloff, M., Jacob, G., et al. (2009). Effect of crosslink density on fracture behavior of model epoxies containing block copolymer nanoparticles. *Polymer*, *50*(19), 4683–4689.

Liu, X., & Wu, Q. (2001). PP/clay nanocomposites prepared by grafting-melt intercalation. *Polymer*, *42*(25), 10013–10019.

Luo, J. -J., & Daniel, I. M. (2003). Characterization and modeling of mechanical behavior of polymer/clay nanocomposites. *Composites Science and Technology*, *63*(11), 1607–1616.

Luo, Y., Feng, C., Wang, Q., Yi, Z., Qiu, Q., Lx, K., et al. (2013). Preparation and characterization of natural rubber/silica nanocomposites using rule of similarity in latex. *Journal of Wuhan University of Technology – Materials Science Edition*, *28*(5), 997–1002.

Ma, C. G., Mai, Y. L., Rong, M. Z., Ruan, W. H., & Zhang, M. Q. (2007). Phase structure and mechanical properties of ternary polypropylene/elastomer/nano-$CaCO_3$ composites. *Composites Science and Technology*, *67*(14), 2997–3005.

Mahrholz, T., Stangle, J., & Sinapius, M. (2009). Quantitation of the reinforcement effect of silica nanoparticles in epoxy resins used in liquid composite moulding processes. *Composites Part A: Applied Science and Manufacturing*, *40*(3), 235–243.

Marouf, B.T., (2009) Effect of microstructure factors on fracture behavior of clay-rubber-epoxy hybrid nanocomposites, Ph.D. Dissertation, Sharif University of Technology: Tehran, Iran.

May, C. (1986). *Epoxy resins: chemistry and technology*: (pp. 1–1288) (2nd ed.). CRC Press.

Miltner, H. E., Assche, G. V., Pozsgay, A., Pukanszky, B., & Mele, B. V. (2006). Restricted chain segment mobility in poly(amide) 6/clay nanocomposites evidenced by quasi-isothermal crystallization. *Polymer*, *47*(3), 826–835.

Mitran, R. -A., Berger, D., Munteanu, C., & Matei, C. (2015). Evaluation of different mesoporous silica supports for energy storage in shape-stabilized phase change materials with dual thermal responses. *The Journal of Physical Chemistry C*, *119*(27), 15177–15184.

Nandiyanto, A. B. D., Iskandar, F., & Okuyama, K. (2008). Nano-sized polymer particle-facilitated preparation of mesoporous silica particles using a spray method. *Chemistry Letters*, *37*(10), 1040–1041.

Nandiyanto, A. B. D., Kim, S. -G., Iskandar, F., & Okuyama, K. (2009). Synthesis of silica nanoparticles with nanometer-size controllable Mesopores and outer diameters. *Microporous and Mesoporous Materials*, *120*(3), 447–453.

Nicolais, N. L., & Narkis, N. M. (1971). Stress-strain behavior of styrene-acrylonitrile/glass bead composites in the glassy region. *Polymer Engineering and Science*, *11*(3), 194–199.

Odegard, G. M., Clancy, T. C., & Gates, T. S. (2005). Modeling of the mechanical properties of nanoparticle/polymer composites. *Polymer*, *46*, 553–562.

Paul, D. R., & Robeson, L. M. (2008). Polymer nanotechnology: nanocomposites. *Polymer*, *49*(15), 3187–3204.

Pearson, R.A. Dittanet, P., Marouf, B.T., Siotong, T., (2010), On the use of different nano-fillers to toughen epoxies, *Polymer Nanocomposites 2010: A New Decade of Opportunities,* 1(1), 1333.

Piggott, M. R., & Chua, P. S. (1987). Recent studies of the glass fiber-polymer interphase. *Industrial & Engineering Chemistry Research*, *26*(4), 672–677.

Preghenella, M., Pegoretti, A., & Migliaresi, C. (2005). Thermo-mechanical characterization of fumed silica-epoxy nanocomposites. *Polymer*, *46*(26), 12065–12072.

Pukanszky, B. (2005). Interfaces and interphases in multicomponent materials: past, present, future. *European Polymer Journal*, *41*(4), 645–662.

Rebizant, V., Abetz, V., Tournilhac, F., Court, F., & Leibler, L. (2003). Reactive tetrablock copolymers containing glycidyl methacrylate. Synthesis and morphology control in epoxy-amine networks. *Macromolecules*, *36*(26), 9889–9896.

Rebizant, V., Venet, A. -S., Tournilhac, F., Girard-Reydet, E., Navarro, C., Pascault, J. -P., et al. (2004). Chemistry and mechanical properties of epoxy-based thermosets reinforced by reactive and nonreactive SBMX block copolymers. *Macromolecules*, *37*(21), 8017–8027.

Reichert, P., Nitz, H., Klinke, S., Brandsch, R., Thomann, R., & Mulhaupt, R. (2000). Poly(propylene)/organoclay nanocomposite formation: influence of compatibilizer functionality and organoclay modification. *Macromolecular Materials and Engineering*, *275*(1), 8–17.

Di Renzo, F., Cambon, H., & Dutartre, R. (1997). A 28-year-old synthesis of micelle-templated mesoporous silica. *Microporous Materials*, *10*(4–6), 283–286.

Reynaud, E., Jouen, T., Gauthier, C., Vigier, G., & Varlet, J. (2001). Nano-fillers in polymeric matrix: a study on silica reinforced PA6. *Polymer*, *42*(21), 8759–8768.

Rothon, R. (2016). Nano-fillers. In *Polymers and polymeric composites: a reference series* (2nd ed., pp. 1–21). Berlin Heindelberg: Springer.

Schadler, L., Brinson, L., & Sawyer, W. (2007). Polymer nanocomposites: a small part of the story. *JOM*, *59*(3), 53–60.

Seltzer, R., Kim, J. -K., & Mai, Y. -W. (2011). Elevated temperature nanoindentation behaviour of polyamide 6. *Polymer International*, *60*(12), 1753–1761.

Shen, L., Wang, L., Liu, T., & He, C. (2006). Nanoindentation and morphological studies of epoxy nanocomposites. *Macromolecular Materials and Engineering*, *291*(11), 1358–1366.

Sheng, N., Boyce, M. C., Parks, D. M., Rutledge, G. C., Abes, J. I., & Cohen, R. E. (2004). Multiscale micromechanical modeling of polymer/clay nanocomposites and the effective clay particle. *Polymer*, *45*(2), 487–506.

Sun, L., Gibson, R. F., Gordaninejad, F., & Suhr, J. (2009). Energy absorption capability of nanocomposites: a review. *Composites Science and Technology*, *69*(14), 2392–2409.

Sun, S., Li, C., Zhang, L., Du, H. L., & Burnell-Gray, J. S. (2006). Effects of surface modification of fumed silica on interfacial structures and mechanical properties of poly(vinyl chloride) composites. *European Polymer Journal*, *42*(7), 1643–1652.

Szazdi, L., Pozsgay, A., & Pukanszky, B. (2007). Factors and processes influencing the reinforcing effect of layered silicates in polymer nanocomposites. *European Polymer Journal*, *43*(2), 345–359.

Tjong, S. C. (2006). Structural and mechanical properties of polymer nanocomposites. *Materials Science Engineering: R: Reports*, *53*(3–4), 73–197.

Trewyn, B. G., Nieweg, J. A., Zhao, Y., & Lin, V. S. -Y. (2007). Biocompatible mesoporous silica nanoparticles with different morphologies for animal cell membrane penetration. *Chemical Engineering Journal*, *137*(1), 23–29.

Trewyn, B. G., Slowing, I. I., Giri, S., Chen, H. T., & Lin, V. S. -Y. (2007). Synthesis and functionalization of a mesoporous silica nanoparticle based on the sol–gel process and applications in controlled release. *Accounts of Chemical Research*, *40*(9), 846–853.

Tsai, J. L., Hsiao, H., & Cheng, Y. L. (2010). Investigating mechanical behaviors of silica nanoparticle reinforced composites. *Journal of Composite Materials*, *44*(4), 505–524.

Uddin, M. F., & Sun, C. T. (2008). Strength of unidirectional glass/epoxy composite with silica nanoparticle-enhanced matrix. *Composites Science and Technology*, *68*(7–8), 1637–1643.

Uddin, M. F., & Sun, C. T. (2010). Improved dispersion and mechanical properties of hybrid nanocomposites. *Composites Science and Technology*, *70*(2), 223–230.

Vaia, R. A., & Wagner, H. D. (2004). Framework for nanocomposites. *Materials Todays*, *7*(11), 32–37.

Wang, G. -T., Liu, H. -Y., Yu, Z. -Z., & Mai, Y. -W. (2009). Evaluation of methods for stiffness predictions of polymer/clay nanocomposites. *Journal of Reinforced Plastics and Composites*, *28*(13), 1625–1649.

Wichmann, M. H. G., Cascione, M., Fiedler, B., Quaresimin, M., & Schulte, K. (2006). Influence of surface treatment on mechanical behaviour of fumed silica/epoxy resin nanocomposites. *Composite Interfaces*, *13*(8–9), 669–715.

Williams, J. G., Donnellan, M. E., James, M. R., & Morris, W. L. (1990). Properties of the interphase inorganic matrix composites. *Materials Science and Engineering: A*, *126*(1–2), 305–312.

Wright, W. W. (1990). The carbon fiber/epoxy resin interphase—a review, part II. *Polymers and Polymer Composites*, *3*(5), 360–401.

Wu, J., Thio, Y. S., & Bates, F. S. (2005). Structure and properties of PBO–PEO diblock copolymer modified epoxy. *Journal of Polymer Science Part B: Polymer Physics*, *43*(15), 1950–1965.

Wu, S., Guo, Q., Kraska, M., Stuhn, B., & Mai, Y. -W. (2013). Toughening epoxy thermosets with block ionomers: the role of phase domain size. *Macromolecules*, *46*(20), 8190–8202.

Wu, S., Guo, Q., Peng, S., Hameed, N., Kraska, M., Stuhn, B., et al. (2012). Toughening epoxy thermosets with block ionomer complexes: a nanostructure–mechanical property correlation. *Macromolecules*, *45*(9), 3829–3840.

Wu, S., Guo, Q., Zhang, T., & Mai, Y. -W. (2013). Phase behavior and nanomechanical mapping of block ionomer complexes. *Soft Matter*, *9*(1), 2662–2672.

Xu, R., Pang, W., Yu, J., Huo, Q., & Chen, J. (2007). *Chemistry of zeolites and related porous materials: synthesis and structure* (1st ed., pp. 1–616) Wiley-Interscience.

Young, R. J., Kinloch, I. A., Gong, L., & Novoselov, K. S. (2012). The mechanics of graphene nanocomposites: a review. *Composites Science and Technology*, *72*(12), 1459–1476.

Zamanian, M., Mortezaei, M., Salehnia, B., & Jam, J. E. (2013). Fracture toughness of epoxy modified with nano-silica particles: particle size effect. *Engineering Fracture Mechanics*, *97*(1), 193–206.

Zhao, D., Feng, J., Huo, Q., Melosh, N., Fredrickson, G. H., Chmelka, B. F., et al. (1998). Triblock copolymer syntheses of mesoporous silica with periodic 50 to 300 angstrom pores. *Science*, *279*(5350), 548–552.

Further reading

Marouf, B. T., Mai, Y. -W., Bagheri, R., & Pearson, R. A. (2016). Toughening of epoxy nanocomposites: nano and hybrid effects. *Polymer Reviews*, *56*(1), 70–112.

Index

Note: Page numbers followed by *f* indicate figures, *t* indicate tables and *s* indicate schemes.

A

Activation energy
- PF, 67, 68*t*
- PF-fsi, 67, 68*t*
- PF-fsi-clay, 67, 68*t*
- PLA-fsi-clay, 102, 103*t*
- PVA, 40, 41*t*, 53, 54*t*
- PVA-fsi-clay, 40, 41*t*
- PVA-si-clay, 53, 54*t*
- ST-co-GMA, 80, 81*t*
- ST-co-GMA-fsi, 80, 81*t*
- ST-co-GMA-fsi-clay, 80, 81*t*

Adsorption isotherm analysis. *See* N_2 adsorption isotherms

Agglomeration, 9, 62, 74–76, 92, 115, 192–193

Alkyl ammonium ions, 113–114

Al_2O_3/SiC system, 206

Aluminum trihydrate, 126

Arrhenius equation, 40, 41*t*, 53, 54*t*, 67, 68*t*, 80, 81*t*, 102, 103*t*

ASTM D5045 standards, 266–267

B

Batch foaming (BF) process, 169

Benzoyl peroxide, 71, 74*f*

Biodegradable polymer-based nanocomposites, 202–203

Bragg's law, 116–117

Brunauer–Emmett–Teller (BET) theory, 29, 93, 271–272

BTO epoxy, 266

C

Carbonaceous char, 126–127

Carbon nanotubes (CNTs), 201–202
- chemical/vapor deposition methods, 208
- graphene cylinders, 208
- reinforced nanocomposite systems, 202–203, 210–211

Carboxyl terminated butadiene acrylonitrile (CTBN), 269

Catalytic decomposition, 210

Cation exchange capacity (CEC), 114, 142–143

Ceramics, 205–206

Ceramic matrix nanocomposites (CMNC), 201–202, 205–206
- catalytic decomposition, 210
- hot pressing, 209
- polymer precursor process, 209
- powder process, 209
- sol-gel process, 209
- solvothermal process, 210
- vapor techniques (*see* Spray pyrolysis)

Chemical/vapor deposition methods, 208, 210

Clay, 1, 3, 33–34, 45–46, 71–72, 141–143, 142*f*
- cation-exchange modification, 5
- minerals, 3–4
- modification
  - by ion-exchange with organic cations, 5
  - by silane reaction, 4–5
- nanocomposites
  - mechanical properties, 12, 145–146
  - morphological properties, 145–146, 147*f*
  - optical properties, 237–238
  - physical properties, 10–11
  - spectral properties, 144–145
  - thermal properties, 13, 144–145
  - UV-Vis absorption, 239*f*
  - UV-Vis spectra, 238*f*
  - water barrier properties, 14

CMNC. *See* Ceramic matrix nanocomposites (CMNC)

CNTs. *See* Carbon nanotubes (CNTs)

Collapse stress. *See* Foam collapse stress

Compatibility, 10, 35–37, 39, 48–50, 61–62, 72, 92–93, 171–173
Condensation polymerization technique
  aniline-formaldehyde copolymers, 15
  PVA-fsi-clay, 26, 27*t*
Conducting polymer nanocomposites, 202–203, 219
  classification, 219–220*t*
  properties, 219–220*t*
Cone calorimetry, 126
Conventional composites, 129–130, 201
Crosslinking reaction, 262
Crystallinity, 120–121, 180, 183*t*

D

DGBEA epoxy, 268–269
Differential scanning calorimetry (DSC), 179–180, 179*f*, 182*f*, 183*t*, 231–232
Diglycidyl ether of bisphenol-A (DGEBA), 234
Dimensional nanoparticles, 221
DMTA. *See* Dynamic mechanical thermal analysis (DMTA)
Dodecylamine (DDA), 4–5
DSC. *See* Differential scanning calorimetry (DSC)
Durability, 144–146, 147*f*
Dynamic mechanical analysis (DMA), 233–234
Dynamic mechanical thermal analysis (DMTA), 180–188, 186*t*, 265

E

Elastic modulus, 174*f*
  foamed composites, 175*f*
  low-density polyethylene, 165*t*, 165*f*
  nanoclays, 224*t*
  silica nanocomposites foams, 165*t*, 165*f*
  solid composites, 175*f*
Elastomeric nanocomposites, 229
Electroless coating, 214
Epoxidized soybean oil (ESO), 12
Epoxy nanocomposites, 125, 264
  fracture behavior, 269–273
  mechanical properties, 264–269
  morphological properties, 269–273
  nano-silica addition, 270*f*
  physical properties, 264–269
  stress–strain curve, 269–270
  tensile properties, 266*t*
Exchangeable cations, 113–114

F

Fiber-reinforced polymer (FRP), 130–131
Fick's law, 102
Fillers, 25, 109
Filler–matrix interaction, 272
Filler-polymer interactions, 166–167
Fire performance index, 127–128
Flame retardance (FR), 126–129
Foam collapse stress, 174–175*f*, 176
Foamed composites, 175*f*
Foamed samples, cellular structure, 194*f*
Fourier transform infrared (FTIR) spectroscopy, 8–10, 28, 160–164, 221–224
  jute cellulose/PE, 162–164, 164*f*
  jute cellulose/PE/clay, 162–164, 164*f*
  jute cellulose/PE/silica, 162–164, 164*f*
  LDPE-SiO_2, 144
  LDPE-ZnO, 144
  PF, 60–61, 61*f*
  PF-fsi, 60–61, 61*f*
  PF-fsi-clay, 60–61, 61*f*
  PLA-fsi-clay, 88–91, 89–90*f*
  pure low-density polyethylene, 161*f*
  PVA, 34–35, 35*f*, 46–48, 47*f*
  PVA-fsi-clay, 34–35, 35*f*
  PVA-si-clay, 46–48, 47*f*
  silica/low-density polyethylene nanocomposite, 162–163*f*
  ST-co-GMA, 72–74, 73*f*
  ST-co-GMA-fsi, 72–74, 73*f*
  ST-co-GMA-fsi-clay, 72–74, 73*f*
Fracture toughness, 124, 140, 261
Free radical polymerization technique
  ST-co-GMA-clay, 7–8
  ST-co-GMA-fsi, 15
  ST-co-GMA-fsi-clay, 27, 27*t*
FTIR. *See* Fourier transform infrared (FTIR) spectroscopy
Fumed silica, 1, 25, 59, 71–72, 87
Fumed silica-epoxy nanocomposites, 159, 266, 266*t*
Functional fiber nanocomposites, 202–203

G

Gasification, 128–129
Glass transition temperature, 125, 183, 233
Glycidyl methacrylate (GMA), 71, 74*f*
GLYMO-modified fumes silica epoxy nanocomposites, 270, 271*f*
Graphene cylinders, 208

H

Heat release rate (HRR), 126
High molecular weight-based nanocomposites, 229–230
Hot pressing method, 209, 214
Hybridization, 145–146

I

Inorganic nanoparticles, 206
In situ intercalative polymerization, 122–123, 217
In situ polymerization, 139, 208, 215–216, 219
Intercalated nanocomposites, 217
Interlayer dimension, 112–113
Interlayer space, 226
Ion-exchange reaction, 113–114, 113*f*
3-Isocyanatopropyltriethoxysilane (IPTES), 87
Izod impact strength, 122–123

J

Jute cellulose/PE/clay nanocomposite (JCPECC), 162–164, 164*f*, 173–174, 178*f*
  scanning electron microscopic images, 192–193, 192*f*
  thermal characteristics, 178*t*
Jute cellulose/PE composite (JCPEC), 162–164, 164*f*, 173–174, 173*f*, 178*f*
  scanning electron microscopic images, 192–193, 192*f*
  thermal characteristics, 178*t*
Jute cellulose/PE/silica nanocomposite (JCPESC), 162–164, 164*f*, 173–174, 178*f*
  scanning electron microscopic images, 192–193, 192*f*
  thermal characteristics, 178*t*
Jute fiber/PE/maleic anhydride/clay nanocomposite (JFPEMACC), 171–173, 172*f*
Jute fiber/PE/maleic anhydride composite (JFPEMAC), 171–173, 172*f*
Jute fiber/PE/maleic anhydride/silica nanocomposite (JFPEMASC), 171–173, 172*f*
Jute fiber/polyethylene (PE)/composite (JFPEC), 171–173, 172*f*

K

Kaolinite, 111
Kelvin equation, 29–30

L

Layered silicates, 110, 216*s*
  biomedical and packaging applications, 129–130
  characteristics, 111–113
  exfoliated systems, 115–116, 116*f*
  high molecular weight, 119
  intercalation, 115–116, 116*f*
  ion-exchange reaction, 113–114, 113*f*
  low molecular weight, 119
  mechanical properties, 117–122
  medium molecular weight, 119
  organic modification, 113–115
  structure, 111–113, 112*f*, 115–117
  transmission electron microscopy, 116–117
  X-ray fluorescence, 116–117
LDPE. *See* Low-density polyethylene (LDPE)
LDPE-EVA. *See* Low density polyethylene-ethylene vinyl acetate (LDPE-EVA)
Linear elastic fracture mechanics (LEFM), 266–267
Linear low density polyethylene (LLDPE), 190
Liquid infiltration, 212
Low-density polyethylene (LDPE), 139, 180
  cellular structure, 194–195, 195*f*
  chemical composition, 160–164
  differential scanning calorimetry, 179–180, 181*t*
  dynamic mechanical thermal analysis, 180–188
  electrical properties, 196

Low-density polyethylene (LDPE) *(Continued)*
- Fourier transform infrared spectroscopy, 160–164, 161*f*
- mechanical properties, 164–176, 170*t*, 171*f*
- modulus of elasticity, 165*t*, 165*f*
- nanosilica particles, 160*t*
- physical properties, 160–164
- scanning electron microscopic images, 188–194
- surface morphologies, 188*f*
- tensile strength, 165*t*, 165*f*
- thermal properties, 176–188
- thermal stability, 177*t*
- thermogravimetric analysis, 176–179

Low density polyethylene-ethylene vinyl acetate (LDPE-EVA), 166–167, 187*f*
- differential scanning calorimetry thermogram, 180, 182*t*, 182*f*
- nanosilica dispersed, 187–188
- scanning electron microscopic images, 190–192
- storage modulus, 186*f*

Low molecular weight-based nanocomposites, 229–230

M

Medium molecular weight-based nanocomposites, 229–230

Mesoporous silica/nano-silica, 263–264

Metal-based nanocomposites, 202–203

Metal matrix nanocomposites (MMNC), 201–202, 206, 211
- chemical processes, 213
- CVD/PVD, 213
- electroless coating, 214
- high energy/ball milling, 212
- hot pressing, 214
- liquid infiltration, 212
- nanoscale dispersion, 214
- PM/infiltration, 214
- rapid solidification process, 212
  - with ultrasonics, 212
- spray pyrolysis, 211

Metalocene, 166, 190

Metal oxide-incorporated polymer nanocomposites, 207

Mica, 111–112

Microcomposites, 210, 260–261

Micron-filled polymers, 159

Microvoids, 122

mLLDPE-SiO_2 nanocomposites, 166, 180–183
- loss modulus, 185*f*
- scanning electron microscopic images, 190*f*
- storage modulus, 184*f*
- tensile properties, 168*t*
- tensile stress–strain curves, 166*f*

Modified layered double hydroxides (MLDH), 10–11

Modulus of elasticity
- low-density polyethylene, 165*t*, 165*f*
- nanoclays, 224*t*
- silica nanocomposites foams, 165*t*, 165*f*

Moisture absorption analysis, 30–31, 102–104, 104*f*
- ST-co-GMA, 80–82, 80*f*
- ST-co-GMA-fsi, 80–82, 80*f*
- ST-co-GMA-fsi-clay, 80–82, 80*f*

Montmorillonite (MMT), 4–6, 126–127, 141–142, 142*f*, 238*f*
- nanocomposites, 228*f*
- UV-Vis absorption, 239*f*

N

N_2 adsorption isotherms
- PF, 64, 64*f*, 65*t*
- PF-fsi, 64, 64*f*, 65*t*
- PF-fsi-clay, 64, 64*f*, 65*t*
- PLA-fsi-clay, 93–96, 97–98*f*, 99*t*
- PVA, 37–38, 37*f*, 38*t*, 50, 50*f*, 51*t*
- PVA-fsi-clay, 37–38, 37*f*, 38*t*
- PVA-si-clay, 50, 50*f*, 51*t*
- ST-co-GMA, 76–77, 76*f*, 77*t*
- ST-co-GMA-fsi, 76–77, 76*f*, 77*t*
- ST-co-GMA-fsi-clay, 76–77, 76*f*, 77*t*

Nanoclays, 45–46
- differential scanning calorimetry, 231–232, 232*f*
- Fourier transform infrared spectroscopy, 222–224*f*
- mechanical properties, 164–166
- modulus of elasticity, 224*t*, 225*f*
- reinforced nanocomposites, 224–226
- surface morphologies, 235–236*f*

tensile strength, 224–226, 224*t*, 225*f*
thermogravimetric analysis, 230*f*
Nanocomposites, 141. *See also specific types of nanocomposites*
adsorption isotherm, 29–30
ceramic matrix nanocomposites, 209–211
clay nanocomposites, 237–238
(*see also* Polypropylene (PP))
conducting polymer nanocomposites, 219
defined, 5, 201
differential scanning calorimetry, 231–232
dynamic mechanical analysis, 233–234
Fourier transform infrared spectroscopy, 28, 221–224
fracture surface, 273–274*f*
fumed silica-epoxy, 266*t*
homogeneous dispersion, 269–270
intercalation, 5–6
layered reinforcement's nanocomposites, 226–230
mechanical properties, 224–230
metal matrix nanocomposites, 211–214
moisture absorption test, 30–31
morphological properties, 234–237
nanoclay reinforced nanocomposites, 224–226
and nanofillers, 261–262
optical properties, 237–238
PF-clay, 7
PF-fumed silica, 7
phase separated, 5–6
physical properties, 221–224
PLA-clay, 8
PLA-fumed silica, 8
polymer matrix nanocomposites, 214–221
preparation, 208–238
prospects and possibilities, 205–208
PVA-clay, 6
PVA-fsi-clay
condensation polymerization technique, 26, 27*t*
by solutionintercalation film-casting technique, 28, 28*t*
solution intercalation technique, 26, 26*t*
PVA-si-clay
solution intercalation technique, 26, 26*t*
PVA-si/fsi, 6
raw materials, 208
scanning electron microscopy analysis, 29, 234–237
ST-co-GMA-clay, 7–8
ST-co-GMA-fsi, 7
ST-co-GMA-fsi-clay
free radical polymerization technique, 27, 27*t*
structure, 5–6
tensile properties, 30, 224–226
thermal properties, 230–234
thermogravimetric analysis, 30, 230–231
two-step chemical method, 234
weight loss, 231*t*
X-ray fluorescence analysis, 28–29
Nanocomposites foams, 196*f*. *See also* Silica nanocomposites foams
Nanofillers, 139–141, 171, 260–262, 260*f*
Nanoparticle-filled polymers, 159
Nanoparticles, 260*f*
Nanoscale dispersion, 214
Nanoscopic reinforcements, 207
Nanosilica (nS), 171*f*, 264, 274
Nanotechnology applications, 147–148
Nonswelling clays, 4
Nylon-6 nanocomposites, 227–230

O

Octahedral sheets, 3–4, 111
1D nanofillers, 260
1D nano-platelets, 260*f*
Onium ions, 113–114
Organic MMT (OMMT), 7, 12
Organoclays, 234, 237
Organo-modified layered silicates (OMLSs), 114, 116–117
Oxirane/glycidyl group, 262

P

PA66 nanocomposites, 118–119
Peak of heat release (PHRR), 126
PE-MA nanocomposites, 119
PF-MMT nanocomposites, 60
Phase-separated composites, 216
Phenol formaldehyde (PF), 2
activation energy, 67, 68*t*
Fourier transform infrared spectroscopy, 60–61, 61*f*
N_2 adsorption isotherms, 64, 64*f*, 65*t*

Phenol formaldehyde (PF) *(Continued)*
oxide contents, 61–62, 62*t*
resins, 59–60
scanning electron microscopy analysis, 62–63, 63*f*
tensile modulus, 65, 66*f*
tensile strength, 65, 66*f*
thermogravimetric analysis, 66–68, 67*f*, 68*t*
X-ray fluorescence analysis, 61–62, 62*t*
Phenol formaldehyde/fumed silica (PF-fsi)
activation energy, 67, 68*t*
Fourier transform infrared spectroscopy, 60–61, 61*f*
N_2 adsorption isotherms, 64, 64*f*, 65*t*
oxide contents, 61–62, 62*t*
scanning electron microscopy analysis, 62–63, 63*f*
tensile modulus, 65, 66*f*
tensile strength, 65, 66*f*
thermogravimetric analysis, 66–68, 67*f*, 68*t*
X-ray fluorescence analysis, 61–62, 62*t*
Phenol formaldehyde/fumed silica/clay (PF-fsi-clay)
activation energy, 67, 68*t*
Fourier transform infrared spectroscopy, 60–61, 61*f*
N_2 adsorption isotherms, 64, 64*f*, 65*t*
oxide contents, 61–62, 62*t*
scanning electron microscopy analysis, 62–63, 63*f*
tensile modulus, 65, 66*f*
tensile strength, 65, 66*f*
thermogravimetric analysis, 66–68, 67*f*, 68*t*
X-ray fluorescence analysis, 61–62, 62*t*
Phyllosilicates, 3–4
PLA-cocamidopropylbetaine (PLACAB), 87–88
PLA-low-density polyethylene (PLA-LDPE), 87
PLA-polycaprolactone-nanoclay (PLA-PCL-nanoclay), 88
Poly(ethylene oxide) (PEO), 10
Poly(methyl methacrylate) (PMMA), 8
Poly-epoxides, 262
Polyethylenes, 139, 180–183
Polyimide polymer matrix nanocomposites, 229
Polylactic acid (PLA), 2, 8
oxide contents, 91, 91*t*
Polylactic acid/fumed silica/clay (PLA-fsi-clay)
activation energy, 102, 103*t*
Fourier transform infrared spectroscopy, 88–91, 89–90*f*
N_2 adsorption isotherms, 93–96, 97–98*f*, 99*t*
scanning electron microscopy analysis, 92–93, 93–96*f*
tensile modulus, 98, 100*f*
tensile strength, 96–98, 99*f*
thermogravimetric analysis, 100–102, 100–101*f*
X-ray fluorescence analysis, 91–92, 91*t*
Polymers, 25
with higher molecular weight, 262
intercalation chemistry, 226
melting properties, 179
nanocomposites, 227, 260–261
condensation polymerization, 15
free radical polymerization, 15
solution intercalation, 15
solution-intercalation film-casting, 15
precursor process, 209
silicate, 216*s*
Polymer-Al/Al_2O_3 nanocomposites, 207
Polymer-based composites, 202–203
Polymer-clay nanocomposites (PCNs), 34
Polymeric epoxy resins, 262–263
Polymerization, 208, 215–216, 218–219
Polymer layered silicate (PLS) nanocomposites, 221, 226–227
Polymer matrix nanocomposites (PMNC), 201–202, 207
dimensional nanoparticles, 221
direct mixing, 218
in situ
intercalative polymerization, 217
polymerization, 218–219
intercalation/prepolymer from solution, 217
melt intercalation, 217
melt mixing, 218
mixing polymerization, 218
properties, 220*t*
sol-gel process, 218
solution mixing, 218
structure, 220*t*
template synthesis, 217

Polymer–nanosilica interaction, 187
Polymer-silica nanocomposite, 46
Polypropylene (PP), 183
  differential scanning calorimetry, 231–232, 232*f*
  Fourier transform infrared spectroscopy, 222–224*f*
  modulus of elasticity, 224*t*, 225*f*
  optical properties, 237–238
  scanning electron microscopy, 188–189, 234
  surface morphologies, 234–236*f*
  tensile strength, 224*t*, 225*f*
  thermogravimetric analysis, 230*f*
  UV-Vis absorption, 239*f*
  UV-Vis spectra, 238*f*
Polyvinyl alcohol (PVA), 1–2, 6, 33, 45
  activation energy, 40, 41*t*, 53, 54*t*
  Fourier transform infrared spectroscopy, 34–35, 35*f*, 46–48, 47*f*
  N_2 adsorption isotherms, 37–38, 37*f*, 38*t*, 50, 50*f*, 51*t*
  oxide contents, 48, 48*t*
  scanning electron microscopy analysis, 35–37, 36*f*, 48–50, 49*f*
  tensile modulus, 39, 39*f*, 52, 52*f*
  tensile strength, 38–39, 38*f*, 51–52, 51*f*
  thermogravimetric analysis, 39–41, 40*f*, 41*t*, 52–54, 53*f*, 54*t*
  X-ray fluorescence analysis, 48, 48*t*
Polyvinyl alcohol/fumed silica/clay (PVA-fsi-clay)
  activation energy, 40, 41*t*
  condensation polymerization technique, 26, 27*t*
  Fourier transform infrared spectroscopy, 34–35, 35*f*
  N_2 adsorption isotherms, 37–38, 37*f*, 38*t*
  scanning electron microscopy analysis, 35–37, 36*f*
  by solution intercalation film-casting technique, 28, 28*t*
  solution intercalation technique, 26, 26*t*
  tensile modulus, 39, 39*f*
  tensile strength, 38–39, 38*f*
  thermogravimetric analysis, 39–41, 40*f*, 41*t*
Polyvinyl alcohol/silica/clay (PVA-si-clay)
  activation energy, 53, 54*t*
  Fourier transform infrared spectroscopy, 46–48, 47*f*
  N_2 adsorption isotherms, 50, 50*f*, 51*t*
  oxide contents, 48, 48*t*
  scanning electron microscopy analysis, 48–50, 49*f*
  solution intercalation technique, 26, 26*t*
  tensile modulus, 52, 52*f*
  tensile strength, 51–52, 51*f*
  thermogravimetric analysis, 52–54, 53*f*, 54*t*
  X-ray fluorescence analysis, 48, 48*t*
Powder process, 209
PP. *See* Polypropylene (PP)
PP-based RTPO/clay nanocomposites, 124–125
PP/PP-MA nanocomposites, 119
Prepolymers, 262
Problem statement, 2–3
Propionic anhydride (PA), 183
Propylene-based nanocomposites, 227–228
Protonated aminododecanoic nanocomposite, 229–230
PU/o-MMT nanocomposites, 120–121, 125–126
PVA-MMT nanocomposites, 46
PVA-silicon dioxide-titanium oxide ($PVA-SiO_2-TiO_2$), 12–13
Pyrophyllite, 111–112

R

Rapid solidification process (RSP), 212
Rice husk ash (RHA), 12
Rubber epoxy nanocomposites, 268

S

Santa Barbara Amorphous (SBA-15), 263–264
Saponite-based nanocomposites, 228*f*
Scanning electron microscopy (SEM) analysis, 29
  PA12 nanocomposites, 122, 123*f*
  PF, 62–63, 63*f*
  PF-fsi, 62–63, 63*f*
  PF-fsi-clay, 62–63, 63*f*
  PLA-fsi-clay, 92–93, 93–96*f*
  PVA, 35–37, 36*f*, 48–50, 49*f*
  PVA-fsi-clay, 35–37, 36*f*
  PVA-si-clay, 48–50, 49*f*

Scanning electron microscopy (SEM) analysis *(Continued)*
ST-co-GMA, 74–76, 75*f*
ST-co-GMA-fsi, 74–76, 75*f*
ST-co-GMA-fsi-clay, 74–76, 75*f*
Semipolymeric epoxy resins, 262–263
Silane coupling agent, 169
Silane reaction, clay modification, 4–5
Silica, 1, 45, 72, 143–144
nanocomposites
mechanical properties, 11–12, 145–146
modulus of elasticity, 224*t*
morphological properties, 145–146, 148*f*
physical properties, 8–10
spectral properties, 144–145
tensile strength, 224*t*
thermal properties, 12–13, 144–145
water barrier properties, 14
Silica epoxy nanocomposites, 267–268
GLYMO-modified fumes, 270, 271*f*
tensile behavior, 269
unmodified fumed, 271*f*
Silica nanocomposites foams
cellular structure, 194–195, 195*f*
chemical composition, 160–164
differential scanning calorimetry, 179–180, 181*t*
dynamic mechanical thermal analysis, 180–188
electrical properties, 196
Fourier transform infrared spectroscopy analysis, 160–164
mechanical properties, 164–176, 170*t*
modulus of elasticity, 165*t*, 165*f*
nanosilica particles, 160*t*
physical properties, 160–164
scanning electron microscopic study, 188–194
tensile strength, 165*t*, 165*f*
thermal properties, 176–188
thermal stability, 177*t*
thermogravimetric analysis, 176–179
weight loss, 177*t*
Silica nanopowder nanocomposites, 273*f*
Smectitic clays, 111–112, 141–142
Sol–gel process, 209–211, 264, 268–269
Solid composites, 175*f*
Solution intercalation technique, 15
PVA-fsi-clay, 26, 26*t*, 28, 28*t*
PVA-si-clay, 26, 26*t*
Solvothermal process, 210
Spray pyrolysis, 210–211
metal matrix nanocomposites, 211
Steel nanocomposites, 211
Stiffness, 120–121
Storage modulus
low density polyethylene-ethylene vinyl acetate, 186*f*
mLLDPE-SiO_2 nanocomposites, 184*f*
SBS-based nanocomposites, 233*f*
vs. temperature, 183–186
zLLDPE-SiO_2 nanocomposites, 184*f*
Strain, 122–126
Structural fiber nanocomposites, 202–203
Styrene (ST), 71
Styrene-butadiene-styrene (SBS) block copolymers, 233–234
Styrene-co-glycidyl methacrylate (ST-co-GMA), 2
activation energy, 80, 81*t*
Fourier transform infrared spectroscopy, 72–74, 73*f*
moisture absorption analysis, 80–82, 80*f*
N_2 adsorption isotherms, 76–77, 76*f*, 77*t*
scanning electron microscopy analysis, 74–76, 75*f*
tensile modulus, 77–78, 78*f*
tensile strength, 77–78, 78*f*
thermogravimetric analysis, 79–80, 79*f*, 81*t*
Styrene-co-glycidyl methacrylate/fumed silica (ST-co-GMA-fsi)
activation energy, 80, 81*t*
Fourier transform infrared spectroscopy, 72–74, 73*f*
free radical polymerization technique, 15
moisture absorption analysis, 80–82, 80*f*
N_2 adsorption isotherms, 76–77, 76*f*, 77*t*
scanning electron microscopy analysis, 74–76, 75*f*
tensile modulus, 77–78, 78*f*
tensile strength, 77–78, 78*f*
thermogravimetric analysis, 79–80, 79*f*, 81*t*
Styrene-co-glycidyl methacrylate/fumed silica/clay (ST-co-GMA-fsi-clay)
activation energy, 80, 81*t*
Fourier transform infrared spectroscopy, 72–74, 73*f*

free radical polymerization technique, 27, 27*t*
moisture absorption analysis, 80–82, 80*f*
N_2 adsorption isotherms, 76–77, 76*f*, 77*t*
scanning electron microscopy analysis, 74–76, 75*f*
schematic reaction, 72, 74*f*
tensile modulus, 77–78, 78*f*
tensile strength, 77–78, 78*f*
thermogravimetric analysis, 79–80, 79*f*, 81*t*
Sustainability, 144–146, 147*f*
Swelling clays, 4

T

Tensile modulus, 11
PA12/clay systems, 118–119
PET nanocomposite, 121–122
PF, 65, 66*f*
PF-fsi, 65, 66*f*
PF-fsi-clay, 65, 66*f*
PLA-fsi-clay, 98, 100*f*
PVA, 39, 39*f*, 52, 52*f*
PVA-fsi-clay, 39, 39*f*
PVA-si-clay, 52, 52*f*
ST-co-GMA, 77–78, 78*f*
ST-co-GMA-fsi, 77–78, 78*f*
ST-co-GMA-fsi-clay, 77–78, 78*f*
Tensile strength (TS), 11, 171–173, 172*f*
fumed silica-epoxy nanocomposites, 266*t*
low-density polyethylene, 165*t*, 165*f*
microcomposite, 227, 229*t*
nanoclay, 224*t*
reinforced nanocomposites, 224–226
nanocomposite, 227, 229*t*
PA12/clay systems, 118–119
PET, 121–122
PF, 65, 66*f*
PF-fsi, 65, 66*f*
PF-fsi-clay, 65, 66*f*
PLA-fsi-clay, 96–98, 99*f*
PP, 122–123
PVA, 38–39, 38*f*, 51–52, 51*f*
PVA-fsi-clay, 38–39, 38*f*
PVA-si-clay, 51–52, 51*f*
silica nanocomposites foams, 165*t*, 165*f*
ST-co-GMA, 77–78, 78*f*
ST-co-GMA-fsi, 77–78, 78*f*
ST-co-GMA-fsi-clay, 77–78, 78*f*
Tetraethylorthosilicate (TEOS), 4–5
Tetrahedral sheets, 3–4, 111
Thermogravimetric analysis (TGA), 30, 176–179, 176*f*, 178*f*, 230–231
PF, 66–68, 67*f*, 68*t*
PF-fsi, 66–68, 67*f*, 68*t*
PF-fsi-clay, 66–68, 67*f*, 68*t*
PLA-fsi-clay, 100–102, 100–101*f*
PVA, 39–41, 40*f*, 41*t*, 52–54, 53*f*, 54*t*
PVA-fsi-clay, 39–41, 40*f*, 41*t*
PVA-si-clay, 52–54, 53*f*, 54*t*
ST-co-GMA, 79–80, 79*f*, 81*t*
ST-co-GMA-fsi, 79–80, 79*f*, 81*t*
ST-co-GMA-fsi-clay, 79–80, 79*f*, 81*t*
Thermoplastics, 227–229, 261
Thermosets, 261
3D nanofillers, 260
Time to ignition (TTI), 126–128
Toughness, 120–126
2D nanofillers, 260
2D nano-tubes, 260*f*
2:1 phyllosilicates, 226, 227*f*

U

Unmodified fumes silica epoxy nanocomposites
dynamic mechanical thermal analysis, 265
TEM analysis, 271*f*
tensile fracture surface, 272*f*

V

Van der Waals forces, 3–4, 142–143

W

Weatherability, 139

X

X-ray fluorescence (XRF) analysis, 10, 28–29
PF, 61–62, 62*t*
PF-fsi, 61–62, 62*t*
PF-fsi-clay, 61–62, 62*t*
PLA-fsi-clay, 91–92, 91*t*
PVA, 48, 48*t*
PVA-si-clay, 48, 48*t*

Y

Young's modulus (YM), 120–121, 145–146, 164–166, 171–173, 172*f*, 227, 267

Z

Ziegler-Natta catalysts, 166, 190
zLLDPE-SiO_2 nanocomposites, 166, 180–183
loss modulus, 185*f*
silica content and, 190*f*
storage modulus, 184*f*
tensile stress–strain curves, 167*f*

CPSIA information can be obtained
at www.ICGtesting.com
Printed in the USA
LVHW06s0952060718
582911LV00005B/16/P

9 780081 021293